FE CIVIL EXAMS

5 PRACTICE EXAMS WITH STEP-BY-STEP SOLUTIONS

MOHAMMAD IQBAL, D.Sc., PE, SE, Esq.
ALI A. IQBAL, PE

Report Errors for This Book

PPI is grateful to every reader who notifies us of a possible error. Your feedback allows us to improve the quality and accuracy of our products. Report errata at **ppi2pass.com**.

FE CIVIL EXAMS:
5 PRACTICE EXAMS WITH STEP-BY-STEP SOLUTIONS

Current release of this edition: 2

Release History

date	edition number	revision number	update
Apr 2021	1	1	New edition.
Mar 2022	1	2	Minor corrections.

Printed in the United States of America.

PPI
ppi2pass.com

ISBN: 978-1-59126-793-5

Table of Contents

Preface and Acknowledgments

Our objective in writing *FE Civil Exams: Five Practice Exams* was to create a book that (1) provides exam-like practice for the NCEES FE Civil exam and (2) familiarizes you with the *NCEES FE Reference Handbook* (*NCEES Handbook*), the only reference you are allowed to use during the exam. We believe this book accomplishes both objectives.

This book originated as a book titled *Civil Engineering FE Exam Preparation Workbook*. We determined that the book would be even more useful to readers as a series of practice exams. For this new book, we have assembled the questions into full-length practice exams with the same difficulty and format as the FE exam. The FE Civil exam has evolved in its format and content, so we've added alternative item types in each exam and written about 70 entirely new questions to better simulate the exam experience. Detailed tables in the introductory portion of this book will help you learn where to look in the *Handbook* to find the data and equations you'll need during the exam.

This book represents a great deal of effort from a great many people. We would like to acknowledge the team at PPI, including

Editorial: Bilal Baqai, Tyler Hayes, Indira Prabhu Kumar, Scott Marley, Grace Wong, Michael Wordelman

Art: Tom Bergstrom

Production: Nikki Capra-McCaffrey, Richard Iriye, Damon Larson, Sean Woznicki, Kimberly Burton-Weisman, Kim Wimpsett

Project Management: Beth Christmas

Content and Product: Nicole Evans, Meghan Finley, Anna Howland, Scott Rutherford, Megan Synnestvedt, Amanda Werts

Publishing Systems: Sam Webster

We'd also like to thank our technical reviewers and calculation checkers: Richard Madsen, PE; David A. Janover, PE, F.NSPE, CFM; Michael Goldrich, PE, M.Eng.; Kennedy Danga, M.Eng., MBA, PMP; and Shady Gergis, EI.

We hope this book helps you with your exam preparation. Though we made every effort to ensure the technical accuracy of our problems, any mistakes you find are ours alone. Please submit suspected errors using PPI's errata reporting website, **ppi2pass.com/errata**.

Mohammad Iqbal, D.Sc., PE, SE, Esq.
Ali A. Iqbal, PE

How to Use This Book

Use this book to practice solving problems by using the *NCEES FE Reference Handbook* as your sole reference. The *Handbook* is the only reference you may use during the NCEES FE Civil exam. During the exam, no external codes and standards are required or permitted, even for code-intensive knowledge areas such as steel and concrete design.

You should be familiar with the *Handbook* before you begin to take the practice exams in this book. Know which exam knowledge areas cover which particular civil engineering concepts and formulas. Then use the review materials of your choice to study those concepts and formulas.

Attempt to solve each problem on your own, using only the *Handbook* as a reference. If you are unsure where to begin on any particular problem, use the tables found in the introductory portion of this book to find the sections in the *Handbook* that each problem references. If you are still unable to solve the problem, review the first few lines of the solution, and see if you can do the rest on your own. If you cannot solve several similar problems, go back and review your study materials, and then retry the problems. Once you feel comfortable solving problems within a given exam, move on to the next exam in the book.

Give yourself plenty of time to study and test yourself; begin at least three months before the exam. The time required to study will vary based on how long you have been out of school and whether your knowledge is general or specialized. Create a study schedule for yourself and stick to it. If you give yourself adequate time to study, if you study hard, and if you work problems until you are sure you understand how to navigate the *Handbook* quickly and efficiently, you will give yourself the best possible chance to pass the NCEES FE Civil exam.

NCEES Handbook Sections by Problem Number

This section shows the relationship between problems in this book and the *NCEES FE Reference Handbook*, version 10.0.1, published for the FE Civil exam. The *Handbook* is the only material allowed as a reference during the exam. The exact titles of chapters and sections from the *Handbook* are given in this section so that they can be used as search terms to find the equations, tables, and other information that you need. If you use these search terms as you study, you will come to learn the ones you use most often, and this will save you time during the exam.

When taking these practice exams, use the *Handbook* as you would for the real exam. Then, while reviewing the solutions, use the tables that follow to find the relevant chapters and sections of the *Handbook*. Be sure to especially familiarize yourself with any problems that you missed.

A word of caution: Subject matter that does *not* appear in the *Handbook* may still appear on the exam. In this section, a *Handbook* chapter without a *Handbook* section indicates that the subject matter as a whole is important for the exam but that there is no specific *Handbook* equation or table that you need for that problem. Familiarize yourself with the exam specifications, in addition to the *Handbook*, to prepare for the sorts of questions you may see.

NCEES Handbook **Sections by Problem Number: Exam 1**

problem	chapter	section
1	Mathematics	Straight Line
2	Mathematics	Trigonometry: Identities
3	Mathematics	Vectors
4	Mathematics	Arithmetic Progression
5	Mathematics	Law of Cosines
6	Mathematics	The Derivative
7	Mathematics	Conic Sections: Ellipse
	Statics	Area and Centroid (Table)
8	Engineering Probability and Statistics	Permutations and Combinations
9	Engineering Probability and Statistics	Dispersion, Mean, Median, and Mode Values
10	Engineering Probability and Statistics	Normal Distribution (Gaussian Distribution)
11	Ethics and Professional Practice	Code of Ethics
12	Ethics and Professional Practice	Code of Ethics
13	Ethics and Professional Practice	Code of Ethics
14	Ethics and Professional Practice	Code of Ethics
15	Ethics and Professional Practice	
16	Engineering Economics	Engineering Economics: Factor Table
17	Engineering Economics	Capitalized Costs
18	Engineering Economics	
19	Engineering Economics	
20	Engineering Economics	Engineering Economics: Factor Table
21	Engineering Economics	
22	Statics	Resolution of a Force
23	Statics	Force
	Units and Conversion Factors	
24	Statics	Resultant (Two Dimensions)
	Mathematics	Law of Sines
25	Statics	Equilibrium Requirements
26	Statics	Systems of Forces Equilibrium Requirements
27	Statics	Centroids of Masses, Areas, Lengths, and Volumes
28	Statics	Area Moment of Inertia (Table)
29	Statics	Friction
30	Dynamics	Particle Rectilinear Motion
31	Dynamics	Torsional Vibration Rigid Body Motion About a Fixed Axis
32	Dynamics	Torsional Vibration Mass Moment of Inertia (Table)
33	Dynamics	Particle Kinetics
34	Dynamics	Potential Energy
35	Mechanics of Materials	Shearing Force and Bending Moment Sign Conventions
36	Statics	Moments (Couples)
	Mechanics of Materials	Shearing Force and Bending Moment Sign Conventions
37	Mechanics of Materials	Uniaxial Loading and Deformation

problem	chapter	section
38	Mechanics of Materials	Shear Stress-Strain Typical Material Properties (Table)
	Units and Conversion Factors	
39	Mechanics of Materials	Composite Sections
40	Mechanics of Materials	Simply Supported Beam Slopes and Deflections
41	Mechanics of Materials	Mohr's Circle – Stress, 2D
42	Mechanics of Materials	Columns
43	Materials Science/Structure of Matter	
44	Materials Science/Structure of Matter	
45	Materials Science/Structure of Matter	
46	Materials Science/Structure of Matter	
47	Materials Science/Structure of Matter	
48	Fluid Mechanics	Minor Losses in Pipe Fittings, Contractions, and Expansions
49	Fluid Mechanics	The Pressure Field in a Static Liquid
50	Fluid Mechanics	
51	Fluid Mechanics	
52	Fluid Mechanics	Reynolds Number
53	Fluid Mechanics	
54	Mathematics	Law of Cosines
55	Mathematics	Law of Sines
56	Mathematics	Regular Polygon (n equal sides)
57	Civil Engineering	Earthwork Formulas
58	Civil Engineering	
59	Civil Engineering	
60	Civil Engineering: Hydrology/Water Resources	NRCS (SCS) Rainfall-Runoff
61	Civil Engineering: Hydrology/Water Resources	
62	Fluid Mechanics	Bernoulli Equation
63	Fluid Mechanics	Pump Power Equation
64	Fluid Mechanics	Head Loss Due to Flow
65	Civil Engineering	
66	Civil Engineering	NRCS (SCS) Rainfall-Runoff
67	Civil Engineering	Darcy's Law
68	Fluid Mechanics	Open-Channel Flow and/or Pipe Flow of Water Flow in Noncircular Conduits
	Civil Engineering	Hydraulic-Elements Graph for Circular Sewers
69	Environmental Engineering	Population Projection Equations: Log Growth = Exponential Growth = Geometric Growth
70	Environmental Engineering	Clarifier
71	Civil Engineering: Structural Analysis	Stability, Determinacy, and Classification of Structures
72	Mechanics of Materials	Simply Supported Beam Slopes and Deflections
73	Civil Engineering: Structural Design	Design of Steel Components: Columns W Shapes Dimensions and Properties (Table)
74	Civil Engineering: Structural Analysis	

problem	chapter	section
75	Civil Engineering: Structural Analysis	Beam Stiffness and Moment Carryover
76	Civil Engineering: Structural Analysis	Beam Stiffness and Moment Carryover
77	Civil Engineering: Structural Analysis	Influence Lines for Beams and Trusses
78	Civil Engineering: Structural Design	W Shapes Dimensions and Properties
79	Civil Engineering: Structural Analysis	
80	Civil Engineering: Structural Design	W Shapes Dimensions and Properties
81	Mechanics of Materials	
82	Mechanics of Materials	Uniaxial Loading and Deformation
83	Civil Engineering: Geotechnical	Phase Relationships
84	Civil Engineering: Geotechnical	Phase Relationships
85	Civil Engineering: Geotechnical	Phase Diagram
86	Civil Engineering: Geotechnical	Phase Relationships
87	Civil Engineering: Geotechnical	Vertical Stress Profiles
88	Civil Engineering: Geotechnical	Horizontal Stress Profiles and Forces
89	Civil Engineering: Geotechnical	
90	Civil Engineering: Geotechnical	
91	Civil Engineering: Geotechnical	Mohr-Coulomb Failure
92	Civil Engineering: Geotechnical	Soil Consolidation Curve Over Consolidated Clay
93	Civil Engineering: Geotechnical	Slope Failure Along Planar Surface
94	Civil Engineering: Transportation	Horizontal Curves
95	Civil Engineering: Transportation	Stopping Sight Distance
96	Civil Engineering: Transportation	
97	Civil Engineering: Transportation	Highway Pavement Design
98	Civil Engineering: Transportation	Basic Freeway Segment Highway Capacity
99	Civil Engineering: Transportation	Traffic Signal Timing
100	Civil Engineering: Transportation	
101	Civil Engineering: Transportation	
102	Civil Engineering: Transportation	Traffic Safety Equations: Crash Rates for Roadway Segments
103	Civil Engineering: Construction	
104	Civil Engineering: Construction	
105	Civil Engineering: Construction	
106	Civil Engineering: Construction	
107	Civil Engineering: Construction	CPM Precedence Relationships
108	Civil Engineering: Construction	
109	Civil Engineering: Construction	
110	Civil Engineering: Construction	

NCEES Handbook **Sections by Problem Number: Exam 2**

problem	chapter	section
111	Mathematics	Straight Line
112	Mathematics	Determinants
113	Mathematics	Integral Calculus Indefinite Integrals Mensuration of Areas and Volumes
114	Mathematics	Quadratic Equation
115	Mathematics	Geometric Progression
116	Mathematics	Derivatives
117	Mathematics	Vectors Vectors: Identities
118	Engineering Probability and Statistics	Permutations and Combinations
119	Engineering Probability and Statistics	Dispersion, Mean, Median, and Mode Values
120	Engineering Probability and Statistics	Dispersion, Mean, Median, and Mode Values
121	Ethics and Professional Practice	Code of Ethics
122	Ethics and Professional Practice	Code of Ethics
123	Ethics and Professional Practice	Model Rules, Section 240.15 Rules of Professional Conduct Model Law, Section 150.10, Grounds for Disciplinary Action–Licensees and Interns
124	Ethics and Professional Practice	Model Rules, Section 240.15 Rules of Professional Conduct
125	Ethics and Professional Practice	
126	Engineering Economics	Engineering Economics: Factor Table Engineering Economics: Nomenclature and Definitions Interest Rate Tables
127	Engineering Economics	Engineering Economics: Factor Table Engineering Economics: Nomenclature and Definitions Interest Rate Tables Non-Annual Compounding
128	Engineering Economics	
129	Mathematics	Derivative: Test for a Maximum
	Engineering Economics	
130	Engineering Economics	Engineering Economics: Factor Table
131	Engineering Economics	Economic Decision Trees
132	Statics	Resultant (Two Dimensions)
133	Statics	Equilibrium Requirements
134	Statics	Plane Truss: Method of Joints
135	Statics	
136	Statics	Resolution of a Force Moments (Couples) Systems of Forces Equilibrium Requirements
137	Statics	Centroids of Masses, Areas, Lengths, and Volumes
138	Statics	Area Moment of Inertia (Table)
139	Statics	Friction
140	Dynamics	Particle Rectilinear Motion

problem	chapter	section
141	Dynamics	Impact
142	Dynamics	Mass Moment of Inertia (Table)
143	Dynamics	Free and Forced Vibration Concept of Weight
144	Dynamics	Plane Circular Motion
145	Mechanics of Materials	Simply Supported Beam Slopes and Deflections
146	Mechanics of Materials	Uniaxial Loading and Deformation
147	Mechanics of Materials	Typical Material Properties (Table) Shear Stress-Strain
	Units and Conversion Factors	Units and Conversion Factors
148	Mechanics of Materials	Shear Stress-Strain
149	Mechanics of Materials	Thermal Deformations
150	Mechanics of Materials	Mohr's Circle – Stress, 2D
151	Mechanics of Materials	Composite Sections
152	Mechanics of Materials	Torsional Strain
153	Materials Science/Structure of Matter	Concrete
154	Materials Science/Structure of Matter	Concrete
155	Materials Science/Structure of Matter	Concrete
156	Materials Science/Structure of Matter	
	Mechanics of Materials	
157	Materials Science/Structure of Matter	Properties of Materials: Mechanical
158	Fluid Mechanics	The Pressure Field in a Static Liquid Density, Specific Volume, Specific Weight, and Specific Gravity
159	Fluid Mechanics	Density, Specific Volume, Specific Weight, and Specific Gravity
160	Fluid Mechanics	Forces on Submerged Surfaces and the Center of Pressure
161	Fluid Mechanics	Forces on Submerged Surfaces and the Center of Pressure
162	Fluid Mechanics	Forces on Submerged Surfaces and the Center of Pressure
	Statics	Area and Centroid (Table)
163	Fluid Mechanics	Similitude
164	Mathematics	Trigonometry
165	Civil Engineering: Transportation	Horizontal Curves
166	Civil Engineering: Transportation	Earthwork Formulas
167	Civil Engineering: Transportation	Latitudes and Departures
168	Civil Engineering: Transportation	
169	Civil Engineering: Transportation	Horizontal Curves
170	Civil Engineering: Hydrology/Water Resources	Rational Formula
171	Fluid Mechanics	Flow in Noncircular Conduits Open-Channel Flow and/or Pipe Flow of Water: Manning's Equation
172	Fluid Mechanics	Pump Power Equation
	Units and Conversion Factors	Units and Conversion Factors
173	Fluid Mechanics	Minor Losses in Pipe Fittings, Contractions, and Expansions Continuity Equation

problem	chapter	section
174	Units and Conversion Factors	Units and Conversion Factors
	Fluid Mechanics	Orifice Discharging Freely into Atmosphere
175	Civil Engineering: Hydrology/Water Resources	
176	Civil Engineering: Hydrology/Water Resources	Darcy's Law
177	Civil Engineering: Hydrology/Water Resources	Dupuit's Formula
178	Environmental Engineering	
179	Environmental Engineering	Clarifier
180	Environmental Engineering	
181	Statics	Equilibrium Requirements
182	Mechanics of Materials	Cantilevered Beam Slopes and Deflections
183	Statics	Radius of Gyration Area Area Moment of Inertia (Table)
184	Civil Engineering: Structural Analysis	Stability, Determinacy, and Classification of Structures Plane Truss
185	Civil Engineering: Structural Analysis	Member Fixed-End Moments (Magnitudes)
186	Civil Engineering: Structural Analysis	Member Fixed-End Moments (Magnitudes)
187	Dynamics	Kinetic Energy
188	Mechanics of Materials	Columns
189	Civil Engineering: Structural Analysis	
	Mechanics of Materials	
190	Civil Engineering: Structural Design	
191	Civil Engineering: Structural Design	
	Mechanics of Materials	
192	Civil Engineering: Structural Design	
193	Civil Engineering: Geotechnical	AASHTO Soil Classification
194	Statics	Equilibrium Requirements
195	Civil Engineering: Geotechnical	Phase Relationships
196	Civil Engineering: Geotechnical	Phase Relationships
197	Civil Engineering: Geotechnical	Phase Relationships
198	Civil Engineering: Geotechnical	
199	Civil Engineering: Geotechnical	Mohr-Coulomb Failure
200	Civil Engineering: Geotechnical	
201	Civil Engineering: Geotechnical	Phase Relationships
202	Civil Engineering: Geotechnical	Phase Relationships
203	Civil Engineering: Geotechnical	
204	Civil Engineering: Transportation	Horizontal Curves
205	Civil Engineering: Transportation	Stopping Sight Distance
	Dynamics	Impulse and Momentum: Friction
206	Civil Engineering: Transportation	Horizontal Curves
207	Civil Engineering: Transportation	
208	Civil Engineering: Transportation	Traffic Signal Timing
209	Civil Engineering: Transportation	Traffic Signal Timing
210	Civil Engineering: Transportation	
211	Civil Engineering: Transportation	Greenshields Model
212	Civil Engineering: Transportation	Traffic Safety Equations: Crash Rates at Intersections

problem	chapter	section
213	Civil Engineering: Construction	
214	Civil Engineering: Construction	
215	Civil Engineering: Construction	
216	Civil Engineering: Construction	
217	Civil Engineering: Construction	Construction: Earned-Value Analysis
218	Civil Engineering: Construction	CPM Precedence Relationships
219	Civil Engineering: Construction	
220	Civil Engineering: Construction	

NCEES Handbook Sections by Problem Number: Exam 3

problem	chapter	section
221	Mathematics	Straight Line
222	Mathematics	Logarithms
223	Mathematics	Integral Calculus
224	Mathematics	Vectors
225	Mathematics	Geometric Progression
226	Mathematics	Curvature in Rectangular Coordinates
227	Engineering Probability and Statistics	Law of Compound or Joint Probability
228	Engineering Probability and Statistics	Dispersion, Mean, Median, and Mode Values
229	Engineering Probability and Statistics	Dispersion, Mean, Median, and Mode Values
230	Engineering Probability and Statistics	Least Squares
231	Ethics and Professional Practice	Model Rules, Section 240.15 Rules of Professional Conduct
232	Ethics and Professional Practice	Model Rules, Section 240.15 Rules of Professional Conduct Model Law, Section 150.10, Grounds for Disciplinary Action–Licensees and Interns
233	Ethics and Professional Practice	
234	Ethics and Professional Practice	Model Rules, Section 240.15 Rules of Professional Conduct
235	Ethics and Professional Practice	
236	Engineering Economics	Engineering Economics: Factor Table Engineering Economics: Nomenclature and Definitions Interest Rate Tables
237	Engineering Economics	
238	Engineering Economics	Economic Decision Trees Interest Rate Tables
239	Engineering Economics	Breakeven Analysis
	Mathematics	Straight Line
240	Engineering Economics	Interest Rate Tables
	Mathematics	Straight Line
241	Engineering Economics	Economic Decision Trees
242	Statics	Resolution of a Force Equilibrium Requirements
243	Statics	Resolution of a Force Equilibrium Requirements
	Mathematics	Trigonometry
244	Statics	
245	Statics	
	Dynamics	Concept of Weight
	Mathematics	Trigonometry
246	Statics	Equilibrium Requirements
247	Statics	Centroids of Masses, Areas, Lengths, and Volumes
248	Statics	Moment of Inertia Parallel Axis Theorem Area Moment of Inertia (Table)
249	Statics	Friction

problem	chapter	section
250	Dynamics	Projectile Motion
251	Dynamics	Dynamics: Radius of Gyration (Table)
252	Dynamics	Kinetic Energy Mass Moment of Inertia (Table)
253	Dynamics	Plane Circular Motion
254	Dynamics	Kinetic Energy
255	Mechanics of Materials: Beams	Shearing Force and Bending Moment Sign Conventions
	Civil Engineering: Structural Design	
256	Mechanics of Materials	
257	Mechanics of Materials	
258	Mechanics of Materials	Torsion Stresses in Beams
259	Mechanics of Materials	Uniaxial Loading and Deformation
260	Mechanics of Materials	Thermal Deformations
261	Mechanics of Materials	Mohr's Circle – Stress, 2D
262	Mechanics of Materials	Stress-Strain Curve for Mild Steel Engineering Strain Typical Material Properties (Table)
263	Material Science/Structure of Matter	Concrete
264	Material Science/Structure of Matter	
265	Material Science/Structure of Matter	Concrete
266	Material Science/Structure of Matter	
267	Material Science/Structure of Matter	
268	Fluid Mechanics	The Pressure Field in a Static Liquid
269	Fluid Mechanics	Density, Specific Volume, Specific Weight, and Specific Gravity The Pressure Field in a Static Liquid
270	Fluid Mechanics	The Pressure Field in a Static Liquid
271	Fluid Mechanics	Impulse-Momentum Principle
272	Fluid Mechanics	Deflectors and Blades
273	Fluid Mechanics	Properties of Water (English Units)
274	Mathematics	
	Civil Engineering: Transportation	Horizontal Curves
275	Civil Engineering: Transportation	Area Formulas
276	Mathematics	Right Circular Cone
	Civil Engineering: Transportation	Earthwork Formulas
277	Civil Engineering: Transportation	Latitudes and Departures
278	Civil Engineering: Transportation	Vertical Curves
279	Civil Engineering: Transportation	Vertical Curves
280	Civil Engineering: Hydrology/Water Resources	Rational Formula
281	Civil Engineering: Hydrology/Water Resources	Manning's Equation Flow in Noncircular Conduits

problem	chapter	section
282	Fluid Mechanics	Bernoulli Equation Turbines Pump Performance Curves Flow Through a Packed Bed
283	Fluid Mechanics	Multipath Pipeline Problems Head Loss Due to Flow
284	Civil Engineering: Hydrology/Water Resources	Weir Formulas: Rectangular
285	Civil Engineering: Hydrology/Water Resources	Darcy's Law
286	Environmental Engineering	
287	Environmental Engineering	Microbial Kinetics: BOD Exertion
288	Environmental Engineering	Clarifier
289	Environmental Engineering	Lime-Soda Softening Equations
290	Environmental Engineering	
291	Civil Engineering: Structural Analysis	
	Statics	Equilibrium Requirements
292	Dynamics	Free and Forced Vibration
	Civil Engineering: Structural Analysis	
293	Civil Engineering: Structural Design	Available Strength in Axial Compression, Kips–W Shapes Design of Steel Components: Columns
	Mechanics of Materials	
294	Civil Engineering: Structural Analysis	Stability, Determinacy, and Classification of Structures Plane Truss
295	Civil Engineering: Structural Analysis	W Shapes Dimensions and Properties Beam Stiffness and Moment Carryover
296	Civil Engineering: Structural Analysis	Member Fixed-End Moments (Magnitudes) Beam Stiffness and Moment Carryover
297	Civil Engineering: Structural Analysis	
	Dynamics	Particle Kinetics
298	Civil Engineering: Structural Design	
299	Civil Engineering: Structural Design	Flat Bars or Angles, Bolted or Welded: Definitions Flat Bars or Angles, Bolted or Welded: Limit States and Available Strengths
300	Civil Engineering: Structural Design	
301	Civil Engineering: Structural Design	Nominal Column Strength Interaction Diagram for Rectangular Section with Bars on End Faces and $\gamma = 0.80$
302	Civil Engineering: Structural Design	Design of Reinforced Concrete Components: Resistance Factors, phi
303	Civil Engineering: Geotechnical	Standard Practice for Classification of Soils for Engineering Purposes (Unified Soil Classification System)
304	Civil Engineering: Geotechnical	Phase Relationships

problem	chapter	section
305	Civil Engineering: Geotechnical	Phase Relationships Phase Diagram
306	Civil Engineering: Geotechnical	Phase Relationships
307	Civil Engineering: Geotechnical	Vertical Stress Profiles
308	Civil Engineering: Geotechnical	Horizontal Stress Profiles and Forces
309	Civil Engineering: Geotechnical	Mohr-Coulomb Failure
310	Civil Engineering: Geotechnical	
311	Civil Engineering: Geotechnical	
312	Civil Engineering: Geotechnical	Horizontal Stress Profiles and Forces
313	Civil Engineering: Geotechnical	
314	Civil Engineering: Transportation	
	Dynamics	Constant Acceleration
315	Civil Engineering: Transportation	Vertical Curves: Sight Distance Related to Curve Length
316	Civil Engineering: Transportation	Vertical Curves: Sight Distance Related to Curve Length
317	Civil Engineering: Transportation	AASHTO Structural Number Equation Highway Pavement Design
318	Civil Engineering: Transportation	Basic Freeway Segment Highway Capacity
319	Civil Engineering: Transportation	
320	Civil Engineering: Transportation	Traffic Safety Equations: Crash Rates at Intersections
321	Civil Engineering: Transportation	Greenshields Model
322	Civil Engineering: Transportation	Greenshields Model
323	Civil Engineering: Construction	
324	Civil Engineering: Construction	
325	Civil Engineering: Construction	
326	Civil Engineering: Construction	
327	Civil Engineering: Construction	Construction: Earned-Value Analysis
328	Civil Engineering: Construction	Construction: Earned-Value Analysis
329	Civil Engineering: Construction	
330	Civil Engineering: Construction	

NCEES Handbook Sections by Problem Number: Exam 4

problem	chapter	section
331	Mathematics	Determinants
332	Mathematics	Logarithms: Identities
333	Mathematics	Quadratic Equation
334	Mathematics	
335	Mathematics	
336	Mathematics	Addition of Two Matrices
337	Mathematics	Curvature of Any Curve
338	Engineering Probability and Statistics	Binomial Distribution
339	Engineering Probability and Statistics	Standard Error of Estimate (S_e^2)
340	Engineering Probability and Statistics	Residual
341	Ethics and Professional Practice	Code of Ethics
342	Ethics and Professional Practice	Model Rules, Section 240.15 Rules of Professional Conduct
343	Ethics and Professional Practice	Code of Ethics Model Law, Section 130.10 General Requirements for Licensure
344	Ethics and Professional Practice	Model Law, Section 150.10, Grounds for Disciplinary Action–Licensees and Interns
345	Ethics and Professional Practice	
346	Engineering Economics	Bonds Interest Rate Tables
347	Engineering Economics	Non-Annual Compounding
348	Engineering Economics	Breakeven Analysis
349	Engineering Economics	Interest Rate Tables
350	Engineering Economics	Economic Decision Trees
351	Engineering Economics	Economic Decision Trees Interest Rate Tables
352	Mechanics of Materials	Torsion Shear Stress-Strain
353	Statics	Resolution of a Force Equilibrium Requirements
354	Statics	Resolution of a Force Systems of Forces Equilibrium Requirements
355	Units and Conversion Factors	
	Mechanics of Materials	Material Properties
356	Statics	Systems of Forces Equilibrium Requirements
357	Statics	Systems of Forces
358	Statics	Area Moment of Inertia (Table)
359	Statics	Moment of Inertia
360	Statics	Friction
361	Statics	Friction
362	Dynamics	Particle Kinetics
363	Dynamics	Mass Moment of Inertia (Table) Equations of Motion
364	Dynamics	Free and Forced Vibration Plane Circular Motion Particle Kinetics
365	Dynamics	Work

problem	chapter	section
366	Mechanics of Materials	Shearing Force and Bending Moment Sign Conventions
367	Mechanics of Materials	Cylindrical Pressure Vessel
368	Mechanics of Materials	Torsional Strain: For Hollow, Thin-Walled Shafts
369	Statics	Plane Truss: Method of Joints Resolution of a Force
370	Mechanics of Materials	Uniaxial Loading and Deformation Columns
371	Statics	Area Moment of Inertia (Table)
372	Statics	Moment of Inertia
373	Material Science/Structure of Matter	Concrete
374	Material Science/Structure of Matter	Concrete
375	Material Science/Structure of Matter	Concrete
376	Material Science/Structure of Matter	Concrete
377	Civil Engineering: Structural Design	Design of Reinforced Concrete Components: Definitions
378	Fluid Mechanics	Continuity Equation
379	Fluid Mechanics	Bernoulli Equation
380	Fluid Mechanics	Stress, Pressure, and Viscosity
381	Fluid Mechanics	Pitot Tube Venturi Meters Orifices
382	Fluid Mechanics	Multipath Pipeline Problems
383	Fluid Mechanics	Forces on Submerged Surfaces and the Center of Pressure
384	Civil Engineering: Transportation	Horizontal Curves
385	Civil Engineering: Transportation	Horizontal Curves
386	Civil Engineering: Transportation	Earthwork Formulas
387	Civil Engineering: Transportation	Latitudes and Departures
388	Civil Engineering: Transportation	Vertical Curves
389	Civil Engineering: Transportation	Vertical Curves
390	Civil Engineering: Hydrology/Water Resources	NRCS (SCS) Rainfall-Runoff
391	Fluid Mechanics	Open-Channel Flow and/or Pipe Flow of Water: Manning's Equation Flow in Noncircular Conduits
392	Fluid Mechanics	Energy Equation Head Loss Due to Flow
393	Fluid Mechanics	Bernoulli Equation Flow in Closed Conduits Head Loss Due to Flow
394	Civil Engineering: Hydrology/Water Resources	Hazen-Williams Equation Values of Hazen-Williams Coefficient C (Table) Flow in Noncircular Conduits
395	Civil Engineering: Hydrology/Water Resources	Darcy's Law
396	Civil Engineering: Hydrology/Water Resources	Darcy's Law
397	Environmental Engineering	Specific Gravity for a Solids Slurry

problem	chapter	section
398	Environmental Engineering	Clarifier
399	Chemical Engineering	
400	Fluid Mechanics	Continuity Equation
401	Civil Engineering: Structural Analysis	Stability, Determinacy, and Classification of Structures
402	Civil Engineering: Structural Analysis	Plane Truss
403	Civil Engineering: Structural Design	Design of Steel Components: Columns
404	Civil Engineering: Structural Analysis	Truss Deflection by Unit Load Method Typical Material Properties (Table)
405	Mechanics of Materials	Simply Supported Beam Slopes and Deflections
406	Statics	Systems of Forces: Equilibrium Requirements
	Mechanics of Materials	Engineering Strain
407	Civil Engineering: Structural Analysis	Influence Lines for Beams and Trusses
408	Civil Engineering: Structural Design	Design of Steel Components: Lateral-Torsional Buckling
409	Civil Engineering: Structural Design	Design of Steel Components: Lateral-Torsional Buckling
410	Civil Engineering: Structural Design	Design of Steel Components: Lateral-Torsional Buckling Values of C_b for Simply Supported Beams
411	Civil Engineering: Structural Design	Beams-Flexure: Singly-Reinforced Beams
412	Civil Engineering: Structural Design	Beams-Flexure: Singly-Reinforced Beams Design of Reinforced Concrete Components: Resistance Factors, phi
413	Civil Engineering: Geotechnical	AASHTO Soil Classification
414	Civil Engineering: Geotechnical	Phase Relationships
415	Civil Engineering: Hydrology/Water Resources	Darcy's Law
416	Civil Engineering: Geotechnical	Vertical Stress Profiles
417	Civil Engineering: Geotechnical	Horizontal Stress Profiles and Forces
418	Civil Engineering: Geotechnical	Mohr-Coulomb Failure
419	Civil Engineering: Geotechnical	Ultimate Bearing Capacity
420	Civil Engineering: Geotechnical	
421	Civil Engineering: Geotechnical	Horizontal Stress Profiles and Forces
422	Civil Engineering: Geotechnical	Horizontal Stress Profiles and Forces
423	Civil Engineering: Geotechnical	Slope Failure Along Planar Surface
424	Civil Engineering: Transportation	Stopping Sight Distance
425	Civil Engineering: Transportation	Vertical Curves: Sight Distance Related to Curve Length Stopping Sight Distance
426	Civil Engineering: Transportation	Stopping Sight Distance
427	Civil Engineering: Transportation	AASHTO Structural Number Equation
428	Civil Engineering: Transportation	Peak Hour Factor
429	Civil Engineering: Transportation	Traffic Signal Timing

problem	chapter	section
430	Civil Engineering: Transportation	Traffic Flow Relationships
431	Civil Engineering: Transportation	Traffic Safety Equations: Crash Rates at Intersections
432	Civil Engineering: Transportation	Traffic Safety Equations: Crash Reduction
433	Civil Engineering: Construction	
434	Civil Engineering: Construction	
435	Civil Engineering: Construction	Construction: Earned-Value Analysis Variances
436	Civil Engineering: Construction	
437	Civil Engineering: Construction	
438	Civil Engineering: Construction	CPM Precedence Relationships
439	Civil Engineering: Construction	
440	Civil Engineering: Construction	

NCEES Handbook **Sections by Problem Number: Exam 5**

problem	chapter	section
441	Mathematics	Quadratic Equation
442	Mathematics	Arithmetic Progression
443	Mathematics	Multiplication of Two Matrices
444	Mathematics	Algebra of Complex Numbers Conic Sections: Circle
445	Mathematics	Integral Calculus
446	Mathematics	Integral Calculus
447	Mathematics	Vectors
448	Engineering Probability and Statistics	Law of Total Probability Law of Compound or Joint Probability
449	Engineering Probability and Statistics	Binomial Distribution
450	Engineering Probability and Statistics	Dispersion, Mean, Median, and Mode Values
451	Ethics and Professional Practice	Model Rules, Section 240.15 Rules of Professional Conduct
452	Ethics and Professional Practice	
453	Ethics and Professional Practice	Model Law, Section 150.10, Grounds for Disciplinary Action–Licensees and Interns
454	Ethics and Professional Practice	
455	Ethics and Professional Practice	
456	Engineering Economics	Engineering Economics: Factor Table
457	Engineering Economics	Breakeven Analysis Engineering Economics: Factor Table
458	Engineering Economics	Breakeven Analysis
459	Engineering Economics	Engineering Economics: Factor Table
460	Engineering Economics	Benefit-Cost Analysis
461	Engineering Economics	Economic Decision Trees
462	Statics	Systems of Forces Equilibrium Requirements
463	Statics	Equilibrium Requirements
464	Statics	Moments (Couples) Moment of Inertia
465	Statics	Plane Truss: Method of Sections
	Mathematics	Law of Sines
466	Civil Engineering: Structural Analysis	Stability, Determinacy, and Classification of Structures
467	Statics	Plane Truss: Method of Joints
468	Statics	Moments (Couples)
469	Statics	Friction
470	Dynamics	Particle Kinetics
471	Dynamics	Constant Acceleration
472	Dynamics	Mass Moment of Inertia (Table) Torsional Vibration
473	Dynamics	Kinetic Energy
474	Dynamics	Kinetic Energy
475	Mechanics of Materials	Shearing Force and Bending Moment Sign Conventions

problem	chapter	section
476	Mechanics of Materials	
477	Mechanics of Materials	Uniaxial Loading and Deformation
478	Mechanics of Materials	Uniaxial Loading and Deformation
479	Mechanics of Materials	Uniaxial Loading and Deformation
480	Mechanics of Materials	Torsional Strength
481	Mechanics of Materials	Elastic Strain Energy
482	Mechanics of Materials	Shearing Force and Bending Moment Sign Conventions Beam Stiffness and Moment Carryover
483	Material Science/Structure of Matter	Concrete
484	Material Science/Structure of Matter	
485	Material Science/Structure of Matter	Concrete
486	Material Science/Structure of Matter	Concrete
487	Material Science/Structure of Matter	Properties of Materials: Mechanical
488	Fluid Mechanics	Reynolds Number
489	Fluid Mechanics	Bernoulli Equation
490	Fluid Mechanics	Deflectors and Blades
491	Fluid Mechanics	Stress, Pressure, and Viscosity
492	Fluid Mechanics	Forces on Submerged Surfaces and the Center of Pressure
493	Fluid Mechanics	Reynolds Number Drag Coefficient for Spheres, Disks, and Cylinders
494	Mathematics	Law of Sines
495	Civil Engineering: Geotechnical	
496	Civil Engineering: Geotechnical	
497	Civil Engineering: Transportation	Horizontal Curves
498	Civil Engineering: Transportation	Vertical Curves
499	Civil Engineering: Transportation	Vertical Curves
500	Civil Engineering: Hydrology/Water Resources	NRCS (SCS) Rainfall-Runoff
501	Civil Engineering: Hydrology/Water Resources	Manning's Equation Flow in Noncircular Conduits
502	Fluid Mechanics	Continuity Equation
503	Fluid Mechanics	Minor Losses in Pipe Fittings, Contractions, and Expansions Energy Equation
504	Fluid Mechanics	Minor Losses in Pipe Fittings, Contractions, and Expansions Properties of Water (English Units)
505	Fluid Mechanics	Multipath Pipeline Problems
506	Civil Engineering: Hydrology/Water Resources	Manning's Equation
507	Safety	Noise Pollution
	Mathematics	Logarithms
508	Civil Engineering: Hydrology/Water Resources	Pan Evaporation
509	Fluid Mechanics	
510	Engineering Probability and Statistics	

problem	chapter	section
511	Civil Engineering: Structural Analysis	Plane Truss: Method of Joints
	Statics	Systems of Forces
512	Mechanics of Materials	Simply Supported Beam Slopes and Deflections
513	Civil Engineering: Structural Design	Design of Steel Components: Columns AISC Table 4-14
514	Mechanics of Materials	Simply Supported Beam Slopes and Deflections Typical Material Properties (Table)
515	Civil Engineering: Structural Analysis	Influence Lines for Beams and Trusses
516	Civil Engineering: Structural Analysis	Influence Lines for Beams and Trusses
517	Civil Engineering: Structural Design	
518	Civil Engineering: Structural Design	Design of Steel Components: Lateral-Torsional Buckling
519	Civil Engineering: Structural Design	
520	Civil Engineering: Structural Design	Design of Reinforced Concrete Components: Definitions
521	Civil Engineering: Structural Design	Beams-Flexure: Singly-Reinforced Beams
522	Civil Engineering: Structural Design	Design of Reinforced Concrete Components: Definitions
523	Civil Engineering: Geotechnical	AASHTO Soil Classification
524	Civil Engineering: Geotechnical	Phase Relationships
525	Civil Engineering: Geotechnical	Phase Relationships Phase Diagram
526	Civil Engineering: Geotechnical	Phase Relationships Phase Diagram
527	Civil Engineering: Hydrology/Water Resources	Darcy's Law
	Civil Engineering: Geotechnical	Vertical Stress Profiles
528	Fluid Mechanics	Reynolds Number
529	Civil Engineering: Geotechnical	
530	Civil Engineering: Geotechnical	AASHTO Soil Classification
531	Civil Engineering: Geotechnical	Mohr-Coulomb Failure
532	Civil Engineering: Geotechnical	
533	Civil Engineering: Geotechnical	
534	Civil Engineering: Transportation	Stopping Sight Distance
535	Civil Engineering: Transportation	Horizontal Curves
536	Civil Engineering: Transportation	Vertical Curves
537	Civil Engineering: Transportation	
538	Civil Engineering: Transportation	Basic Freeway Segment Highway Capacity
539	Civil Engineering: Transportation	Traffic Signal Timing
540	Civil Engineering: Transportation	Highway Pavement Design
541	Civil Engineering: Transportation	Highway Pavement Design
542	Civil Engineering: Transportation	Basic Freeway Segment Highway Capacity
543	Civil Engineering: Construction	CPM Precedence Relationships
544	Civil Engineering: Construction	Earthwork Formulas
545	Civil Engineering: Construction	
546	Civil Engineering: Construction	

problem	chapter	section
547	Civil Engineering: Construction	
548	Civil Engineering: Construction	Construction: Earned-Value Analysis
549	Civil Engineering: Construction	
550	Civil Engineering: Construction	

Exam 1 Instructions

In accordance with the rules established by your state, you may use any approved battery- or solar-powered, silent calculator to work this examination. However, no blank papers, writing tablets, unbound scratch paper, or loose notes are permitted. Sufficient paper will be provided. The *NCEES FE Reference Handbook* is the only reference you are allowed to use during this exam.

You are not permitted to share or exchange materials with other examinees.

You will have six hours in which to work this session of the examination. Your score will be determined by the number of questions that you answer correctly. There is a total of 110 questions. All 110 questions must be worked correctly in order to receive full credit on the exam. There are no optional questions. Each question is worth 1 point. The maximum possible score for this section of the examination is 110 points.

Partial credit is not available. No credit will be given for methodology, assumptions, or work written on scratch paper.

Record all of your answers on the Answer Sheet. Mark your answers with a no. 2 pencil. Answers marked in pen may not be graded correctly. Marks must be dark and must completely fill the bubbles. Record only one answer per question. If you mark more than one answer, you will not receive credit for the question. If you change an answer, be sure the old bubble is erased completely; incomplete erasures may be misinterpreted as answers.

If you finish early, check your work and make sure that you have followed all instructions. After checking your answers, you may submit your answers and leave the examination room. Once you leave, you will not be permitted to return to work or change your answers.

When permission has been given by your proctor, you may begin your examination.

Name: ______________________________________
Last First Middle Initial

Examinee number: ______________

Examination Booklet number: ______________

Fundamentals of Engineering Examination

Exam 1

Exam 1 Answer Sheet

1. (A) (B) (C) (D)
2. (A) (B) (C) (D)
3. ____________
4. ____________
5. (A) (B) (C) (D)
6. ____________
7. (A) (B) (C) (D)
8. ____________
9. (A) (B) (C) (D)
10. (A) (B) (C) (D)
11. (A) (B) (C) (D) (E)
12. (A) (B) (C) (D)
13. (A) (B) (C) (D) (E)
14. (A) (B) (C) (D) (E)
15. (A) (B) (C) (D)
16. (A) (B) (C) (D)
17. (A) (B) (C) (D)
18. (A) (B) (C) (D)
19. (A) (B) (C) (D)
20. (A) (B) (C) (D)
21. (A) (B) (C) (D)
22. (A) (B) (C) (D)
23. (A) (B) (C) (D)
24. (A) (B) (C) (D)
25. ____________
26. (A) (B) (C) (D)
27. (A) (B) (C) (D)
28. (A) (B) (C) (D)
29. (A) (B) (C) (D)
30. (A) (B) (C) (D)
31. (A) (B) (C) (D)
32. (A) (B) (C) (D)
33. (A) (B) (C) (D)
34. (A) (B) (C) (D)
35. (A) (B) (C) (D)
36. (A) (B) (C) (D)
37. (A) (B) (C) (D)
38. (A) (B) (C) (D)
39. (A) (B) (C) (D)
40. (A) (B) (C) (D)
41. (A) (B) (C) (D)
42. (A) (B) (C) (D)
43. (A) (B) (C) (D)
44. (A) (B) (C) (D)
45. (A) (B) (C) (D)
46. (A) (B) (C) (D)
47. ____________
48. (A) (B) (C) (D)
49. (A) (B) (C) (D)
50. (A) (B) (C) (D)
51. (A) (B) (C) (D)
52. (A) (B) (C) (D)
53. (A) (B) (C) (D)
54. (A) (B) (C) (D)
55. (A) (B) (C) (D)
56. (A) (B) (C) (D)
57. (A) (B) (C) (D)
58. (A) (B) (C) (D)
59. (A) (B) (C) (D)
60. (A) (B) (C) (D)
61. (A) (B) (C) (D)
62. (A) (B) (C) (D)
63. (A) (B) (C) (D)
64. (A) (B) (C) (D)
65. (A) (B) (C) (D)
66. (A) (B) (C) (D)
67. (A) (B) (C) (D)
68. (A) (B) (C) (D)
69. (A) (B) (C) (D)
70. (A) (B) (C) (D)
71. (A) (B) (C) (D)
72. (A) (B) (C) (D)
73. (A) (B) (C) (D)
74. (A) (B) (C) (D)
75. (A) (B) (C) (D)
76. (A) (B) (C) (D)
77. (A) (B) (C) (D)
78. (A) (B) (C) (D)
79. (A) (B) (C) (D)
80. (A) (B) (C) (D)
81. (A) (B) (C) (D)
82. (A) (B) (C) (D)
83. (A) (B) (C) (D)
84. (A) (B) (C) (D)
85. (A) (B) (C) (D)
86. (A) (B) (C) (D)
87. (A) (B) (C) (D)
88. (A) (B) (C) (D)
89. (A) (B) (C) (D)
90. (A) (B) (C) (D)
91. (A) (B) (C) (D)
92. (A) (B) (C) (D)
93. (A) (B) (C) (D)
94. (A) (B) (C) (D)
95. (A) (B) (C) (D)
96. (A) (B) (C) (D)
97. (A) (B) (C) (D)
98. (A) (B) (C) (D)
99. (A) (B) (C) (D)
100. (A) (B) (C) (D)
101. (A) (B) (C) (D)
102. (A) (B) (C) (D)
103. (A) (B) (C) (D)
104. (A) (B) (C) (D)
105. (A) (B) (C) (D)
106. (A) (B) (C) (D)
107. (A) (B) (C) (D)
108. (A) (B) (C) (D)
109. (A) (B) (C) (D)
110. (A) (B) (C) (D)

Exam 1

1. The equation of a line passing through points (1, 2) and (5, 0) is

(A) $2y + x = 5$

(B) $y + 2x = 5$

(C) $y + 3x = 5$

(D) $3y + x = 5$

2. The relation $\tan\theta + \cot\theta$ can be written as

(A) $2\tan^2\theta$

(B) $2\csc^2\theta$

(C) $2\csc 2\theta$

(D) $\csc 2\theta$

3. The area of a parallelogram bounded by vectors from the origin to points (0, 5) and (4, 5) is ______. (Fill in the blank.)

4. A mason builds a wall using bricks and mortar. The bottom row has 30 bricks. Each row has one brick less than the row below it. The wall has 16 rows. The number of bricks used to erect the wall is ______. (Fill in the blank.)

5. A surveyor measures two segments AB and AC to be 120 m and 140 m long, respectively. The angle between them is 36°30′. The perimeter of the triangular lot ABC is most nearly

(A) 104 m

(B) 344 m

(C) 355 m

(D) 7250 m

6. Consider the following limit.

$$\lim_{x\to 1}\frac{x^2+2x-3}{x^2+x-2}$$

The limit equals ______. (Fill in the blank.)

7. The area of the region inside the circle $x^2 + y^2 = 16$ and outside the ellipse $x^2/16 + y^2/9 = 1$ is

(A) 3π

(B) 4π

(C) 9π

(D) 16π

8. A three-digit combination is used to open a safe. The combination is unknown. The lock keypad shows numbers from 0 to 9. The number of possible selections to open the combination lock is ______. (Fill in the blank.)

9. A 500,000 ft^2 area is excavated to construct a shopping mall. The foundation is expected to be partially on sandy silt and partially on rock. One geotechnical report estimates that the foundation is 25% sandy silt, and another report estimates that the foundation is 55% sandy silt. The unit cost to excavate sandy-silt-type material is $100/$ft^2$, and it is $300/$ft^2$ to excavate the rock to the specified depth. The expected foundation cost is most nearly

(A) $100 million

(B) $107 million

(C) $110 million

(D) $115 million

10. A city's rainfall is modeled as a continuous random variable, x. Its probability, using the unit normal distribution, is described using the probability density function as

$$f(x) = \frac{1}{\sqrt{2\pi}}e^{-\frac{1}{2}(x-15)^2}$$

The fraction of the year with rainfall between 14 in and 16 in is most nearly

(A) 0.33

(B) 0.50

(C) 0.68

(D) 0.97

11. Which three statements define the Code of Engineering Ethics?

(A) a set of guidelines that describe how a licensed engineer should behave professionally

(B) a set of aspirations that describe how a licensed engineer should behave professionally

(C) a set of rules that describe a licensed engineer's responsibilities to the public, clients and other licensees

(D) a set of laws that describe how a licensed engineer should behave professionally

(E) ethics are a set of rules that incorporate criminal penalties

12. A licensed engineer's first and foremost responsibility in performance of professional services is to the

(A) clients

(B) employers

(C) customers

(D) public welfare

13. An engineer designed a floor for a client under a written contract. After the floor was constructed, a person tripped and fell on the floor and was injured. The injured person filed a claim against the engineer, alleging that the engineer is liable for plaintiff's injuries. Out of the following, the two statements that are likely to be most correct are

(A) The engineer cannot be sued because he or she did not have a direct contract with the injured person.

(B) The engineer can be sued even though he or she did not have a direct contract with the injured person.

(C) The engineer can only be sued if another engineer issues a certificate stating that the design engineer was at fault.

(D) The engineer can be found liable only if another design firm issues a certificate that the design engineer was at fault.

(E) The engineer can be liable if he or she is found negligent by a court of proper jurisdiction.

14. A state board of licensed engineers found that a licensee violated the state's professional code of ethics. Which one of the following penalties may NOT be imposed on a licensee found violating the code of ethics?

(A) censure and reprimand

(B) license suspension

(C) attainment of additional education or training

(D) $1000 fine for each violation

(E) requirement to move to another state

15. A project was awarded at a lump sum price of $2 million. The retainage was set at 10%. The construction is 95% complete, and the engineer notified the owner that it is substantially complete. The amount the owner should retain at this stage is most nearly

(A) $0

(B) $100,000

(C) $190,000

(D) $200,000

16. A community sold a property for $24 in 1626. If the proceeds were invested at 7% interest compounded yearly, the value of the investment at the end of 2016 would have been most nearly

(A) $70 million

(B) $700 million

(C) $7 billion

(D) $7 trillion

17. An engineer owns a design firm that generates $1 million per year in gross fees. The engineer calculates that the firm receive $100,000 per year in profits after paying all employee salaries, benefits, and other expenses. Assume zero inflation and no growth. If the expected rate of return is 10%, the fair market price of the firm is most nearly

(A) $600,000

(B) $1,000,000

(C) $1,600,000

(D) $6,000,000

18. Which statement about sunk cost is NOT correct?

(A) A company spends $50,000 on a marketing study to see if its new testing tool will succeed in the marketplace. The study concludes that the tool will not be profitable. At this point, the $50,000 is a sunk cost.

(B) A company pays a new recruit $15,000 to join the organization. The individual proves to be unreliable and the company is considering terminating the individual's employment. The $15,000 payment is considered a sunk cost.

(C) A sunk cost is a cost that an entity has incurred and that it can no longer recover.

(D) Sunk costs should be considered when deciding whether or not to continue investing in an ongoing project.

19. The method of considering more than the initial cost of a facility is called

(A) green construction or sustainability costing

(B) modified accelerated recovery

(C) life-cycle costing

(D) carbon emission average costing

20. A machine is covered by the manufacturer's warranty for the first year. After this, the repair cost is expected to be as follows.

cost	amount
at the end of year 2	$1000
at the end of year 3	$2000
at the end of year 4	$3000
at the end of year 5	$4000

If the interest rate is 8%, the annual equivalent repair cost is most nearly

(A) $1850

(B) $2150

(C) $2250

(D) $2450

21. Your client, Gr8 Oil, must decide whether to drill for oil at a particular location for which it has drilling rights. There are only two possible outcomes: striking oil and not striking oil. If the decision is made to drill and Gr8 strikes oil, the oil would be worth $13,000,000. The cost of drilling would be $3,000,000. At present, Gr8 can sell the drilling rights to a prospector for $1,000,000, with a contingency to receive another $1,000,000 if the prospector strikes the oil. Based on the geological assessment, the probability of striking the oil at the location is 0.30. If Gr8 decides to drill, the maximum payoff to Gr8 is most nearly

(A) $3,000,000

(B) $7,000,000

(C) $10,000,000

(D) $13,000,000

22. A bolt is being pulled by two forces, A and B, as shown.

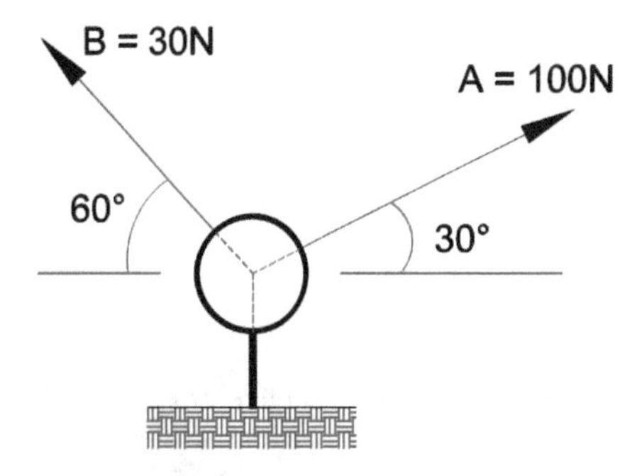

Which of the following diagrams correctly graphically represents the resultant **R**?

(A) 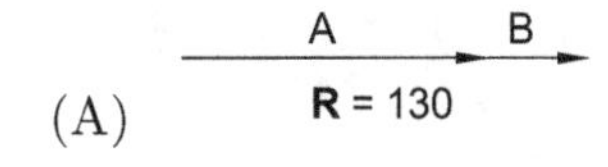

(B) 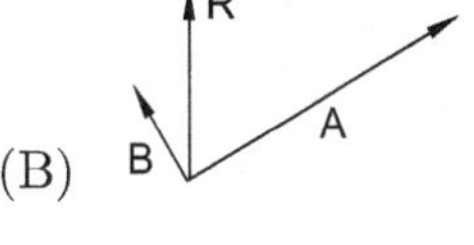

(C) 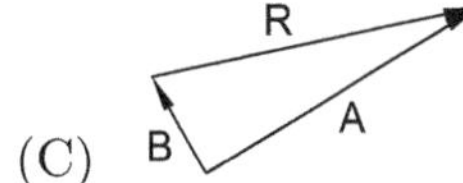

(D) 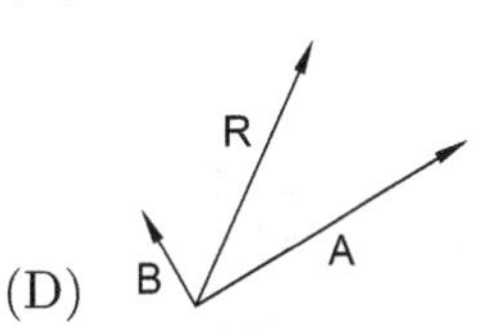

23. A pizza that has a mass of 1 kg is sitting still before it is sliced. The gravitational force the pizza exerts on the pan it sits on is

(A) 0.1 N

(B) 1 N

(C) 10 N

(D) 30 N

24. Two tugboats, A and B, are pulling a barge, as shown.

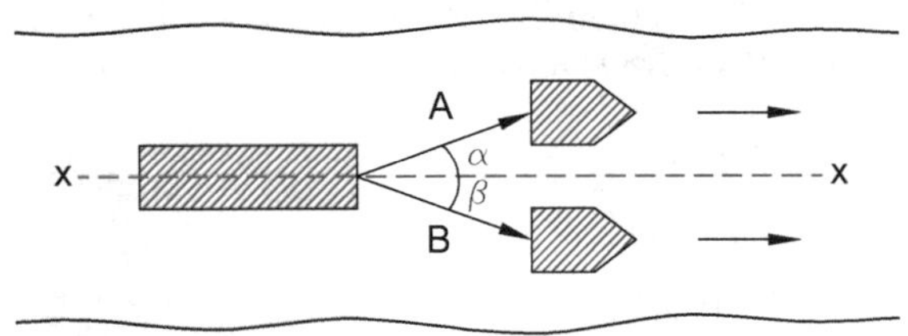

The rope tension for tugboat B is required to be two-thirds of the tension for tugboat A. If rope A makes an angle of 30 degrees with respect to the x-axis, the angle rope B should have to keep the barge moving along the x-axis is most nearly

(A) 27°

(B) 31°

(C) 45°

(D) 49°

25. A two-member truss is carrying a 4 kN load, as shown.

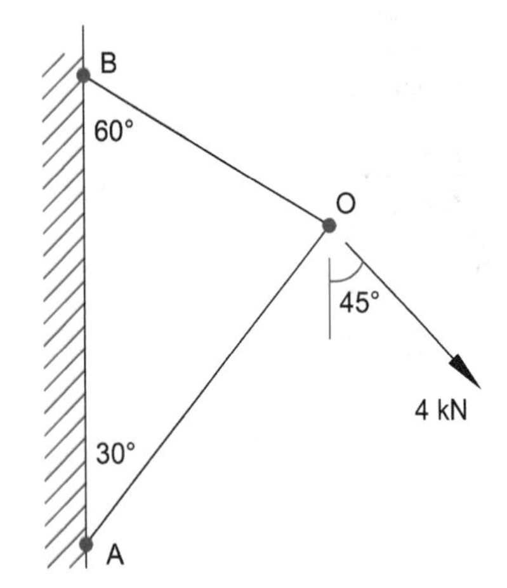

Determine whether the members are in tension or compression, and find the member forces F_{AO} and F_{BO}. Fill in the table from the options given below. (Not all options will be used.)

- tension
- compression
- 1.04 kN
- 3.86 kN
- 4.90 kN

	member	
	AO	BO
tension or compression		
force magnitude (kN)		

26. Cantilever AB is being propped by cable BC, as shown.

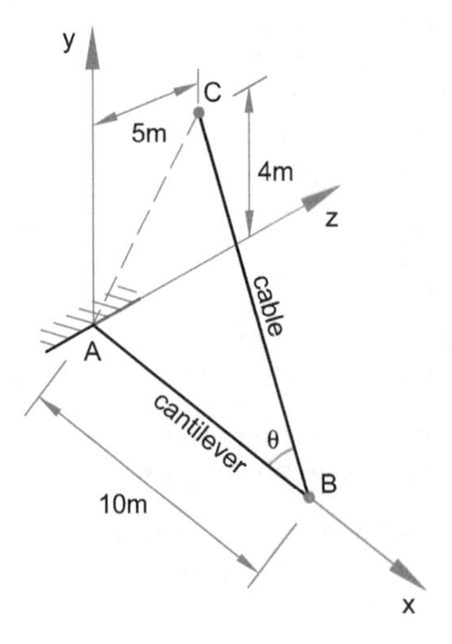

The tension in cable BC is 9 kN. If counterclockwise moment is considered positive, the moment the force exerts about support A is most nearly

(A) –81 kN-m

(B) –49 kN-m

(C) 49 kN-m

(D) 81 kN-m

27. An L-shaped section ABCDEF is shown.

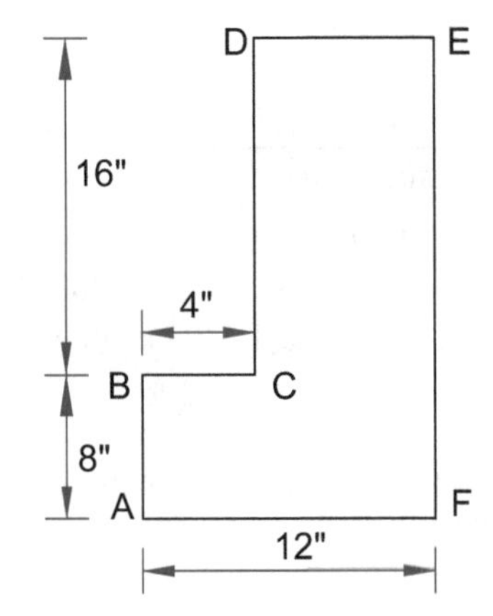

The shortest distance between the centroid and side AF is most nearly

(A) 10 in

(B) 11 in

(C) 12 in

(D) 13 in

28. A steel girder is made from three welded plates, as shown.

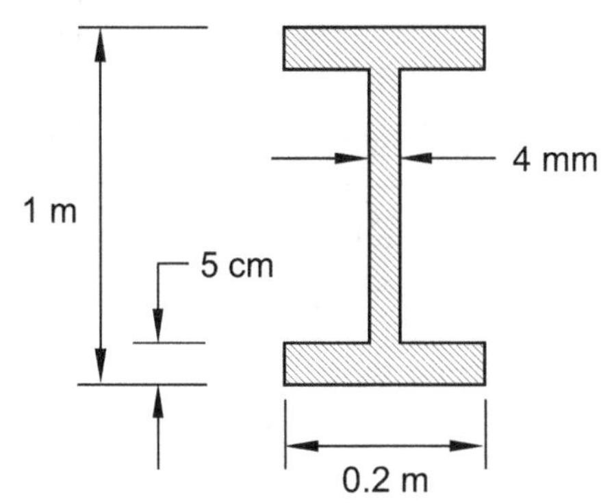

The moment of inertia about the major axis of the girder is most nearly

(A) 5×10^{-4} m^4

(B) 33×10^{-4} m^4

(C) 48×10^{-4} m^4

(D) 33×10^{-3} m^4

29. A 100 kg block is placed on a flat plane. The friction coefficient at the interface between the block and the plane surface is 0.3. The limiting friction is most nearly

(A) 3.5 N

(B) 29 N

(C) 32 N

(D) 290 N

30. A ball is thrown straight upward from a height of 5 ft above the ground with an initial speed of 60 ft/sec. The maximum height the ball reaches above the ground is most nearly

(A) 56 ft

(B) 61 ft

(C) 110 ft

(D) 120 ft

31. A constant torque of 0.1 N·m is applied to an initially stationary flywheel with a moment of inertia of 3.14 kg·m^2. At the end of its tenth revolution, the rotational speed of the flywheel is most nearly

(A) 0.20 rad/s

(B) 2.0 rad/s

(C) 3.1 rad/s

(D) 31 rad/s

32. A circular disk weighs 50 lbf and has a diameter of 18 in. A rod of negligible mass is used to support the disk. The system's period of torsional vibration is 1 sec. The torsional stiffness of the rod is most nearly

(A) 1.8 ft-lbf/rad

(B) 3.2 ft-lbf/rad

(C) 8.5 ft-lbf/rad

(D) 17 ft-lbf/rad

33. An elevator is designed to carry a 1000 lbf load, including its self-weight. The elevator accelerates upward at a constant rate of 10 ft/sec^2. If the elevator is hung by a single cable, the tension load in the cable is most nearly

(A) 1000 lbf

(B) 1320 lbf

(C) 1550 lbf

(D) 1750 lbf

34. A spring is 12 in long. A 1 kip force is used to compress it by 4 in. Assume that the spring acts linearly. The work done in compressing the spring an additional 4 in is most nearly

(A) 2 in-kips

(B) 4 in-kips

(C) 6 in-kips

(D) 8 in-kips

35. A simply supported beam is subjected to two point-loads, as shown.

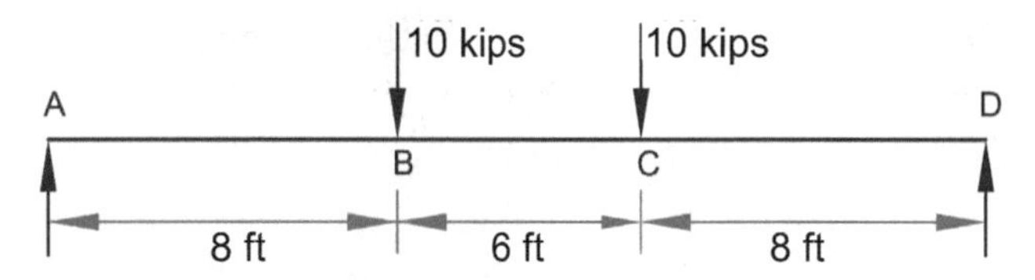

The shear force at the midspan of the beam is

(A) 0 kips

(B) 10 kips

(C) 15 kips

(D) 20 kips

36. Consider the frame shown.

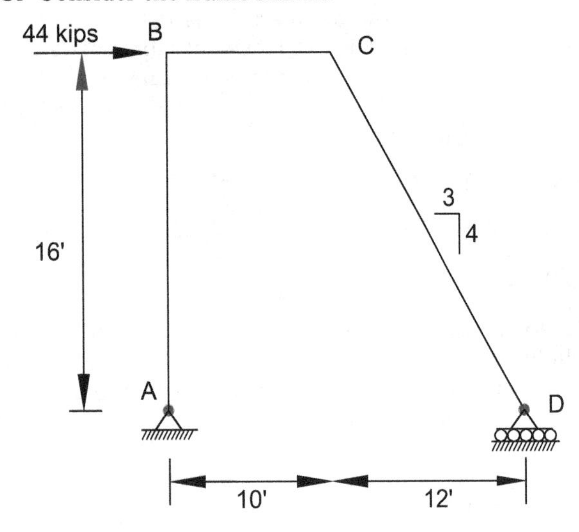

The bending moment at point C is

(A) 0 ft-kips

(B) 32 ft-kips

(C) 384 ft-kips

(D) 704 ft-kips

37. A steel bar is pulled by a 300 kN axial force. If the bar is 1 m long and has a diameter of 3 cm, its elongation is most nearly

(A) 1 mm

(B) 2 mm

(C) 10 mm

(D) 27 mm

38. 1 m long, 20 cm wide, 20 mm thick aluminum bar is pulled by a 300 kN axial force. The decrease in bar width is most nearly

(A) 50 μm

(B) 70 μm

(C) 90 μm

(D) 110 μm

39. Consider a composite steel-concrete beam. The steel is grade 60, and its modulus of elasticity is 29×10^6 psi. The properties of the concrete are as follows.

property type	value
compressive strength (psi)	4000
modulus of elasticity (ksi)	3600
modulus of rupture (psi)	450

The modular ratio of the materials is most nearly

(A) 8

(B) 31

(C) 3600

(D) 8000

40. A hollow steel shaft has an outer diameter of 400 mm and a wall thickness of 20 mm. The shaft is carrying a uniformly distributed load of 2.5 kN/m over its span of 12 m. The shaft is simply supported at its ends. Neglecting self-weight, the shaft's maximum deflection is most nearly

(A) 6 mm

(B) 8 mm

(C) 10 mm

(D) 12 mm

41. A point is subjected to a 5 ksi biaxial load in tension. The diameter of Mohr's circle for the state of stress is most nearly

(A) 0 ksi

(B) 5 ksi

(C) $5\sqrt{2}$ ksi

(D) 10 ksi

42. A 6 cm square, 2 m long steel bar is loaded concentrically. If the ends are pinned, the buckling capacity is most nearly

(A) 50 kN

(B) 150 kN

(C) 500 kN

(D) 5000 kN

43. An asphalt concrete mix weighs 152 lbf/ft^3. The asphalt binder is 5% by weight, the binder density is 64 lbf/ft^3, and the binder absorption is zero. The aggregate density is 172 lbf/ft^3. The percentage of the binder in the mix by volume is most nearly

(A) 2%

(B) 7%

(C) 12%

(D) 95%

44. An asphalt concrete mix weighs 152 lbf/ft^3. The asphalt binder is 5% by weight, the binder density is 64 lbf/ft^3, the binder absorption is zero, and the binder volume is 10%. The aggregate density is 172 lbf/ft^3. The percentage by volume of the air voids in the mix is most nearly

(A) 0%

(B) 6%

(C) 12%

(D) 84%

45. A concrete mix has a water-cement ratio of 0.4. The volume of water needed per sack of cement, weighing 94 lbf each, is most nearly

(A) 2.5 gal

(B) 4.5 gal

(C) 7.0 gal

(D) 37 gal

46. Three 6 in diameter concrete cylinders were tested and failed at an average compressive load of 255 kips. The concrete average compressive strength is most nearly

(A) 1500 psi

(B) 2500 psi

(C) 6000 psi

(D) 9000 psi

47. A stress-strain diagram of a material is shown.

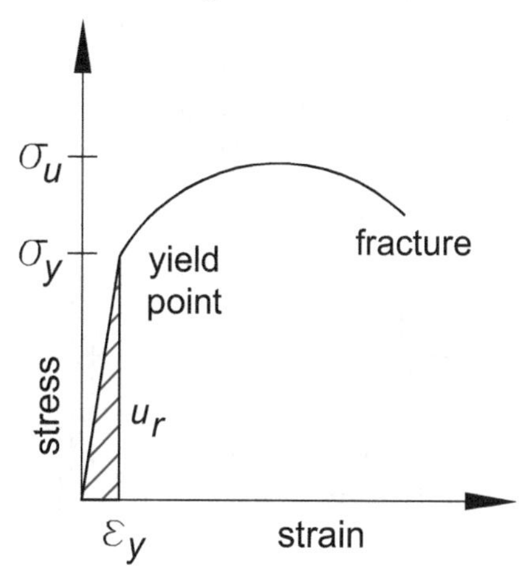

Point to the stress at which a material will experience permanent deformation.

48. Steady incompressible fluid flow in conduits and pipes is expressed by Bernoulli's energy equation.

$$\frac{p_1}{\gamma} + \frac{v_1^2}{2g} + z_1 = \frac{p_2}{\gamma} + \frac{v_2^2}{2g} + z_2 + h_f + h_{f,\text{fitting}}$$

Which of the following statements regarding the term $h_{f,\text{fitting}}$ is correct?

(A) It can be determined using the Moody diagram.

(B) It represents the head loss due to pipe friction.

(C) Manning's equation can also be used to determine its value.

(D) It is generally expressed as $C\frac{\text{v}^2}{2g}$.

49. A dam is 100 ft long and 13 ft from the base. The water behind the dam is 12 ft deep. The total lateral force caused by the hydrostatic pressure at the base of the dam is most nearly

(A) 75 kips

(B) 85 kips

(C) 450 kips

(D) 550 kips

50. Water flows from tank A to tank B using the pipe system shown.

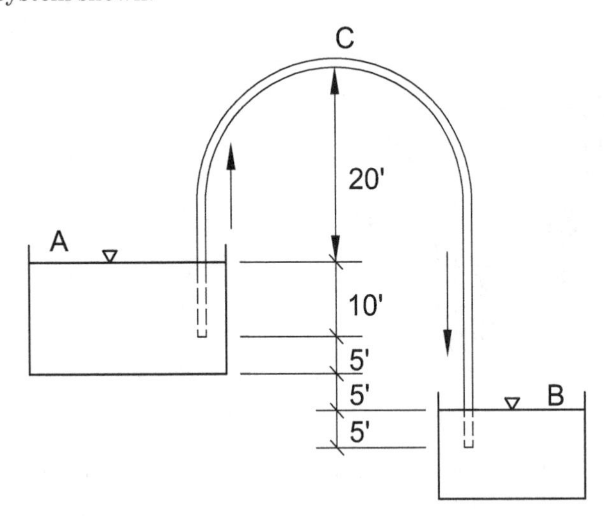

The difference in the elevation heads between the tanks is

(A) 15 ft

(B) 20 ft

(C) 40 ft

(D) 45 ft

51. A dam cross section is shown.

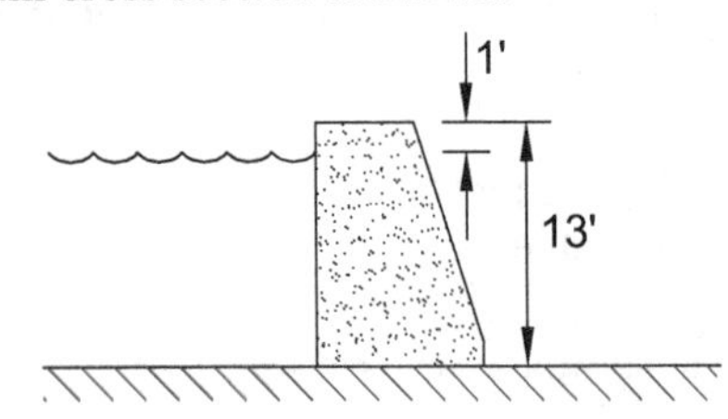

The maximum overturning moment over its 100 ft length is most nearly

(A) 450 ft-kips

(B) 1800 ft-kips

(C) 2200 ft-kips

(D) 3400 ft-kips

52. Oil flows through a pipe at a velocity of 10 m/s. The specific gravity is 2.0, and the kinematic viscosity is 0.0003 m^2/s. For a 2.54 cm diameter pipe, the Reynolds number of the oil is

(A) 423

(B) 846

(C) 1690

(D) 33,300

53. A 12 in diameter pipe carries oil flowing at a rate of 3 MGD. The oil's average velocity is most nearly

(A) 1.5 ft/sec

(B) 6.0 ft/sec

(C) 24 ft/sec

(D) 72 ft/sec

54. Two sides of a triangular-shaped parcel are 80 ft and 100 ft with a 60° angle between them. The length of the third side of the parcel is most nearly

(A) 81 ft

(B) 86 ft

(C) 92 ft

(D) 100 ft

55. Two sides of a triangular-shaped parcel are 80 ft long and 100 ft long, with a 60° angle between them. The third side is 91.65 ft long. The largest angle of the parcel is most nearly

(A) 60°

(B) 71°

(C) 75°

(D) 79°

56. A closed traverse has six segments. Its four interior angles measure 90° each. The sum of the remaining interior angles is most nearly

(A) 0°

(B) 90°

(C) 180°

(D) 360°

57. A pile is made of recycled aggregate material. The pile has a diameter of 100 ft at the base and 25 ft at the top. It is 25 ft high. The volume of piled material is most nearly

(A) 2576 yd^3

(B) 3860 yd^3

(C) 5787 yd^3

(D) 105,000 yd^3

58. What is the southern azimuth of a line with a bearing of S12°34′56″E?

(A) 12°34′56″

(B) 77°25′04″

(C) 167°25′04″

(D) 347°25′04″

59. Which of the following statements is NOT correct?

(A) A leveling staff is a crew assigned to a route-surveying task under the supervision of a licensed professional surveyor.

(B) A leveling staff is a rod used in surveying.

(C) A level line is one where all points are normal to the direction of the force of gravity.

(D) A freely suspended plumb bob shows the direction of the gravitational force.

60. 5 in of precipitation fall on a 300 ft wide, 400 ft long open field in 24 hours. The drainage area has a curve number of 60. The runoff volume expected from the event is most nearly

(A) 3000 ft^3

(B) 13,000 ft^3

(C) 30,000 ft^3

(D) 50,000 ft^3

61. Which of the following statements regarding a unit hydrograph is NOT correct?

(A) A unit hydrograph represents 1 in (or 1 mm) of direct runoff from one unit of effective rainfall occurring uniformly in space over a unit period of time.

(B) The unit hydrographs can be superimposed, and linearity of the relation applies.

(C) A unit hydrograph reflects all features of the watershed basin.

(D) By using a unit hydrograph, the properties of a watershed basin can be altered.

62. A 12 in diameter pipe is used to carry water at a flow rate of 3 ft/sec from reservoir A to reservoir B, as shown. Separation of dissolved gases occurs at an absolute pressure of 8 ft of water, and the change in the water level in each tank is negligible. Assume full pipe flow, and use the following factors: an entry loss of 1.0 ft, an exit loss of 0.1 ft. The pipe friction loss between points A and C is 15 ft, and the barometric pressure is 30 in of mercury (or 34 ft of water).

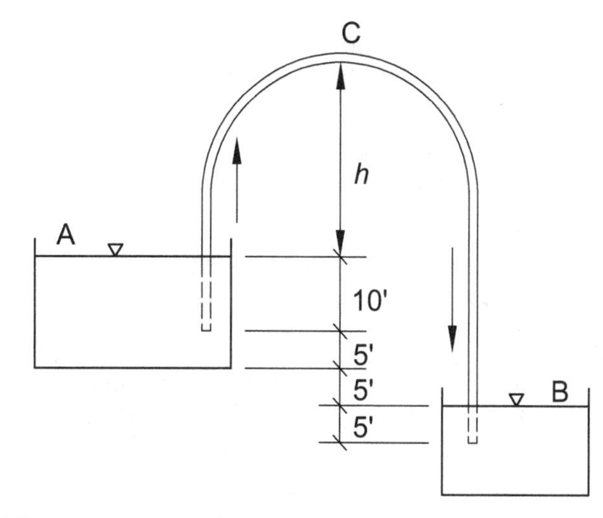

The maximum height, h, that can be used for siphoning is most nearly

(A) 10 ft

(B) 12 ft

(C) 15 ft

(D) 30 ft

63. A 6 in diameter, 1 mi long steel pipeline is used to move water. The elevation difference between the upper and the lower reservoir is 100 ft. The pump is 80% efficient. The friction factor is 0.028. Neglecting minor losses, the pumping horsepower required to maintain a discharge of 0.5 ft^3/sec is most nearly

(A) 2 hp

(B) 6 hp

(C) 9 hp

(D) 20 hp

64. A 10 cm diameter hydraulically smooth pipe carries water flowing at a velocity of 2 m/s. The pipe is 100 m long. The water temperature is 90°C. The head loss in the smooth pipe due to friction is most nearly

(A) 0.6 m

(B) 1.3 m

(C) 2.5 m

(D) 9.5 m

65. Which of following statements regarding a hydraulic jump is correct?

(A) For a hydraulic jump to occur, the depth upstream must be at or more than the critical depth.

(B) For a hydraulic jump to occur, the depth upstream must be less than the critical depth.

(C) For a hydraulic jump to occur, the specific energy upstream must be at the maximum.

(D) There is no loss of energy at the hydraulic jump.

66. The area in a drainage basin is evaluated using the NRCS rainfall-runoff method. The area is found to have a curve number of 100. Which of the following statements regarding the soil in the area is correct?

(A) The soil is either saturated or impervious.

(B) The soil is dry to the wilting point.

(C) The soil can fully pass through a number 100 sieve.

(D) The soil is fully covered with grass.

67. A stratum consists of two layers of thicknesses, b_1 and b_2, and hydraulic conductivities, K_1 and K_2, as shown.

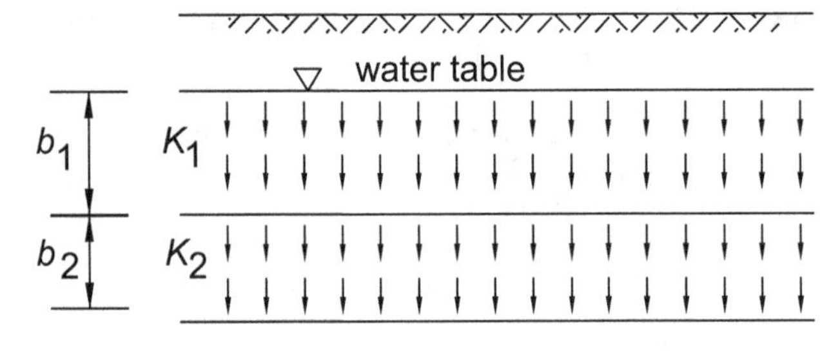

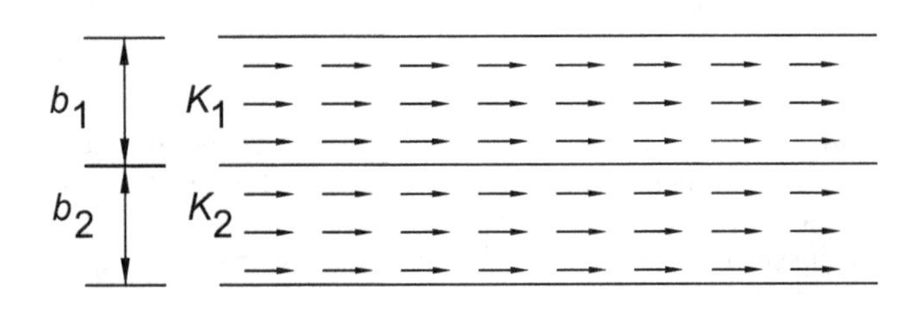

Which of the following statements is correct?

(A) Transmissivity is the product of permeability and the thickness of the aquifer.

(B) If the flow is perpendicular to stratification, the resultant hydraulic conductivity, $\overline{K}$, is

$$\overline{K} = \frac{b_1 K_1 + b_2 K_2}{b_1 + b_2}$$

(C) If the flow is parallel to stratification, the resultant hydraulic conductivity, $\overline{K}$, is

$$\frac{b_1 + b_2}{\overline{K}} = \frac{b_1}{K_1} + \frac{b_2}{K_2}$$

(D) A confined aquifer is an aquifer in which groundwater is under greater-than-atmospheric pressure.

68. A 36 in diameter sewer pipe is running three-quarters full. Its hydraulic radius is most nearly

(A) 8 in

(B) 11 in

(C) 15 in

(D) 18 in

69. The initial density of bacteria is 1000 cells per liter. After 20 generations of exponential growth during the growth phase, without any decay or microbial death, the number of bacteria in a liter is most nearly

(A) 20,000 cells/L

(B) 40,000 cells/L

(C) 1,000,000 cells/L

(D) 1,000,000,000 cells/L

70. A wastewater plant intends to use horizontal flow grit chambers as pretreatment. A chamber is 6 ft wide and 8 ft deep. The design flow rate is 20 ft^3/sec. The approach velocity in the chamber is most nearly

(A) 0.42 ft/sec

(B) 0.84 ft/sec

(C) 1.3 ft/sec

(D) 2.0 ft/sec

71. Examine the arch shown.

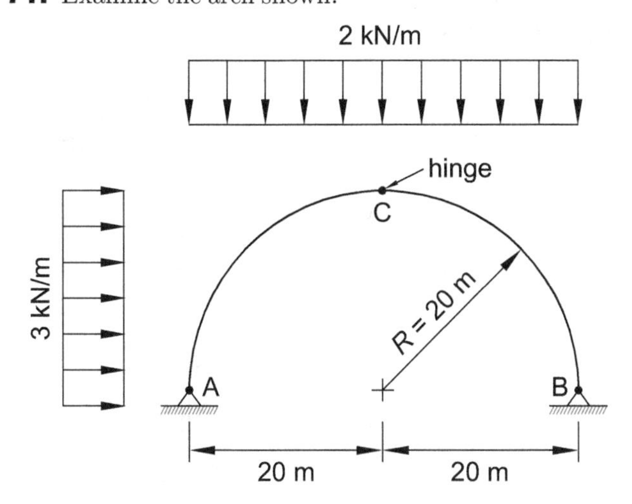

The arch is

(A) unstable

(B) stable and statically determinate

(C) stable and statically indeterminate

(D) redundant

72. A steel beam is carrying a uniformly distributed load of 2 kips/ft.

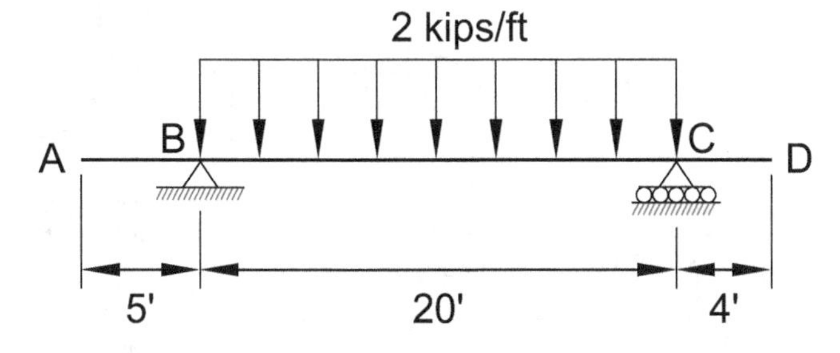

The moment of the inertia of the beam is 500 in^4. Its vertical deflection at point A is most nearly

(A) 0.048 in

(B) 0.40 in

(C) 7.7 in

(D) 9.6 in

73. A W12 × 79, grade 50 steel column is shown.

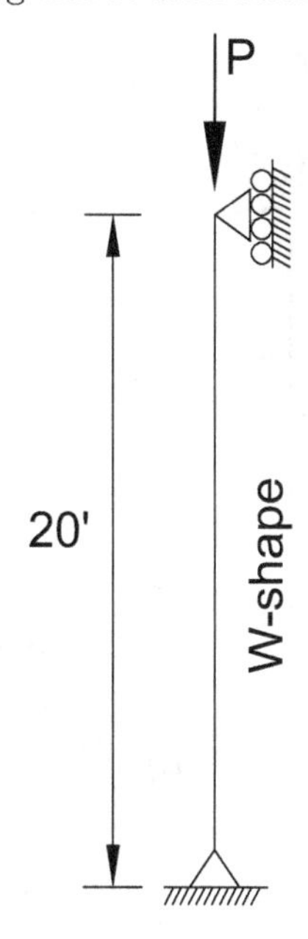

Using AISC standards, the buckling capacity of the column is most nearly

(A) 28.5 kips

(B) 330 kips

(C) 660 kips

(D) 730 kips

74. Which of the following statements regarding planar trusses is correct?

(A) Planar truss members are subjected to combined axial loads, shear forces, and bending moments.

(B) Planar truss members must have full moment connections at the joints.

(C) For trusses to be stable and statically determinate, all of the truss supports should have zero translation but be capable of rotation.

(D) Planar truss members are subjected to axial loads only.

75. A frame is carrying a 24 kip point load.

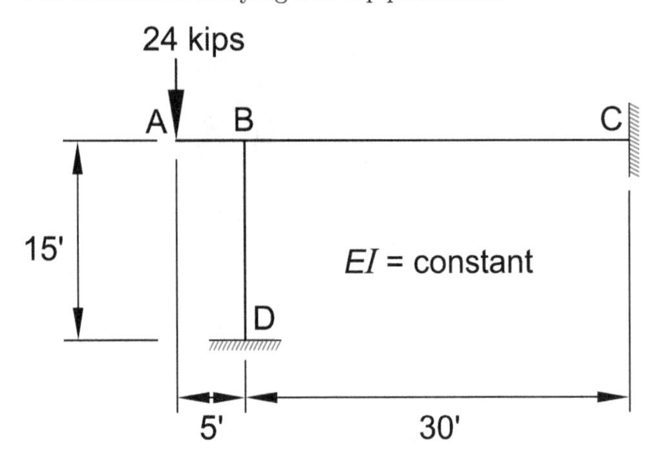

All members have equivalent Young's moduli and moments of inertia. The carryover moment at support C is most nearly

(A) 0 ft-kips

(B) 20 ft-kips

(C) 30 ft-kips

(D) 60 ft-kips

76. A moment of 100 ft-kips is applied at the non-fixed end of a propped cantilever. The induced moment at the other end is most nearly

(A) 0 ft-kips

(B) 50 ft-kips

(C) 100 ft-kips

(D) 200 ft-kips

77. A unit load moves horizontally from point C to point G over frame ABCDEFG.

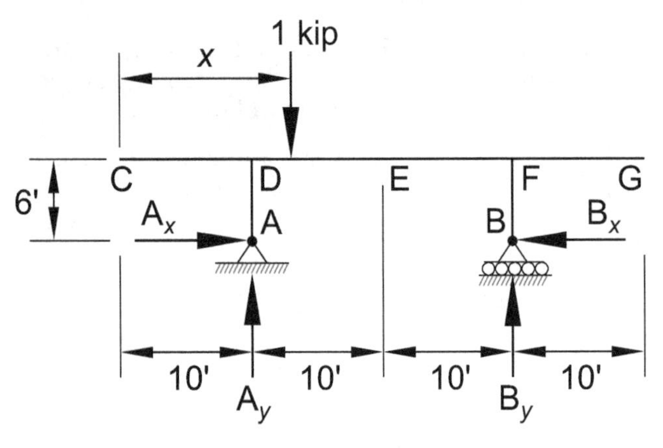

Which of the following is the influence line diagram for the vertical reaction at support A?

(A)

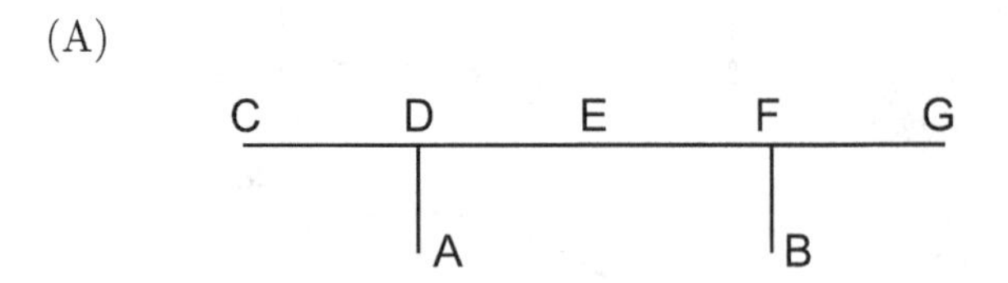

(B)

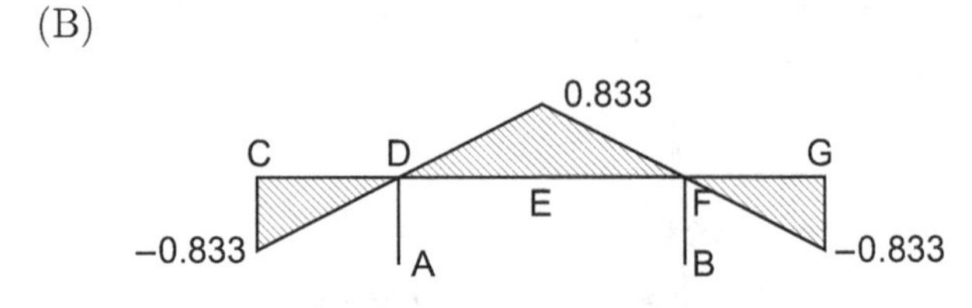

(C)

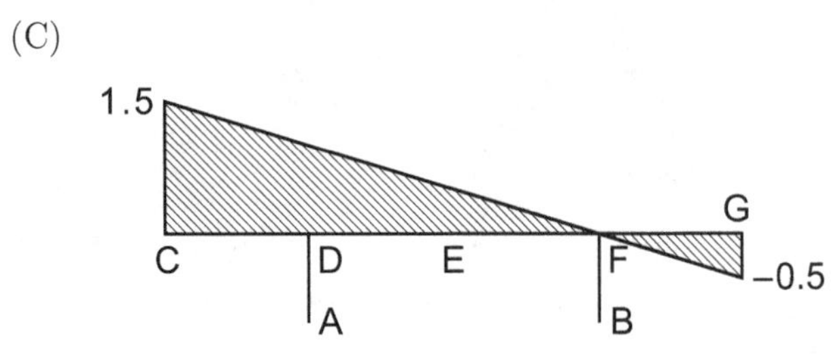

(D)

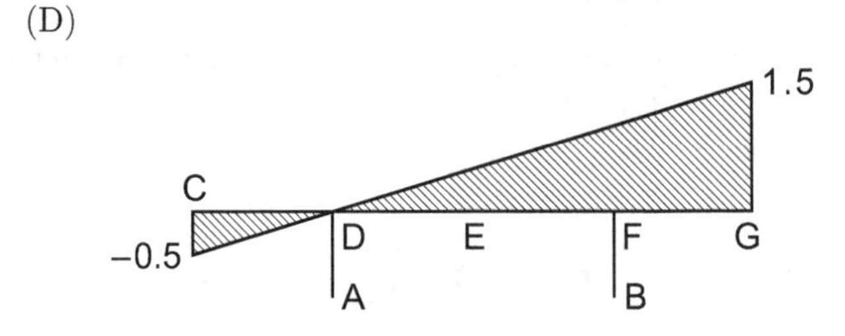

78. Consider a W21 × 44, grade 50 steel shape. The beam spans 30 ft, as shown.

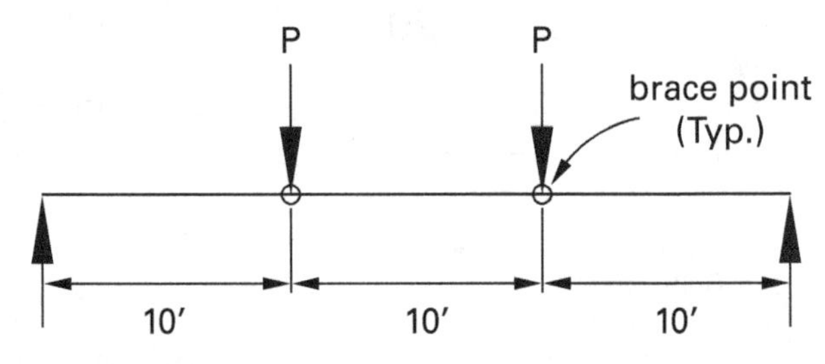

The beam carries two point-loads and is braced at the load points. Its moment capacity about its major axis is most nearly

(A) 250 ft-kips

(B) 280 ft-kips

(C) 308 ft-kips

(D) 358 ft-kips

79. A concrete slab is cast on top of a steel beam. No shear studs or other connectors were used in the construction.

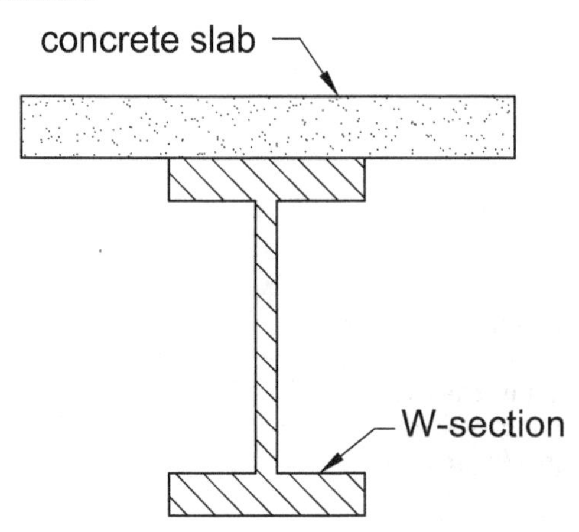

Which of the following statements is correct?

(A) There is no composite action, and the beam is unbraced.

(B) There is no composite action, and the beam is fully braced.

(C) The composite action is partial, and the beam is partially braced.

(D) There is full composite action, and the beam is fully braced.

80. A frame with a diagonal brace is subjected to a wind load, P.

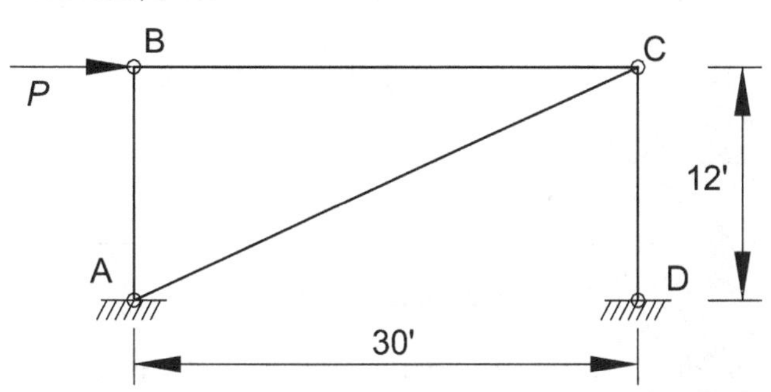

The brace consists of a W10 × 45 steel section with a yield strength of 50 ksi, pinned at both ends. The maximum wind force the frame can resist at yield is most nearly

(A) 0 kips

(B) 520 kips

(C) 560 kips

(D) 675 kips

81. An L-beam carries a 50 kip load at the midpoint of its 6 in wide ledge, as shown.

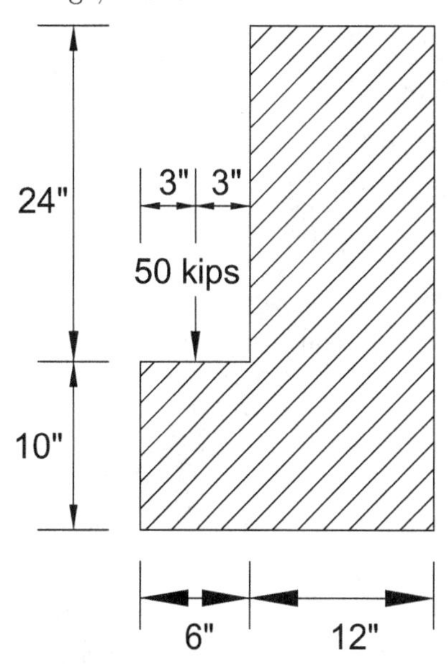

The torsion on the beam is most nearly

(A) 0 in-kips

(B) 130 in-kips

(C) 390 in-kips

(D) 740 in-kips

82. A high-strength steel rod with a cross-sectional area of 4 in^2 passes through a circular concrete element with a diameter of 24 in. The diameter of the opening that the steel rod passes through is 4 in. The rod is fitted at each end with a steel cap plate, a washer, and a nut.

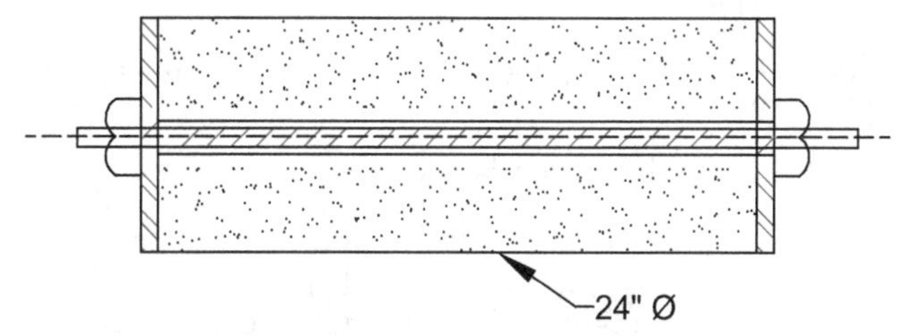

The nuts are tightened until the tensile stress in the rod reaches 55 ksi. Assuming the modular ratio of steel to concrete is 8, the stress in the concrete section is most nearly

(A) 0.125 ksi

(B) 0.5 ksi

(C) 6.00 ksi

(D) 6.875 ksi

83. The liquid limit of a silty soil is 52 and its plastic limit is 35. The soil type and plasticity index are most nearly

(A) CH and 35

(B) CH and 17

(C) OH and 52

(D) MH and 17

84. Which of the following statements regarding permeability through soil is NOT correct?

(A) The coefficient of permeability has the same units as velocity.

(B) The velocity head plays a significant role and cannot be ignored when determining permeability.

(C) Head loss occurs because water molecules, while traveling from upstream to downstream, expend energy in overcoming the frictional resistance provided by the soil.

(D) The hydraulic gradient is defined as the head loss per unit length.

85. Which of the following statements regarding the phase diagram of soils is NOT correct?

(A) It shows three parts: solids at the bottom, water above it, and the air on the top.

(B) The three parts in the diagram depict the conditions at the bottom of a water channel where the soil is at the bed, water is above it, and the atmospheric air at the top.

(C) It shows the weight-volume relationship of a soil mass.

(D) The weight of air is taken as zero.

86. A soil sample weighs 222 g and its volume is 101 cm^3. The sample is oven dried and its weight is 185 g. The moisture content of the soil is most nearly

(A) 20%

(B) 30%

(C) 45%

(D) 55%

87. A footing is placed 2 ft below a riverbed. The river is 20 ft deep. The soil density is 120 pcf. The soil effective vertical stress at the footing's bottom level is most nearly

(A) 125 psf

(B) 240 psf

(C) 1360 psf

(D) 1490 psf

88. A retaining wall is 12 ft high. The backfill weighs 120 pcf and is saturated below 5 ft from the grade. The angle of internal friction for the backfill is 30°. The resultant active lateral force behind the wall per unit wall length is most nearly

(A) 2900 lbf/ft

(B) 3900 lbf/ft

(C) 4900 lbf/ft

(D) 8800 lbf/ft

89. A dam's cross section is shown.

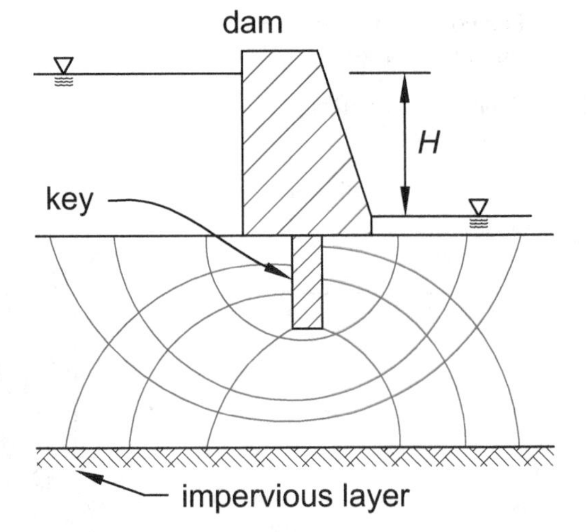

The dam's head drop from upstream to downstream is H. The dam has a key, and it rests on a permeable layer of soil with an impermeable layer underneath it. Which statement regarding the key is NOT correct?

(A) The key increases resistance to dam sliding.

(B) The key increases resistance to dam overturning.

(C) The flow in the permeable bed decreases as the key depth is increased.

(D) The head drop, H, increases as the key depth is increased.

90. A 3 in diameter clay soil sample with a 6 in height is tested under uniaxial loading. The clay's ultimate strength is 499 lbf. Using the Mohr-Coulomb failure criterion, its cohesion is most nearly

(A) 25 psi

(B) 35 psi

(C) 50 psi

(D) 70 psi

91. A 4 ft diameter straight caisson is drilled 40 ft deep to rest on fractured rock. The allowable bearing capacity of the rock is 20 ksf with a factor of safety of 3. Neglecting skin friction, the caisson's ultimate capacity is most nearly

(A) 61 kips

(B) 83 kips

(C) 250 kips

(D) 750 kips

92. For soil consolidation curves such as the one shown in the *NCEES Handbook*, which statement is NOT correct?

(A) The range of recompression applies to overconsolidated soils.

(B) The recompression index is also known as the swelling index.

(C) Point P_C is the past maximum consolidation stress.

(D) As the pressure is increases on a soil sample, the void ratio, e, decreases linearly.

93. An embankment slopes at 40° from horizontal and is 15 ft high. The soil is silty sand that weighs 120 pcf. Its angle of internal friction is 30°, its cohesion coefficient is 300 psf, and its slip surface angle is 20°. The factor of safety against slope stability is most nearly

(A) 0.3

(B) 1.4

(C) 2.4

(D) 3.4

94. A roadway has a design speed of 55 mph on a superelevated horizontal curve of 1000 ft radius. The coefficient of friction is 0.10. The superelevation is most nearly

(A) 0.01%

(B) 0.1%

(C) 1.0%

(D) 10%

95. A highway has a design speed of 55 mph and slopes downward at a grade of 4%. Assuming that a driver's deceleration is 10 ft/sec^2 and that the reaction time is 2.5 sec, the stopping sight distance at the design speed is nearly

(A) 375 ft

(B) 575 ft

(C) 700 ft

(D) 825 ft

96. Which of the following is an INCORRECT statement about roadway curves?

(A) A vehicle at the bottom of a sag curve may bottom out.

(B) A vehicle approaching the top of a crest curve may become airborne at the crest of the curve.

(C) The posted design speed is increased as the grade change becomes longer and larger.

(D) The higher the design speed for a road, the larger the superelevation.

97. A truck is carrying gravel from a stone pit. The truck has three axles. The axle loads are as follows.

	front axle (single)	middle axles (tandem)	rear axle (single)
fully loaded truck	18,000 lbf	30,000 lbf each	20,000 lbf
empty truck	10,000 lbf	12,000 lbf	8000 lbf

The truck factor for the fully loaded truck is most nearly

(A) 1.0

(B) 3.2

(C) 3.8

(D) 7.0

98. Level of service F denotes

(A) free flow

(B) fluid flow

(C) forced flow

(D) Fibonacci flow

99. An intersection is 60 ft wide and has a posted speed of 35 mph. The vehicle design length is 19 ft, and the driver's reaction time is assumed to be 2.3 sec. The length of the red clearance interval for the intersection should be most nearly

(A) 1.0 sec

(B) 1.5 sec

(C) 2.3 sec

(D) 4.0 sec

100. During a traffic study, 15 vehicles pass at a point on a highway in a span of 10 minutes. The traffic flow is most nearly

(A) 15 vehicles per hour

(B) 90 vehicles per hour

(C) 900 vehicles per hour

(D) 2160 vehicles per hour

101. During a traffic study, it is observed that six vehicles are between two points 200 ft apart, as shown.

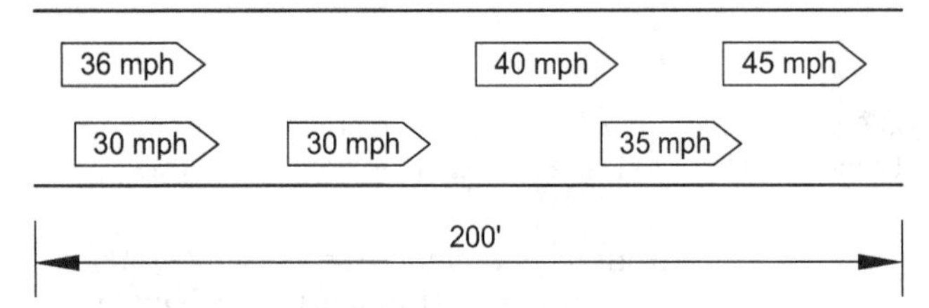

The roadway density is most nearly

(A) 6 vehicles per mile

(B) 160 vehicles per mile

(C) 206 vehicles per mile

(D) 880 vehicles per mile

102. Which of the following statements is INCORRECT?

(A) Average daily traffic (ADT) is the average of 24 hr count collected over a number of days more than 1 but less than 365.

(B) Average annual daily traffic (AADT) is the average of 24 hr count collected over every day in a year.

(C) Vehicle miles of travel (VMT) is a measure of travel by an average motorist for business use during business hours.

(D) Peak hour volume (PHV) is the maximum number of vehicles that pass a point during a period of 60 consecutive minutes.

103. The parties in any construction bonding arrangement are

(A) the owner, contractor, and subcontractors

(B) the contractor, architect, and owner

(C) the construction manager, surety, and engineer

(D) the owner, contractor, and surety

104. A project was awarded at a lump sum price of $2 million. The retainage was set in the contract as 10% of the work completed. The construction is 95% complete and only 5% is left to complete. At this time, the owner is holding $190,000 as retainage. The amount the owner should additionally retain at this stage is most nearly

(A) $0

(B) $10,000

(C) $100,000

(D) $200,000

105. Guardrails have been a preferred and recommended means of fall protection for a long time. Under OSHA regulations, the required height for the top rail of a guardrail is most nearly

(A) 25 in

(B) 30 in

(C) 35 in

(D) 45 in

106. An engineer can help save on formwork costs by taking all the following measures EXCEPT

(A) using commercially available formwork sizes

(B) permitting maximum reuse of formwork by minimizing the number of sizes of beams, columns, and footings

(C) considering available sizes of dressed lumber, plywood, and other ready-made formwork components

(D) developing a special formwork size for a project

107. Which statement regarding activities in a critical path method diagram is correct?

(A) The float is always positive for activities on the critical path.

(B) The critical path is the longest path that connects activities from start to finish.

(C) The critical path is the shortest path that connects activities from start to finish.

(D) The terms *activity* and *event* are synonymous.

108. A 100 ft long concrete retaining wall is being planned to be built. The wall will be 10 ft tall and 10 in thick. The cost of ready-mix concrete delivered on site is \$100 per yd^3. Assume 10% waste in concrete. The cost of the concrete needed to build the wall is most nearly

(A) \$3100

(B) \$3400

(C) \$83,000

(D) \$92,000

109. A single-story building has 100 columns. Each column has a diameter of 24 in and has eight #9 vertical steel rebars. The rebar length is 12 ft 9 in.

The column ties are #3 at 12 in on center with 12 in lap.

Concrete cover to tie centerline is 2 in.

The weight of vertical reinforcement is most nearly

(A) 10.2 tons

(B) 17.3 tons

(C) 34.7 tons

(D) 92.2 tons

110. A truck hauling fill material takes 20 min to load, haul, dump, and return. The delay time is 10 min per hour. The number of trips the truck can make in an 8 hr day is most nearly

(A) 16

(B) 20

(C) 24

(D) 48

STOP!

DO NOT CONTINUE!

This concludes the examination. If you finish early, check your work and make sure that you have followed all instructions. After checking your answers, you may turn in your examination booklet and answer sheet and leave the examination room. Once you leave, you will not be permitted to return to work or change your answers.

Exam 2 Instructions

In accordance with the rules established by your state, you may use any approved battery- or solar-powered, silent calculator to work this examination. However, no blank papers, writing tablets, unbound scratch paper, or loose notes are permitted. Sufficient paper will be provided. The *NCEES FE Reference Handbook* is the only reference you are allowed to use during this exam.

You are not permitted to share or exchange materials with other examinees.

You will have six hours in which to work this session of the examination. Your score will be determined by the number of questions that you answer correctly. There is a total of 110 questions. All 110 questions must be worked correctly in order to receive full credit on the exam. There are no optional questions. Each question is worth 1 point. The maximum possible score for this section of the examination is 110 points.

Partial credit is not available. No credit will be given for methodology, assumptions, or work written on scratch paper.

Record all of your answers on the Answer Sheet. Mark your answers with a no. 2 pencil. Answers marked in pen may not be graded correctly. Marks must be dark and must completely fill the bubbles. Record only one answer per question. If you mark more than one answer, you will not receive credit for the question. If you change an answer, be sure the old bubble is erased completely; incomplete erasures may be misinterpreted as answers.

If you finish early, check your work and make sure that you have followed all instructions. After checking your answers, you may submit your answers and leave the examination room. Once you leave, you will not be permitted to return to work or change your answers.

When permission has been given by your proctor, you may begin your examination.

Name: ______________________________
Last First Middle Initial

Examinee number: ______________

Examination Booklet number: ______________

Fundamentals of Engineering Examination

Exam 2

Exam 2 Answer Sheet

111. ________________
112. ________________
113. (A) (B) (C) (D)
114. ________________
115. ________________
116. (A) (B) (C) (D)
117. ________________
118. ________________
119. (A) (B) (C) (D)
120. (A) (B) (C) (D)
121. (A) (B) (C) (D) (E)
122. (A) (B) (C) (D)
123. (A) (B) (C) (D) (E)
124. (A) (B) (C) (D)
125. (A) (B) (C) (D)
126. (A) (B) (C) (D)
127. (A) (B) (C) (D)
128. (A) (B) (C) (D)
129. (A) (B) (C) (D)
130. (A) (B) (C) (D)
131. ________________
132. ________________
133. (A) (B) (C) (D)
134. ________________
135. (A) (B) (C) (D)
136. ________________
137. ________________
138. (A) (B) (C) (D)
139. (A) (B) (C) (D)
140. (A) (B) (C) (D)
141. (A) (B) (C) (D) (E)
142. (A) (B) (C) (D)
143. (A) (B) (C) (D)
144. (A) (B) (C) (D)
145. ________________
146. (A) (B) (C) (D)
147. (A) (B) (C) (D)
148. (A) (B) (C) (D)
149. (A) (B) (C) (D)
150. (A) (B) (C) (D)
151. (A) (B) (C) (D)
152. (A) (B) (C) (D)
153. (A) (B) (C) (D)
154. (A) (B) (C) (D) (E)
155. (A) (B) (C) (D)
156. (A) (B) (C) (D)
157. (A) (B) (C) (D)
158. (A) (B) (C) (D)
159. (A) (B) (C) (D)
160. (A) (B) (C) (D)
161. (A) (B) (C) (D)
162. (A) (B) (C) (D)
163. (A) (B) (C) (D)
164. ________________
165. (A) (B) (C) (D)
166. (A) (B) (C) (D)
167. (A) (B) (C) (D)
168. (A) (B) (C) (D)
169. (A) (B) (C) (D)
170. (A) (B) (C) (D)
171. (A) (B) (C) (D)
172. (A) (B) (C) (D)
173. (A) (B) (C) (D)
174. (A) (B) (C) (D)
175. (A) (B) (C) (D)
176. (A) (B) (C) (D)
177. (A) (B) (C) (D)
178. (A) (B) (C) (D)
179. (A) (B) (C) (D)
180. (A) (B) (C) (D)
181. (A) (B) (C) (D)
182. (A) (B) (C) (D)
183. (A) (B) (C) (D)
184. (A) (B) (C) (D)
185. (A) (B) (C) (D)
186. (A) (B) (C) (D)
187. (A) (B) (C) (D)
188. (A) (B) (C) (D)
189. (A) (B) (C) (D)
190. (A) (B) (C) (D)
191. (A) (B) (C) (D)
192. ________________
193. (A) (B) (C) (D)
194. (A) (B) (C) (D)
195. (A) (B) (C) (D)
196. (A) (B) (C) (D)
197. (A) (B) (C) (D)
198. (A) (B) (C) (D)
199. (A) (B) (C) (D)
200. (A) (B) (C) (D)
201. (A) (B) (C) (D)
202. (A) (B) (C) (D)
203. (A) (B) (C) (D)
204. (A) (B) (C) (D)
205. (A) (B) (C) (D)
206. (A) (B) (C) (D)
207. (A) (B) (C) (D)
208. (A) (B) (C) (D)
209. (A) (B) (C) (D)
210. (A) (B) (C) (D)
211. (A) (B) (C) (D)
212. ________________
213. (A) (B) (C) (D)
214. (A) (B) (C) (D)
215. (A) (B) (C) (D)
216. (A) (B) (C) (D)
217. ________________
218. (A) (B) (C) (D)
219. (A) (B) (C) (D)
220. (A) (B) (C) (D)

Exam 2

111. For a geothermal project, the temperature increases from 20°C at the surface of the earth to 90°C at a depth of 2 km below the ground. Assuming a linear temperature gradient, the underground temperature at 3800 m below the ground surface is ______. (Fill in the blank.)

112. A system of three simultaneous equations is given.

$$x + 2y - z = -7$$

$$2x + 3y + 2z = -3$$

$$x - 2y - 2z = 3$$

If the determinant of matrix of the equations is 17, the value of x is ______. (Fill in the blank.)

113. Line $y = 2x$ is rotated about the y-axis. The volume between $y = 0$ and $y = 10$ is

(A) $50\pi/3$

(B) 50π

(C) $250\pi/3$

(D) 100π

114. Consider a 50 m by 30 m parking lot. The owner wants to double its size by increasing both width and length equally. The minimum additional width rounded to the nearest meter length is most nearly ______. (Fill in the blank.)

115. A series is given.

$$1 - \frac{1}{3} + \frac{1}{9} - \frac{1}{27} + \ldots$$

The sum of the series is ______. (Fill in the blank.)

116. A particle moves along a line so that its position at time t is

$$s(t) = \frac{t^2 + 1}{t + 1}$$

The velocity at $t = 2.5$ is most nearly

(A) 0.78

(B) 0.84

(C) 0.90

(D) 2.07

117. The volume of a parallelepiped bounded by vectors originating from the origin to points (0, 4, 0), (4, 8, 0), and (1, 2, 3) is ______. (Fill in the blank.)

118. Out of eight male and 12 female workers, a crew of five workers is formed. If the crew is composed of one man and four women, how many ways can the crew be formed? ______. (Fill in the blank.)

119. A factory employs 100 workers. Their wages vary according to the tasks they perform, as shown.

no. of workers	hourly wage
10	\$10.00
10	\$30.00
15	\$25.00
20	\$12.00
20	\$15.00
25	\$20.00

The average wage of the employees is most nearly

(A) \$16/hr

(B) \$18/hr

(C) \$20/hr

(D) \$21/hr

120. If the mean one-year rainfall is 30 in and the standard deviation is 6.6, the coefficient of variation is most nearly

(A) 0.22%

(B) 4.5%

(C) 22%

(D) 98%

121. Which of the following statements regarding acceptable ethical behavior are correct? (Select all that apply.)

(A) Ethics vary with time.

(B) Ethics vary with location.

(C) Ethics vary with culture.

(D) Ethics vary with time, location, and culture.

(E) Ethics norms are universal and do not vary with time, location, or culture.

122. An owner enters into an oral agreement with a licensed engineer to perform a condition appraisal of an old building. The engineer determines that the building is too dangerous for human occupancy. The engineer suggests that the repair work would be too expensive and the building should be demolished. The engineer submits the findings and invoices the owner. The owner refuses to pay. State law permits entering into an oral contract. Under contract law, which of the following options describes the fairest result?

(A) The engineer is not entitled to the fees because there is no written agreement between the parties.

(B) The engineer is not entitled to the fees because the engineer's report caused the owner's building to shut down and the owner lost rental income.

(C) The engineer acted unethically when recommending the demolition of the client's property.

(D) The engineer is entitled to the fees because the engineer has a valid agreement and performed services properly.

123. According to engineering ethics, a licensee shall sign and seal only those design documents that do which of the following? (Select all that apply.)

(A) meet the minimum requirements of a building code

(B) safeguard the life, health, property, and welfare of the public

(C) serve the economic goals of the client

(D) require all green or environmentally friendly materials

(E) have been prepared by the licensee or under the licensee's responsible charge

124. Which of the following methods of advertising is most likely to violate an ethical standard for engineering design firms?

(A) social media advertising

(B) publishing and distributing calendars showing photos of projects the firm has completed

(C) inviting a prospective client to the Super Bowl and paying all expenses to attend the event

(D) using a firm brochure with self-laudatory language

125. A homeowner asked a contractor for a quote to repair a leaky basement wall. The contractor replied that she can do the work for $5000. The homeowner did not respond to the quote. After two days, the contractor changed her mind and told the homeowner that she was withdrawing her quote. Which of the following statements is correct under contract law?

(A) The contractor can walk away because the offer was verbal and thus invalid.

(B) The contractor can walk away because her offer was not accepted in time.

(C) The contractor must perform the work for $5000.

(D) The contractor is in breach of contract and must pay the homeowner $5000.

126. A tree is worth \$100 today. If its value increases 10% per year, the tree's worth after 25 years is most nearly

(A) \$450

(B) \$970

(C) \$1100

(D) \$1300

127. A car is on sale for \$30,000. A dealer asks for a 20% down payment now, then 60 monthly payments of \$599. The first monthly payment would be due one month after the transaction closes. The effective annual interest rate being charged is most nearly

(A) 12%

(B) 13%

(C) 16%

(D) 19%

128. A concrete mix requires two admixtures, a minimum of 3 doses but no more than 9 doses of admixture A, and a minimum of 8 doses of admixture B. The maximum sum of both doses is 16. The cost of admixture A and admixture B is \$4 and \$12 per dose, respectively. The minimum cost of admixtures to produce the mix is most nearly

(A) \$96

(B) \$108

(C) \$192

(D) \$216

129. A tollway authority at present charges \$3.00 as a toll fee. A total of 20,000 motorists use the road daily. The authority wants to increase the toll fee. It is estimated that for each quarter dollar increase in the toll, 1000 fewer motorists will use the toll road. The toll fee increase that will maximize the authority's income is most nearly

(A) \$1.00

(B) \$1.50

(C) \$2.00

(D) \$2.50

130. A piece of equipment is valued at \$100,000. If the value is depreciated at a rate of 10% per year, its book value after 10 years will be most nearly

(A) \$0

(B) \$35,000

(C) \$65,000

(D) \$250,000

131. Your client, Gr8 Oil, must decide whether or not to drill for oil at a particular location for which Gr8 has drilling rights. There are two possible outcomes: striking oil and not striking oil. If the decision is made to drill and Gr8 strikes oil, the oil would be worth \$13,000,000. The cost of drilling would be \$3,000,000. At present, Gr8 can sell the drilling rights to a wildcatter or prospector for \$1,000,000, with a contingency to receive another \$1,000,000 if the wildcatter strikes the oil. Based on the geological assessment, the probability of striking the oil at the location is 0.30. If Gr8 decides not to drill, the expected maximum payoff to Gr8, in whole millions of dollars, is ______. (Fill in the blank.)

132. A bolt is being pulled by two forces A and B as shown.

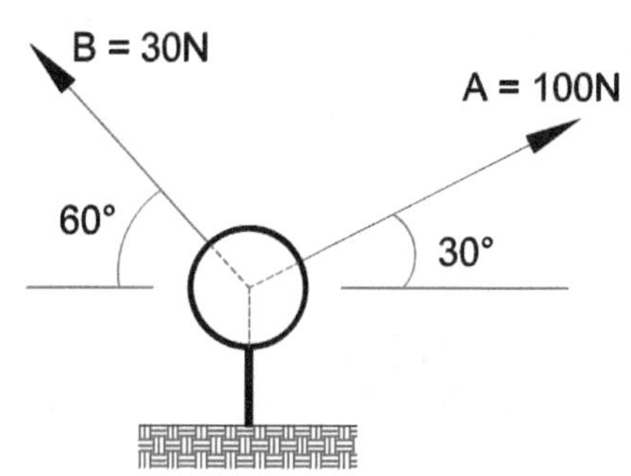

The following options are given for the resultant force.

104 N

115 N

∡ 45° from x-axis

∡ 47° from x-axis

∡ 135° from x-axis

Fill in the table with the attributes most nearly representing the resultant.

resultant force (N)	angle (clockwise from x-axis)

133. A concrete block weighing 1200 lbf needs leveling on one edge. A crowbar is used to lift the edge, as shown in the illustration.

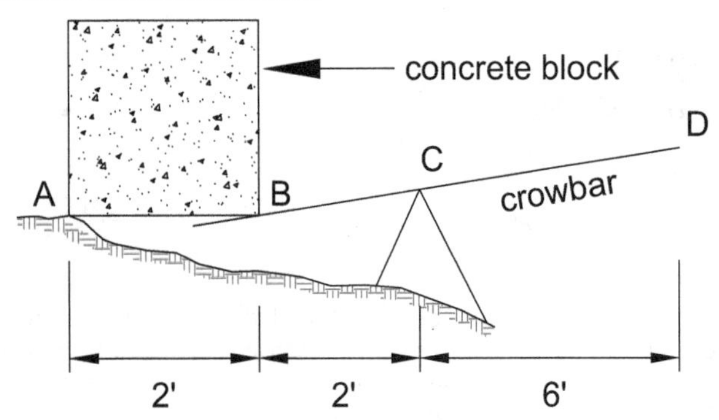

The downward force required at point D to lift the block is most nearly

(A) 120 lbf

(B) 200 lbf

(C) 800 lbf

(D) 1200 lbf

134. Four truss elements are connected at a node as shown in the illustration. Force F_1 is 10 kN (compressive), and force F_4 is 20 kN (compressive). To determine the forces F_2 and F_3, the following options are provided.

- tension
- compression
- 6 kN
- 8 kN
- 10 kN

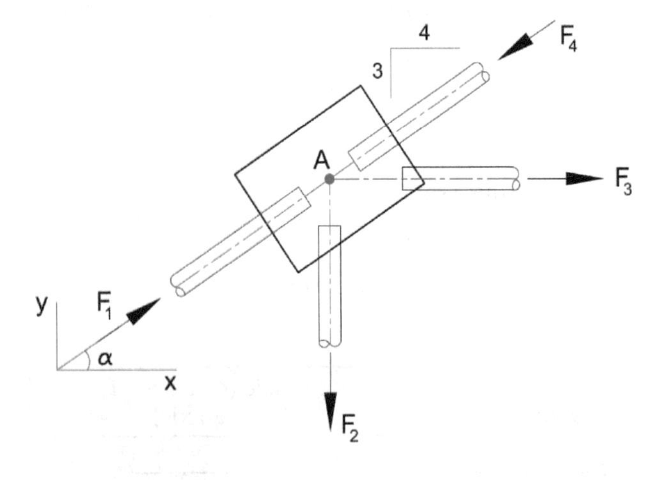

Fill in the table with the values most nearly representing forces F_2 and F_3.

	force F_2 (kN)	force F_3 (kN)
tension or compression?		
magnitude		

135. A billboard is 2 m wide and 1 m tall, mounted on top of a 5 m long pole.

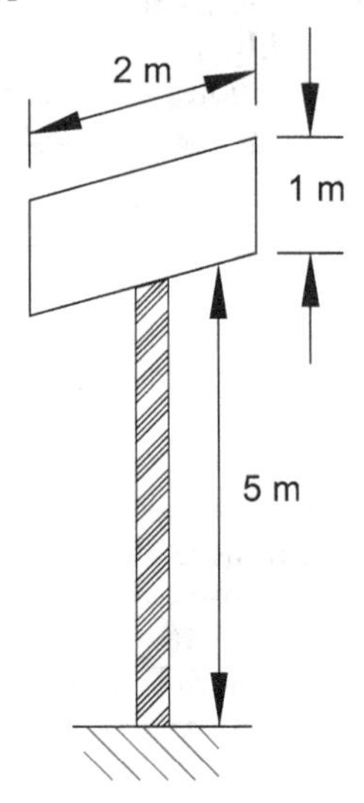

The total wind force acting on the billboard is 200 N. The resulting moment at the base of the pole is most nearly

(A) 0.10 kN·m

(B) 0.50 kN·m

(C) 1.1 kN·m

(D) 1.6 kN·m

136. A cantilever is carrying a point load, as shown in the illustration.

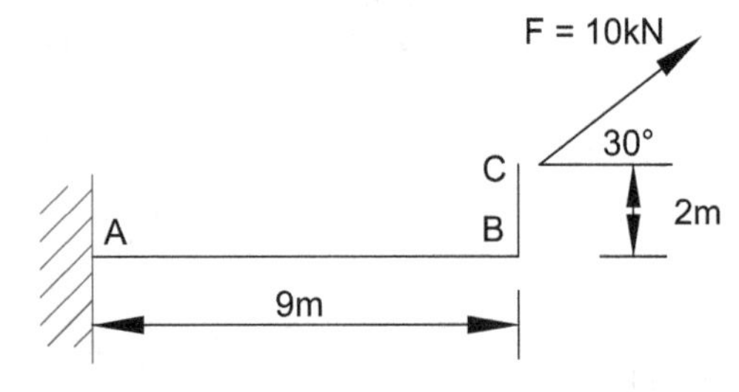

To determine the moment at support A, the following options are provided.

- clockwise
- counterclockwise
- 28 kN·m

- 62 kN·m
- 98 kN·m

Fill in the table using the given options.

	moment
clockwise or counterclockwise	
magnitude (k·Nm)	

137. A composite beam-slab section is shown.

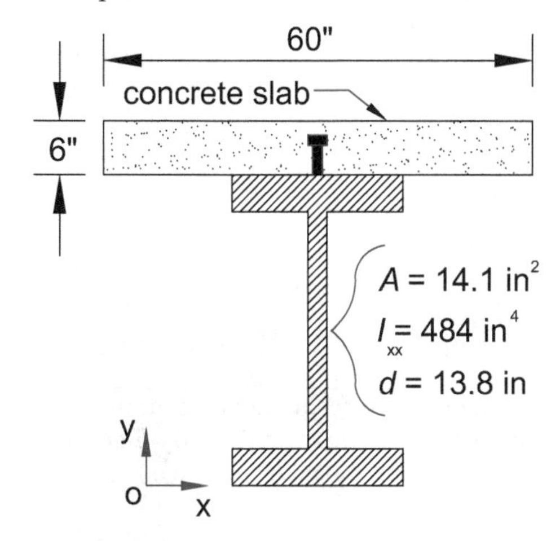

The concrete is 5000 psi and the steel section is a W14 × 48 beam. The force, F_y, is 60 ksi. The distance in inches from the bottom flange to the centroid of the composite section is most nearly ______. (Fill in the blank.)

138. Two identical rectangular steel bars are welded together, one on the top of the other. The ratio of the moment of inertia (MOI) of the welded bars to the MOI of a single bar is

(A) 2

(B) 4

(C) 8

(D) 10

139. A 100 kg block is placed on a sloping plane. The friction coefficient at the interface between the block and the sloping plane surface is 0.3. The sloping angle from the horizontal line at which the block would start sliding down is most nearly

(A) 17°

(B) 26°

(C) 33°

(D) 45°

140. In order to determine the height of a cliff, a stone is dropped from the cliff top. The stone is seen hitting the ground 2.9 s later. Neglecting air resistance to the stone, the cliff height above the ground is most nearly

(A) 10 m

(B) 30 m

(C) 40 m

(D) 90 m

141. A car has a mass of 6000 lbm and travels at 30 mph eastbound when it collides head-on with a 15,000 lbm truck traveling westbound at the same speed. After the collision, the vehicles stick together. If momentum is conserved during the collision, which two of the following options most nearly represent the vehicles' common velocity after the crash?

(A) 10 mph

(B) 13 mph

(C) 14 mph

(D) westbound

(E) eastbound

142. A flywheel weighs 1000 lbf. The radius of the flywheel is 18 in. Assuming that the mass is well distributed, the mass moment of inertia of the flywheel about the x-axis is most nearly

(A) 0.25 slugs-ft^2

(B) 35 slugs-ft^2

(C) 2100 slugs-ft^2

(D) 8500 slugs-ft^2

143. A 4 lbf weight stretches a spring by 6 in from its natural length. The undamped natural frequency of vibration of the spring is most nearly

(A) 3 rad/sec

(B) 6 rad/sec

(C) 8 rad/sec

(D) 10 rad/sec

144. A crankshaft is 50 cm long and rotates about point A, as shown.

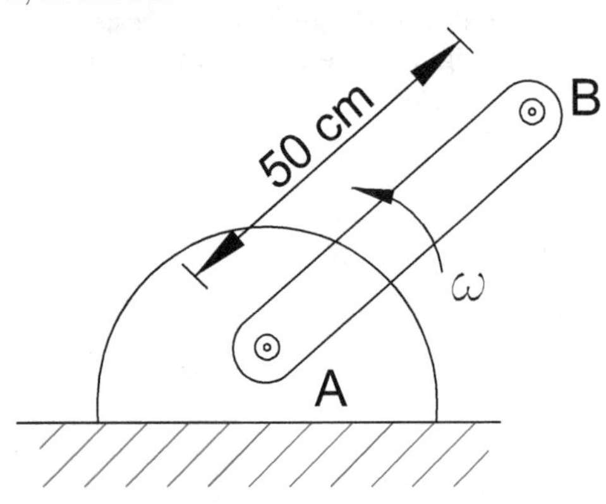

If the shaft is rotating at a constant angular velocity of 20 rad/s, the acceleration at shaft tip B is most nearly

(A) 0 m/s^2

(B) 10 m/s^2

(C) 100 m/s^2

(D) 200 m/s^2

145. A beam is subjected to two point loads, as shown. Draw a circle on the region in the illustration, where the beam bending moment is the maximum.

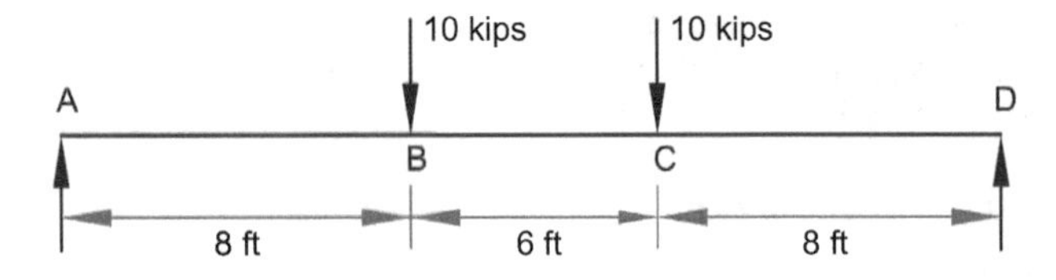

146. A 1 in diameter bolt resists a 16 kip force, as shown in the illustration.

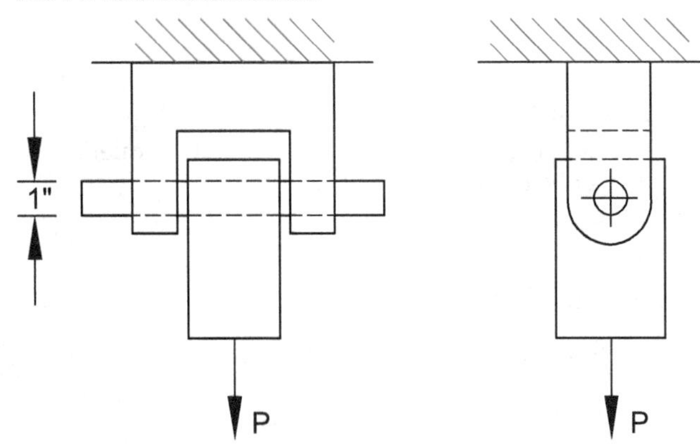

The shear stress in the bolt is most nearly

(A) 5 ksi

(B) 10 ksi

(C) 20 ksi

(D) 30 ksi

147. A 12 in × 12 in × 1 in steel plate is fixed to a surface along its bottom length, as shown.

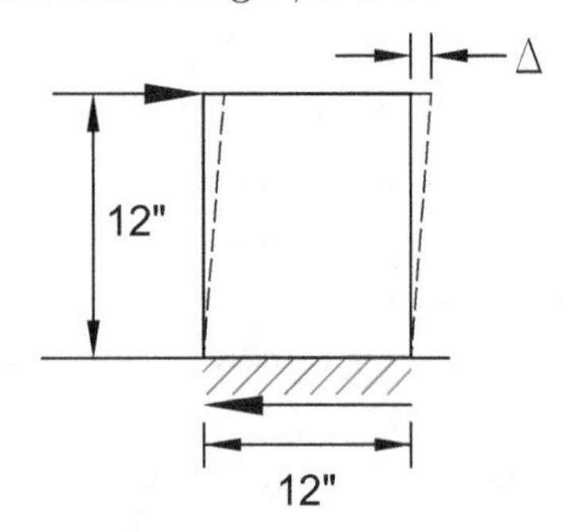

If a load of 24 kips is applied at its top surface and the bending deformation is negligible, the maximum deflection under the load is most nearly

(A) 0.075×10^{-3} in

(B) 0.18×10^{-3} in

(C) 0.82×10^{-3} in

(D) 2.1×10^{-3} in

148. The modulus of elasticity of a material is 200 GPa, and the Poisson ratio is 0.3. The modulus of rigidity of the material is most nearly

(A) 65 GPa

(B) 77 GPa

(C) 150 GPa

(D) 210 GPa

149. A metallic surveying tape was calibrated at 68°F. The tape was used in the field when the temperature was 98°F. The modulus of thermal expansion of the tape material is 6.5×10^{-6} in/in per °F. The measured distance was 3001.20 ft. The true distance was

(A) 2999.32 ft

(B) 3000.61 ft

(C) 3001.79 ft

(D) 3003.08 ft

150. A steel sphere is pressurized so that it is subjected to 5 ksi biaxial tensile stress. The maximum shear stress in the sphere for the state of stress is most nearly

(A) 0 ksi

(B) 5 ksi

(C) 7 ksi

(D) 10 ksi

151. Consider a composite steel-concrete member. The modular ratio of the member's materials is defined as

(A) $\dfrac{\text{modulus of elasticity of concrete}}{\text{modulus of elasticity of steel}}$

(B) $\dfrac{\text{modulus of elasticity of steel}}{\text{modulus of elasticity of concrete}}$

(C) $\dfrac{\text{shear modulus of concrete}}{\text{shear modulus of steel}}$

(D) $\dfrac{\text{shear modulus of steel}}{\text{shear modulus of concrete}}$

152. A steel shaft is subjected to a set of torsional forces as shown.

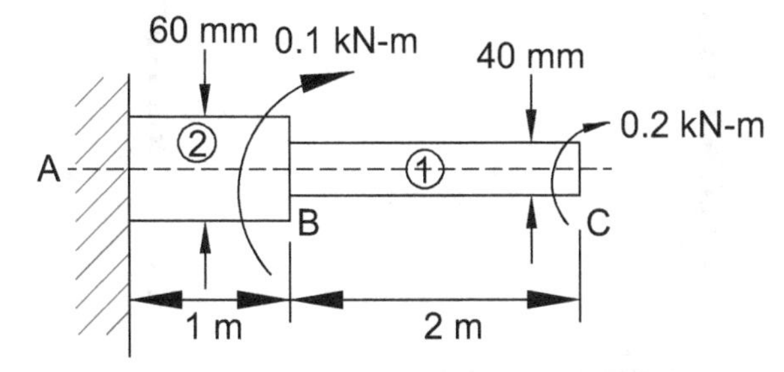

The shaft's total rotation at point C is most nearly

(A) 0.0235°

(B) 0.176°

(C) 1.14°

(D) 1.31°

153. A concrete mixture is to be used for a parking garage exposed to freeze and thaw cycles. Which type of admixture should be considered for use?

(A) entrained air

(B) entrapped air

(C) retarder

(D) accelerator

154. Portland cement concrete mix is generally described as 1:x:y, where x and y are variables (e.g., 1:2:3). Out of the following statements, select the two that correctly describe the mix designation.

(A) The mix contains one part of water.

(B) The mix contains x parts of cement.

(C) The mix contains x parts of fine aggregate.

(D) The mix contains x parts of coarse aggregate.

(E) The mix contains y parts of coarse aggregate.

155. A concrete mix has a ratio of 1:2:3. The water-cement ratio and the unit weight of the mix are 0.4 and 145 lbm/ft^3, respectively. Assuming that a sack of cement weighs 94 lbm, the number of sacks of cement needed to produce 1 yd^3 of mix is most nearly

(A) 5 sacks/yd^3

(B) 6 sacks/yd^3

(C) 7 sacks/yd^3

(D) 10 sacks/yd^3

156. Grade 60 steel reinforcement is used in a concrete column. For concrete having $f_c' = 5000$ psi, the ratio of steel strength to concrete strength in the column is most nearly

(A) 0.10

(B) 1.0

(C) 12

(D) 120

157. The ability of a material to undergo deformation without failure under high tensile stress is called

(A) hardness

(B) resilience

(C) ductility

(D) fracture toughness

158. Two pressure gauges are attached on the inside of a tank filled with a liquid. The first gauge, located 5 m from the bottom of the tank, reads 80 kPa. The second gauge, located 8 m from the bottom, reads 57.4 kPa. Neglect friction and fitting losses. The unit weight of the liquid is most nearly

(A) 393 kg/m^3

(B) 768 kg/m^3

(C) 1180 kg/m^3

(D) 7530 kg/m^3

159. The specific gravity of a liquid is 2.7 compared against a water density of 1000 kg/m^3. The specific weight of the liquid is most nearly

(A) 160 $\text{kg/m}^2\text{·s}^2$

(B) 270 $\text{kg/m}^2\text{·s}^2$

(C) 2700 $\text{kg/m}^2\text{·s}^2$

(D) 27,000 $\text{kg/m}^2\text{·s}^2$

160. The wall of a circular water tank has a hole in it. The hole is under a water pressure head of 10 ft, as shown. A steel plate is installed outside the wall at the opening and welded all around.

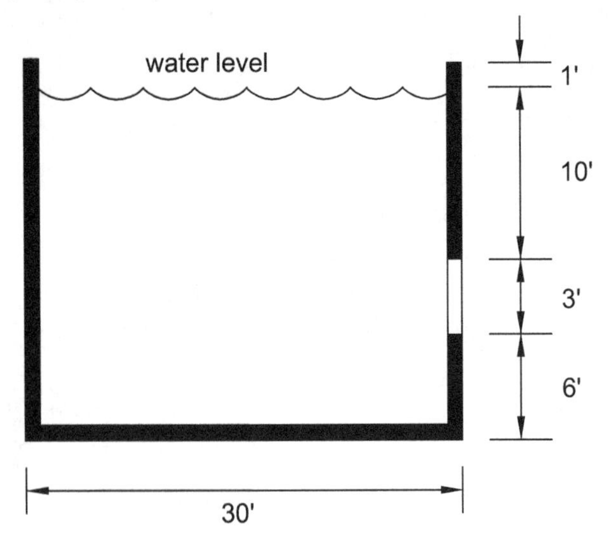

The type of force on the weld is

(A) null

(B) tensile

(C) compressive

(D) combined bending and torsion

161. A 3 in wide and 3 ft tall strip is removed from the wall of a steel tank that contains water under a pressure head of 10 ft, as shown.

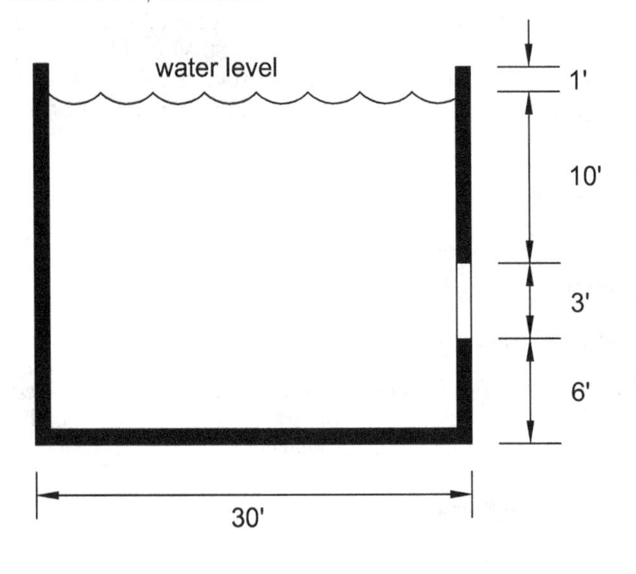

A steel plate is installed outside the wall at the opening and welded all around. Assuming the weld is adequate, the total force on the weld is most nearly

(A) 538 lbf

(B) 624 lbf

(C) 811 lbf

(D) 1070 lbf

162. A plate was used in plugging the hole in the water tank wall as shown. The pressure distribution diagram is also shown.

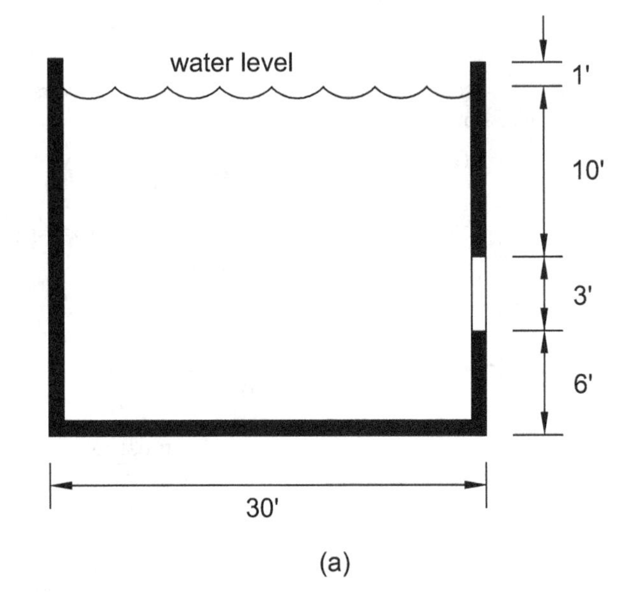

(a)

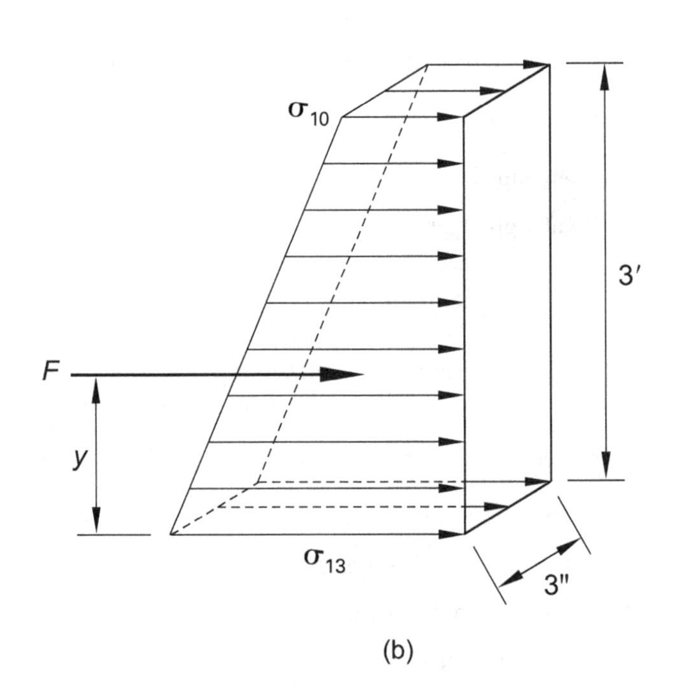

(b)

The location of the center of hydrostatic pressure against the plate located above the plate base is most nearly

(A) 17 in

(B) 18 in

(C) 19 in

(D) 139 in

163. In model similarity, if gravitational and inertial forces are the only contributing forces, which number is kept constant in similitude?

(A) Cauchy number

(B) Froude number

(C) Reynolds number

(D) Weber number

164. In an equilateral triangle-shaped parcel, the height (altitude) of the triangle is 5 ft less than its side length. The side length, to the nearest one-tenth of a foot, of the triangle is equal to ______. (Fill in the blank.)

165. A horizontal curve has a diameter of 500 m. The intersection angle is 60°.

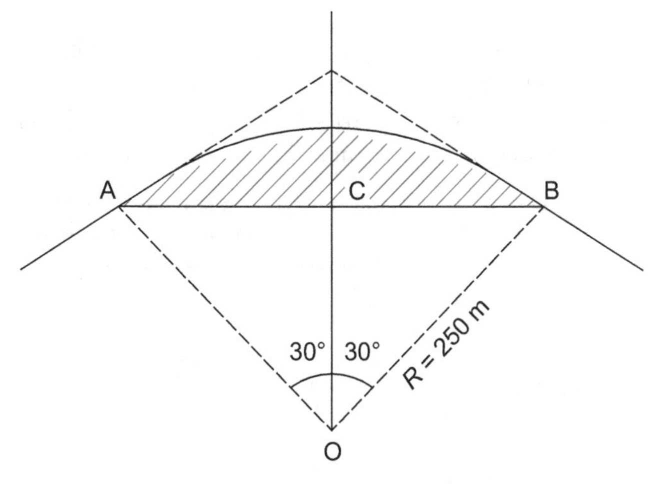

The shaded area of the curve shown in the illustration is most nearly

(A) 5662 m^2

(B) 6201 m^2

(C) 27,060 m^2

(D) 32,720 m^2

166. Two cross sections of a proposed roadway are located at station 1+65.00 and station 3+50.00. One cross section needs 200 ft^2 of cut and the other needs 125 ft^2 of fill. The net excavation required between the two sections is most nearly

(A) 257 yd^3

(B) 1120 yd^3

(C) 13,800 yd^3

(D) 32,100 yd^3

167. Line AB bears N12°34′56″E, and line AC bears S12°34′56″E. The deflection angle between the lines is

(A) straight east or 90°

(B) 25°09′52″ (left)

(C) 167°25′04″ (left)

(D) 154°50′08″ (right)

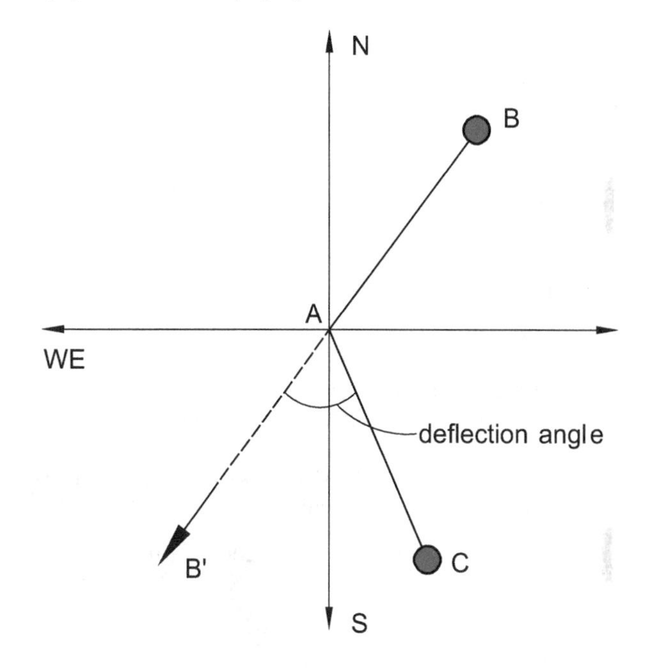

168. A surveyor takes several leveling readings. The instrument is on a known point of elevation of 123.45 ft, and the height of the instrument (HI) is 5.15 ft. To determine the elevation of the underside of a beam, an inverted sight reading of 3.13 ft is obtained. To determine the elevation of a point on a slope, a reading of 4.32 ft is obtained. The elevations of the underside of the beam and the point on the slope are respectively

(A) 128.60 ft, 126.58 ft

(B) 131.73 ft, 124.28 ft

(C) 145.47 ft, 132.92 ft

(D) 151.43 ft, 113.98 ft

169. For a horizontal curve of radius 250 ft and an intersection angle of 66°, the curve length is most nearly

(A) 272 ft

(B) 288 ft

(C) 785 ft

(D) 2880 ft

170. An urban watershed area consists of four types of surfaces, as tabulated below.

surface type	area (acres)	coefficient (C)
asphalt pavement	24	0.73
concrete pavement	16	0.85
lawns (heavy soil)	35	0.22
roofs	24	0.90

The composite runoff coefficient of the area is most nearly

(A) 0.6

(B) 0.7

(C) 0.9

(D) 2.7

171. Which of the following statements regarding the hydraulic radius of a channel is INCORRECT?

(A) Hydraulic radius is the cross-sectional area divided by wetted perimeter.

(B) Hydraulic radius is also called hydraulic mean depth.

(C) Hydraulic radius can be used for both circular and noncircular shapes.

(D) Hydraulic radius of a channel has no effect on water discharge.

172. A pump is used to lift water to a 100 ft high water tank and deliver 5 cfs. The pump is 85% efficient. The total head loss in the suction and discharge pipes, including other minor losses, is 48 ft. The specific weight of the water is 62.42 lbf/ft^3. The horsepower needed to pump the water is most nearly

(A) 5 hp

(B) 50 hp

(C) 60 hp

(D) 100 hp

173. A 1200 ft long galvanized steel pipe of 1 ft outside diameter and a 1 in thick wall connects two reservoirs. The water discharge rate through the pipe is 2 ft^3/sec. Assume the pipe has a round entrance and a sharp exit for the fittings. The total fitting loss for the pipe is most nearly

(A) 0 ft

(B) 0.025 ft

(C) 0.23 ft

(D) 2.0 ft

174. A 2 in diameter, sharp-edged orifice discharges freely into the atmosphere from a tank. The water level is maintained at 10 ft above the center of the orifice. The orifice discharge rate is most nearly

(A) 0.340 gps

(B) 2.55 gps

(C) 149 gps

(D) 366 gps

175. A drainage channel is being considered along a highway to prevent erosion. It may be necessary to provide protective linings on the bottom and along the side of the channel. Which of the following statements regarding the design of the linings in the drainage channels is correct?

(A) Rigid linings are necessary for proper design of the channel.

(B) The preferred method to design the linings is based on the maximum shear force on the lining.

(C) The preferred method to design the linings is based on the bearing capacity of the lining determined using the California Bearing Test (CBT).

(D) Rock riprap is an example of rigid lining.

176. The specific discharge, q, (or Darcy velocity, v) is expressed in various ways. Which of the following equations for specific discharge is INCORRECT?

(A) $q = -K\left[\frac{dh}{dx}\right]$

(B) $q = -K\left[\frac{h_2 - h_1}{L}\right]$

(C) $q = -K\left[\frac{\Delta h}{L}\right]$

(D) $q = K\left[\frac{h_1 + h_2}{2L}\right]$

177. A well draws from an aquifer with a coefficient of permeability of 0.001 ft/sec.

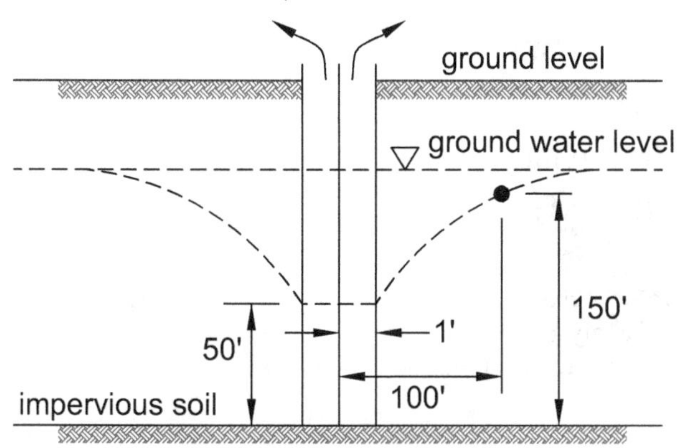

The well discharge rate is most nearly

(A) 11.8 ft^3/sec

(B) 13.6 ft^3/sec

(C) 31.4 ft^3/sec

(D) 42.7 ft^3/sec

178. Which of the following statements is correct?

(A) The more biochemical oxygen demand (BOD) a water sample has, the healthier the water is.

(B) The more dissolved oxygen (DO) the sample has, the healthier the water is.

(C) Where wastewater is discharged into a clean water stream, both BOD and DO concentrations decrease.

(D) The abbreviation BOD_5 denotes an average of five samples.

179. A wastewater plant intends to use horizontal flow grit chambers as pretreatment. The design flow rate is 20 ft^3/sec. The chamber is 6 ft wide and 8 ft deep. Assume no friction loss or head loss. If the grit chamber is 100 ft long, the amount of time a particle would remain in the chamber is most nearly

(A) 1 min

(B) 2 min

(C) 3 min

(D) 4 min

180. Which of the following statements is true?

(A) Water hardness is defined as the sum of all monovalent cations.

(B) Water hardness is defined as the sum of all divalent cations.

(C) Water hardness is defined as the sum of all polyvalent cations.

(D) Water hardness is defined as the sum of all cations and anions.

181. The arch shown is subjected to both uniformly distributed horizontal and vertical loads.

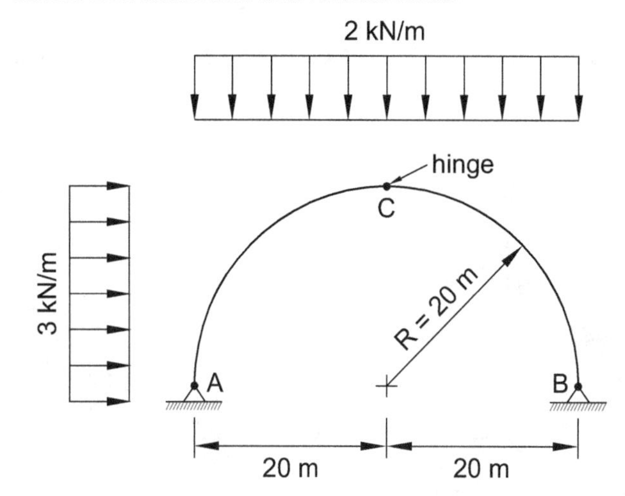

The vertical reaction at support A is

(A) 25 kN ↑

(B) 40 kN ↑

(C) 55 kN ↑

(D) 60 kN ↓

182. A 16 ft steel cantilever is made of a W16 × 57 steel section and carries a uniformly distributed dead load of 1.0 kip/ft and a live load of 0.8 kip/ft over its entire span. The maximum deflection about the major axis at the ultimate load of the cantilever is most nearly

(A) 0.0015 in

(B) 0.012 in

(C) 1.1 in

(D) 1.6 in

183. A concrete rectangular column is shown.

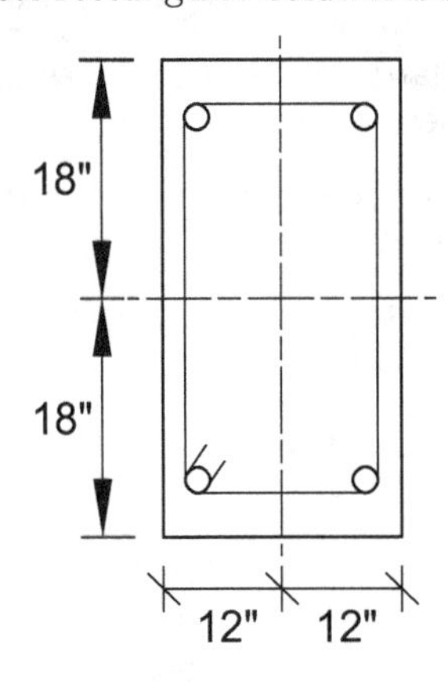

What is most nearly the controlling radius of gyration of the column?

(A) 7 in

(B) 9 in

(C) 10 in

(D) 20 in

184. A truss is shown.

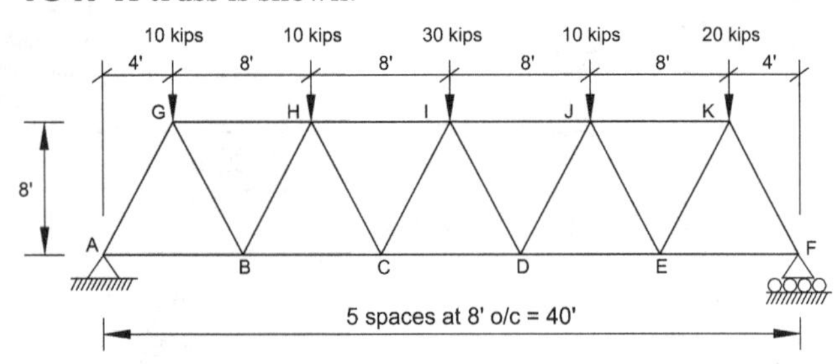

This truss is

(A) unstable

(B) statically determinate

(C) statically indeterminate to the first degree

(D) statically indeterminate to the second degree

185. A beam carries a point load, as shown.

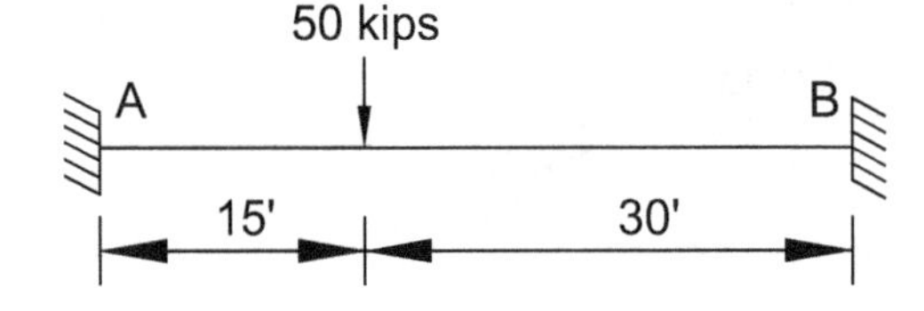

Assuming clockwise moment is positive, the moments at ends A and B, respectively, are most nearly

(A) −167 ft-kips, 333 ft-kips

(B) −333 ft-kips, 167 ft-kips

(C) 167 ft-kips, −333 ft-kips

(D) 333 ft-kips, −167 ft-kips

186. A fixed-end beam spans 30 ft. The beam carries a uniformly distributed load (UDL) of 1 kip/ft and a 10 kip concentrated load at its midspan.

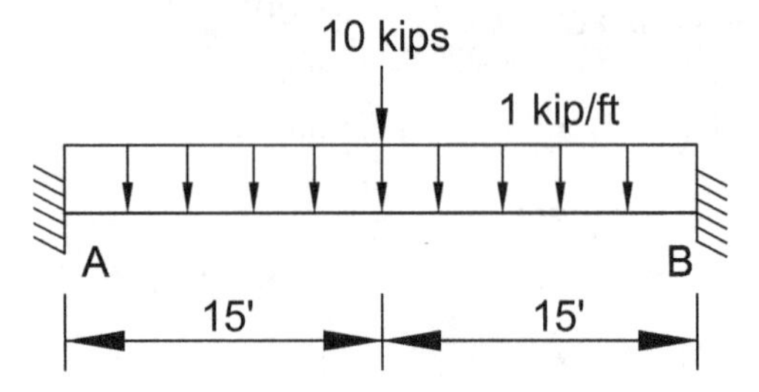

Assuming counterclockwise moment is positive, the bending moment at support A is most nearly

(A) −112.5 ft-kips

(B) −37.50 ft-kips

(C) 37.50 ft-kips

(D) 112.5 ft-kips

187. A car weighing 6000 lbf travels at 35 mph when it collides with a rigid barrier. The car's front end is crushed, buckled, and bent under impact. At a result, the car, which is 18 ft in length before the impact, is 16 ft in length after the impact. The impact force on the barrier is most nearly

(A) 6200 lbf

(B) 15,000 lbf

(C) 26,000 lbf

(D) 120,000 lbf

188. A frame with a diagonal brace is subjected to a wind load, P, as shown. The brace consists of a W10 × 54 steel section pinned at both ends. The steel yield stress, F_y, is 50 ksi.

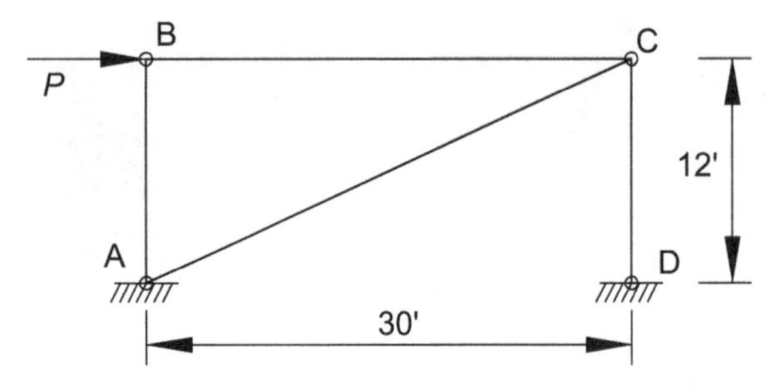

Assuming that the diagonal brace capacity is the controlling factor for the frame stability, the critical force, P, that the frame can resist is most nearly

(A) 0 kips
(B) 144 kips
(C) 155 kips
(D) 550 kips

189. Four steel sections are considered for a beam design. The properties of the sections are shown.

section properties	A	B	C	D
area (in^2)	21.1	17.9	16.8	14.7
moment of inertia (in^4)	597	640	758	800

The section that would deflect the least is

(A) section A
(B) section B
(C) section C
(D) section D

190. The LRFD method is used in steel design. The abbreviation LRFD denotes

(A) limiting resistance and failure design
(B) load and resistance force design
(C) load and resistance factor design
(D) long-span roof and floor design

191. A reinforced concrete cantilever beam is designed to carry a uniformly distributed downward load. The beam needs eight #9 bars as flexural reinforcement. Out of the bar patterns shown, which bar pattern should be used to place the reinforcement in the beam?

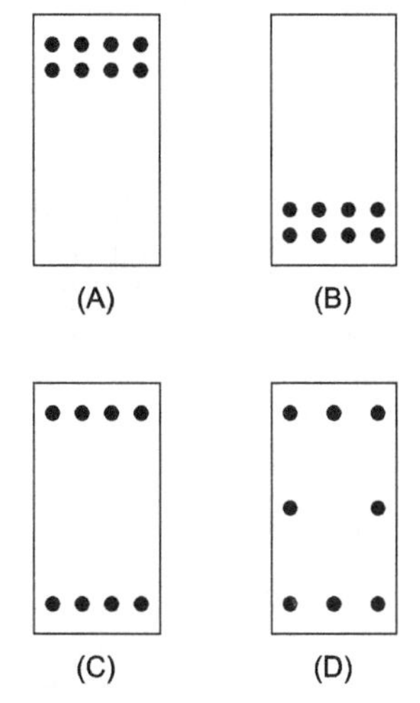

192. Consider the reinforced concrete column interaction diagram as shown. Two points, X and Y, are shown in the illustration. Draw a circle on one of the points that pertains to safe loading.

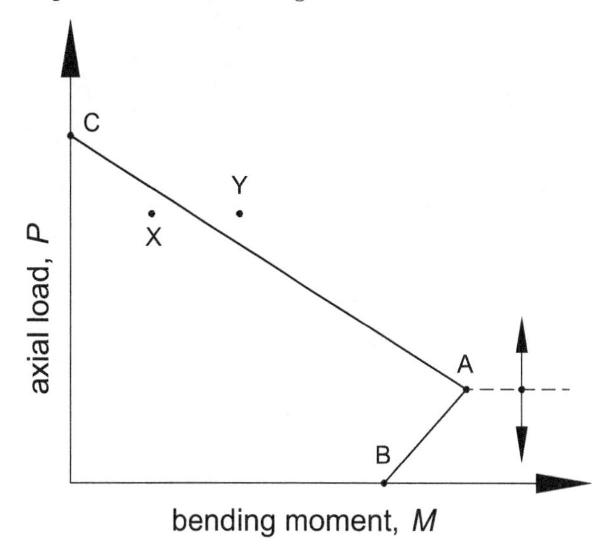

193. The liquid limit of an elastic silty soil is 52. The plastic limit is 35. The soil has the grain-size characteristics shown.

sieve size	percent finer
2.0 (no. 10)	100
0.075 (no. 200)	50
0.050	40
0.005	30
0.002	20

The AASHTO soil classification is

(A) A-7-5 (6)
(B) A-7-6 (6)
(C) A-2-7 (8)
(D) A-5 (8)

194. A rectangular combined footing supports two columns, as shown.

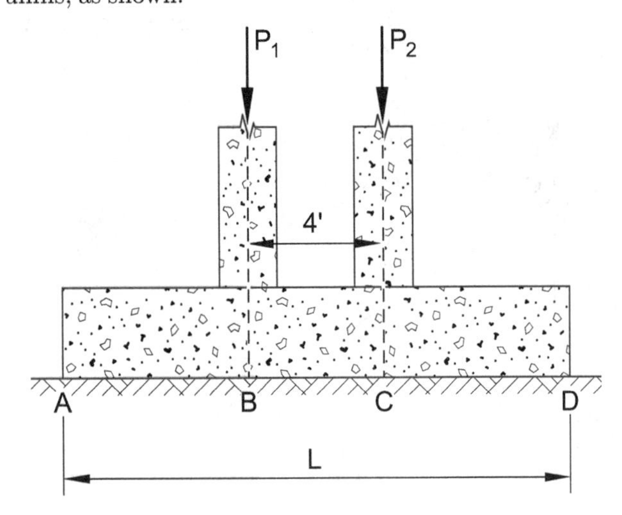

The design data is

$$\begin{aligned} \text{column 1 load} &= 200 \text{ kips} \\ \text{column 2 load} &= 400 \text{ kips} \\ \text{allowable soil pressure} &= 6 \text{ ksf} \\ \text{footing size} &= 7 \text{ ft} \times 15 \text{ ft} \end{aligned}$$

It is required to uniformly distribute the loads to achieve a uniform pressure on the soil below the footing. Assuming the footing is rigid, the distance AB is most nearly

(A) 4 ft 10 in

(B) 5 ft 6 in

(C) 7 ft 2 in

(D) 7 ft 6 in

195. A soil sample weighing 222 g is tested. To determine the soil volume, it is placed in a 500 cm^3 container. 390 cm^3 of water is needed to fill the container. The unit weight of the soil is most nearly

(A) 2.02 g/cm^3

(B) 62.4 g/cm^3

(C) 112 g/cm^3

(D) 126 g/cm^3

196. A sample of sandy soil is tested. The minimum and maximum dry unit weights are 110 pcf and 140 pcf, respectively. The project specifications require that the soil in the field should be compacted to at least 70% of relative density level. The minimum acceptable dry density of the compacted soil is most nearly

(A) 120 pcf

(B) 125 pcf

(C) 130 pcf

(D) 135 pcf

197. A light-concrete 3 ft cube is placed at the bed of a pond that is 3 ft deep. The concrete density is 120 pcf. Assuming the block base is in full contact with the bed and exerts a uniform pressure on the bed, the contact pressure the bed is subjected to is most nearly

(A) 57.4 psf

(B) 172 psf

(C) 1550 psf

(D) 3240 psf

198. A 12 ft high retaining wall and its lateral soil pressure diagrams are shown.

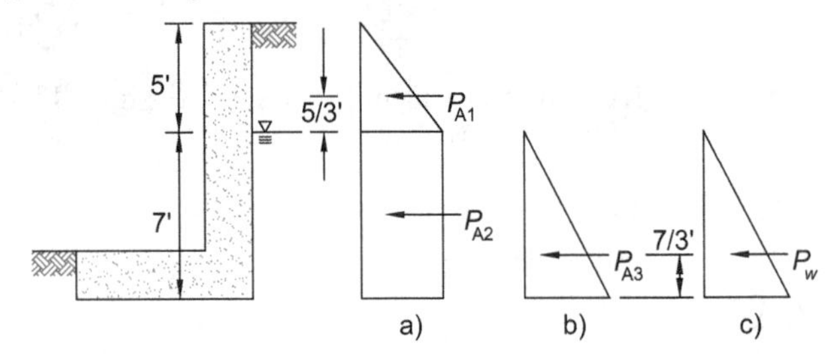

If the pressures P_{A1}, P_{A2}, P_{A3}, and P_w are 500, 1400, 470, and 1529 lbf, respectively, the overturning moment at the base of the wall per unit wall length is most nearly

(A) 5200 ft-lbf

(B) 8500 ft-lbf

(C) 11,000 ft-lbf

(D) 14,000 ft-lbf

199. A sand layer is 20 ft thick and weighs 115 pcf. Its angle of internal friction is 35°. Using the Mohr-Coulomb criteria, the soil shear strength at the bottom of the layer is most nearly

(A) 0 psi

(B) 10 psi

(C) 15 psi

(D) 1610 psi

200. An end-bearing concrete pier is designed to carry 600 kips working load. The allowable bearing capacity of rock is 12 ksf. Considering that the pier is short and of adequate concrete strength, the pier diameter is most nearly

(A) 4 ft

(B) 8 ft

(C) 10 ft

(D) 12 ft

201. A sample taken from a normally loaded clay is tested. The sample's liquid limit, LL, is 40, and the effective stress is 1500 psi. The vertical stress on the sample is increased to 2500 psi. The corresponding change in void ratio is most nearly

(A) 0.06

(B) 0.10

(C) 0.20

(D) 0.60

202. A 10 ft thick clay layer is bounded by two layers of sandy soil. The clay is fully saturated and normally loaded with an initial void ratio of 0.70. The compression index is 0.18, and the effective stress is increased from 1500 psi to 2500 psi. The total settlement of the clay layer is most nearly

(A) 0.5 in

(B) 1.0 in

(C) 3.0 in

(D) 6.0 in

203. The LEAST effective natural method of soil erosion control is

(A) surface treatment using a compost blanket

(B) a vegetative filler strip planted parallel to the runoff flow

(C) erosion control blankets fabricated from natural and synthetic materials

(D) seeding and fertilizing

204. A freeway has a posted speed limit of 55 mph. An existing horizontal curve along the freeway has a radius of 950 ft. The superelevation slope is 8%, and the side friction factor is 0.12. The maximum safe speed that should be posted for the curved region of the freeway is most nearly

(A) 40 mph

(B) 45 mph

(C) 50 mph

(D) 55 mph

205. A car was traveling up a 3% incline on a road with a posted speed limit of 40 mph. The driver applied the brakes to avoid hitting a deer on the road. The vehicle left skid marks of 80 ft in length. The surface friction was 0.6. Which of the following statements is correct?

(A) The car was traveling above the posted speed limit when the brakes were applied.

(B) The car was traveling within the posted speed limit when the brakes were applied.

(C) The driver's reaction/perception time is needed to determine the speed the car was traveling when the brakes were applied.

(D) The driver's weight and the weight of the car are needed to determine the speed the car was traveling when the brakes were applied.

206. Which of the following statements about the superelevation of a roadway is INCORRECT?

(A) The purpose of providing superelevation is to counterbalance centripetal forces.

(B) For roads, the centerline is maintained at grade, the inner edge is lowered, and outer edge is raised.

(C) For railroads, the inner edge is maintained at grade, the centerline is raised by half of the superelevation, and the outer edge is raised by the full amount of superelevation.

(D) Higher superelevation rates are easier to attain in urban settings than in rural areas.

207. Which of the following statements about AASHTO pavement design method is INCORRECT?

(A) The design method is used for a full range of ESALs.

(B) The structural number, SN, is determined based on the projected design ESAL.

(C) The design method is limited to flexible pavements.

(D) The engineer can select the type of surface used, which can be either a layer of asphalt concrete, a single surface treatment, or a double-surface treatment.

208. A pedestrian-crossing green light signal interval is being programmed. The crosswalk is 105 ft long and its width is less than or equal to 10 ft. If 10 pedestrians are expected to cross during the interval, the minimum green light time interval for the pedestrian crossing is most nearly

(A) 23 sec

(B) 32 sec

(C) 36 sec

(D) 41 sec

209. A traffic intersection is 60 ft wide, and the posted speed is 40 mph. Traffic signal design for the intersection assumes that the deceleration is 11 ft/sec^2 and the driver's reaction time is 2.3 sec. The length of the yellow light time interval at the intersection should be most nearly

(A) 2.0 sec

(B) 3.9 sec

(C) 5.0 sec

(D) 7.3 sec

210. During a traffic study, six vehicles are observed between two points 200 ft apart, as shown.

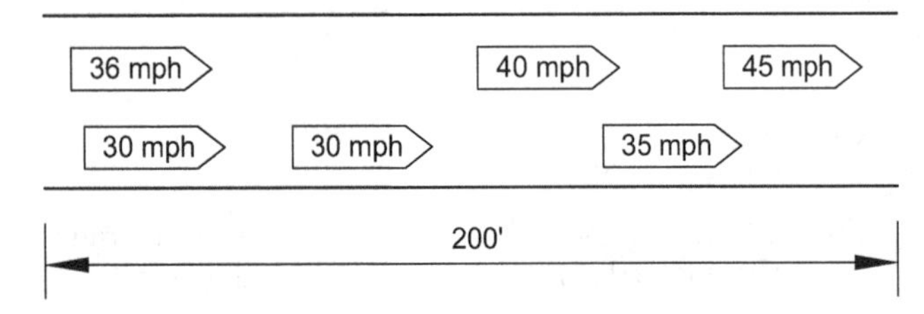

The time mean speed on the roadway is most nearly

(A) 32 mph

(B) 36 mph

(C) 43 mph

(D) 48 mph

211. The traffic flow on a roadway was studied, and the following relationship was established.

$$2\text{v} + k - 120 = 0$$

v = speed in miles per hour

k = density in vehicles per mile

The density (vehicles/mile) at which the road traffic will jam is most nearly

(A) 0 vehicles/mi

(B) 60 vehicles/mi

(C) 120 vehicles/mi

(D) 240 vehicles/mi

212. At an intersection, 20 crashes were recorded last year, and the 24 hr vehicle volume entering from all approaches was 10,000. The number of crashes per million entering vehicles, rounded to two decimal places, is ______. (Fill in the blank.)

213. Two unit-item bids, in dollars per cubic yard, are given.

work item	bid A	bid B
10,000 yd^3 earth excavation	\$8	\$4.40
1000 yd^3 rock excavation	\$15	\$50

The earth and rock excavation quantities are estimated based on the geotechnical engineering report; however, due to differing site conditions, the quantities may change significantly.

As a project engineer, you should recommend to the owner to

(A) accept bid B

(B) accept bid B for the soil excavation work item and reject the rock excavation work item

(C) accept bid A for the rock excavation work item and reject the soil excavation work item

(D) reject bid B

214. An owner has a budget of \$100,000 to route and seal concrete floor cracks in a parking garage. The cracks are estimated to total 25,000 linear feet, and the estimated unit cost is \$4 per feet. The owner signs a unit price contract with the contractor at a unit repair cost of \$5 per feet. The price includes 10% net profit for the contractor. At completion, the cracks are a total of 30,000 linear ft. The owner's obligation to pay the contractor for completing the project is most nearly

(A) \$100,000

(B) \$110,000

(C) \$150,000

(D) \$165,000

215. A free-standing concrete wall 10 ft tall and 10 in thick is to be built. The unit weight of the concrete is 145 lbf/ft^3 in flowable phase. The concrete has a high slump. The space between the wall formwork is filled rapidly before stiffening of the concrete takes place. The maximum lateral force that the newly placed concrete exerts against 1 ft width of the formwork is most nearly

(A) 5000 lbf

(B) 6000 lbf

(C) 7000 lbf

(D) 8000 lbf

216. To avoid "struck-by" personal injuries and fatalities from rolling objects, which of the following measure is INCORRECT?

(A) Train workers for safety.

(B) Install traffic signs alerting vehicle drivers of presence of workers.

(C) Install traffic signals around construction areas at potentially hazardous points.

(D) Increase compressed air pressure used for cleaning to above 30 psi.

217. A contractor signed a contract to haul and compact 100,000 yd^3 of fill material in 10 months at a lump sum price of $1.1 million, which includes 10% profit. The contractor allocated uniform resources for the construction period and projected uniform productivity. After two months, the contractor has hauled 15,000 yd^3 and spent $210,000. The contractor's budgeted cost of the work performed is ______. (Fill in the blank.)

218. An arrow diagram created using the critical path method (CPM) is shown.

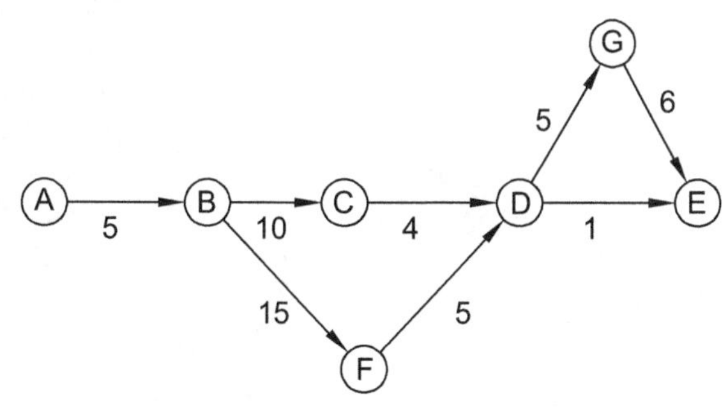

The minimum numbers of days to complete the project is

(A) 20

(B) 30

(C) 36

(D) 51

219. A 100 ft long concrete retaining wall, 10 ft tall and 10 in thick, is being planned to be built. The cost of leasing the forms is $2 per ft^2 for a single use. The labor cost of erecting and stripping the forms is $5 and $2 per ft^2, respectively. The total cost of the formwork to build the wall is most nearly

(A) $4000

(B) $10,000

(C) $12,000

(D) $18,000

220. A single-story building has 100 columns. The column properties are

- column diameter, 24 in
- vertical steel rebars, 8 #9
- rebar length, 12 ft 9 in
- column steel ties, #3 @ 12 in on center with 12 in lap
- concrete cover to tie centerline, 2 in

The weight of column ties is most nearly

(A) 1.38 tons

(B) 1.65 tons

(C) 1.78 tons

(D) 2.92 tons

STOP!

DO NOT CONTINUE!

This concludes the examination. If you finish early, check your work and make sure that you have followed all instructions. After checking your answers, you may turn in your examination booklet and answer sheet and leave the examination room. Once you leave, you will not be permitted to return to work or change your answers.

Exam 3 Instructions

In accordance with the rules established by your state, you may use any approved battery- or solar-powered, silent calculator to work this examination. However, no blank papers, writing tablets, unbound scratch paper, or loose notes are permitted. Sufficient paper will be provided. The *NCEES FE Reference Handbook* is the only reference you are allowed to use during this exam.

You are not permitted to share or exchange materials with other examinees.

You will have six hours in which to work this session of the examination. Your score will be determined by the number of questions that you answer correctly. There is a total of 110 questions. All 110 questions must be worked correctly in order to receive full credit on the exam. There are no optional questions. Each question is worth 1 point. The maximum possible score for this section of the examination is 110 points.

Partial credit is not available. No credit will be given for methodology, assumptions, or work written on scratch paper.

Record all of your answers on the Answer Sheet. Mark your answers with a no. 2 pencil. Answers marked in pen may not be graded correctly. Marks must be dark and must completely fill the bubbles. Record only one answer per question. If you mark more than one answer, you will not receive credit for the question. If you change an answer, be sure the old bubble is erased completely; incomplete erasures may be misinterpreted as answers.

If you finish early, check your work and make sure that you have followed all instructions. After checking your answers, you may submit your answers and leave the examination room. Once you leave, you will not be permitted to return to work or change your answers.

When permission has been given by your proctor, you may begin your examination.

Name: ____

Last First Middle Initial

Examinee number: ______________

Examination Booklet number: ______________

Fundamentals of Engineering Examination

Exam 3

Exam 3 Answer Sheet

221. (A) (B) (C) (D)
222. ____________
223. ____________
224. ____________
225. ____________
226. ____________
227. ____________
228. ____________
229. ____________
230. ____________
231. (A) (B) (C) (D) (E)
232. (A) (B) (C) (D)
233. (A) (B) (C) (D)
234. (A) (B) (C) (D)
235. (A) (B) (C) (D)
236. ____________
237. ____________
238. (A) (B) (C) (D) (E)
239. (A) (B) (C) (D)
240. ____________
241. (A) (B) (C) (D)
242. (A) (B) (C) (D) (E)
243. (A) (B) (C) (D)
244. (A) (B) (C) (D) (E)
245. (A) (B) (C) (D)
246. (A) (B) (C) (D) (E)
247. (A) (B) (C) (D)
248. (A) (B) (C) (D)
249. (A) (B) (C) (D)
250. (A) (B) (C) (D)
251. (A) (B) (C) (D)
252. (A) (B) (C) (D)
253. (A) (B) (C) (D)
254. (A) (B) (C) (D)
255. (A) (B) (C) (D)
256. ____________
257. ____________
258. (A) (B) (C) (D)
259. (A) (B) (C) (D)
260. ____________
261. (A) (B) (C) (D)
262. (A) (B) (C) (D)
263. (A) (B) (C) (D) (E)
264. (A) (B) (C) (D)
265. (A) (B) (C) (D)
266. (A) (B) (C) (D)
267. (A) (B) (C) (D)
268. (A) (B) (C) (D)
269. (A) (B) (C) (D)
270. (A) (B) (C) (D)
271. (A) (B) (C) (D)
272. (A) (B) (C) (D)
273. ____________
274. (A) (B) (C) (D) (E)
275. (A) (B) (C) (D)
276. (A) (B) (C) (D)
277. (A) (B) (C) (D)
278. (A) (B) (C) (D)
279. (A) (B) (C) (D)
280. (A) (B) (C) (D)
281. (A) (B) (C) (D)
282. (A) (B) (C) (D) (E)
283. (A) (B) (C) (D)
284. (A) (B) (C) (D)
285. (A) (B) (C) (D)
286. (A) (B) (C) (D) (E)
287. (A) (B) (C) (D)
288. (A) (B) (C) (D)
289. (A) (B) (C) (D)
290. (A) (B) (C) (D)
291. (A) (B) (C) (D) (E)
292. (A) (B) (C) (D)
293. (A) (B) (C) (D)
294. (A) (B) (C) (D)
295. (A) (B) (C) (D)
296. (A) (B) (C) (D)
297. (A) (B) (C) (D)
298. (A) (B) (C) (D)
299. (A) (B) (C) (D)
300. (A) (B) (C) (D)
301. (A) (B) (C) (D)
302. (A) (B) (C) (D)
303. (A) (B) (C) (D)
304. (A) (B) (C) (D)
305. (A) (B) (C) (D)
306. (A) (B) (C) (D)
307. (A) (B) (C) (D)
308. (A) (B) (C) (D)
309. (A) (B) (C) (D)
310. (A) (B) (C) (D)
311. (A) (B) (C) (D)
312. (A) (B) (C) (D)
313. (A) (B) (C) (D)
314. (A) (B) (C) (D)
315. (A) (B) (C) (D)
316. (A) (B) (C) (D)
317. (A) (B) (C) (D)
318. (A) (B) (C) (D)
319. (A) (B) (C) (D)
320. (A) (B) (C) (D)
321. (A) (B) (C) (D)
322. (A) (B) (C) (D)
323. (A) (B) (C) (D)
324. (A) (B) (C) (D)
325. (A) (B) (C) (D)
326. (A) (B) (C) (D)
327. (A) (B) (C) (D)
328. (A) (B) (C) (D)
329. (A) (B) (C) (D)
330. (A) (B) (C) (D)

Exam 3

221. A line passes through (2, 5) and is perpendicular to the line $3x - 5y = 7$. The equation of the line is

(A) $-5x + 3y = 5$

(B) $3x + 2y = 16$

(C) $3x + 5y = 31$

(D) $5x + 3y = 25$

222. Earthquake intensities are measured using a base-10 logarithmic scale called the Richter scale. The Richter scale measurement gives the exponent of the relative intensity of the earthquake. The 1906 San Francisco earthquake is estimated to be 8.3 on the Richter scale. An Alaskan earthquake measured 8.4. The ratio of intensities of the Alaskan earthquake to the 1906 San Francisco earthquake to the second decimal place is ______. (Fill in the blank.)

223. A parabola described by the equation $y = 2x^2$ is rotated about its y-axis. The volume between $y = 0$ and $y = 10$ is ______. (Fill in the blank.)

224. Two vectors are

$$\mathbf{A} = 2\mathbf{i} + 2\mathbf{j} + \mathbf{k}$$

$$\mathbf{B} = 6\mathbf{i} - 3\mathbf{j} - 2\mathbf{k}$$

The angle between the vectors, to two decimal places, is ______. (Fill in the blank.)

225. A construction crew initially places 100 yd^3 of concrete per unit time. The crew undergoes a series of productivity training sessions. After each productivity training session, the rate of placement of concrete increases by 5% from the previous period. The volume of concrete the crew can place in a day after 10 training sessions, in whole cubic yards, is most nearly ______. (Fill in the blank.)

226. The equation for a parabola is shown.

$$y = x - \frac{1}{6}x^2$$

The parabola's curvature at $x = 3$ is ______. (Fill in the blank.)

227. A state lottery requires a winning player to match 3 numbers drawn from an urn containing 10 balls, numbered 0 to 9. The three numbers are drawn randomly from the urn, one at a time, without replacing the picked ball. The probability of one entry winning the lottery is most nearly ______. (Fill in the blank.)

228. A 500,000 ft^2 area is excavated to construct a shopping mall. The foundation is expected to be partially on sandy silt and partially on rock. One geotechnical report estimates that the foundation is 25% sandy silt, and the other report estimates that the foundation is 55% sandy silt. The unit cost to excavate sandy-silt-type material is $\$100/ft^2$, and it costs $\$300/ft^2$ to excavate the rock to the specified depth. The owner gives the first report double the weight of the second report. The expected foundation cost (in millions of dollars) is ______. (Fill in the blank.)

229. If the mean 1 year rainfall in an area is 30 in and the standard deviation is 6.6, the coefficient of variation is most nearly ______%. (Fill in the blank.)

230. A data set consists of four points, as shown.

x	y
2	9
3	11
5	15
9	22

Using least square regression, the line equation that best fits the data form is $y =$ ______ + ______ x. (Fill in the blanks.)

231. Which of the following statements about the professional code of ethics is (are) INCORRECT, if any?

(A) Bribery is defined as remuneration for performance of an act that is consistent with the work one is hired to perform.

(B) Bribery is defined as remuneration for performance of an act that is inconsistent with the project contract.

(C) A gift is defined as something given after an act is done in accordance with one's obligation, without prior knowledge or expectation.

(D) If a gift is perceived as a bribe, then it is a bribe.

(E) It is always ethical for a licensee to make a political contribution.

232. Which of the following is NOT an example of professional misconduct?

(A) fraud

(B) gross incompetence

(C) gross negligence

(D) running a website denying climate change

233. When providing design services for a construction project, which standard of care do the courts traditionally assign to the engineer in responsible charge?

(A) The engineer should produce a set of perfect drawings.

(B) The engineer should meet the minimum requirements of the governing building code and no more.

(C) The engineer should use state-of-the-art methods that a reasonably prudent engineer would follow when designing similar projects in similar localities.

(D) The engineer is responsible for the job site safety of workers.

234. A licensee makes a professional judgment that is overruled. The licensee believes that the life, health, property, or welfare of the public is endangered. Who should the licensee inform?

(A) the licensee's employer

(B) the licensee's client

(C) other authorities as may be appropriate

(D) A or B, and C

235. An owner issues a request for proposal (RFP) for the construction of an office building per drawings and specifications. The RFP requests potential bidders submit their bids. A contractor submits a bid to do the work for \$1.1M. After two months, the owner asks the contractor to reduce the price and do the work for \$1M. Under contract law, which statement is the fairest?

(A) The contractor may walk away and refuse to do the job for the reduced sum.

(B) The contractor can't walk away and must do the work for \$1M.

(C) The contractor is entitled to demand that the owner accept its original \$1.1M bid.

(D) The contract may declare the original bid expired since the owner took too long to call the contractor back.

236. A design firm anticipates that it would need to upgrade its computers and drafting system in 5 years. The upgrade will cost \$500,000. The firm makes deposits every month into an account that earns 0.5% interest compounded monthly. The minimum amount the firm should deposit every month to save for the anticipated cost, to the nearest whole dollar, is ______. (Fill in the blank.)

237. A prospective project is expected to generate the annual cash flows shown. The inflation rate is expected to be 10%, which is accounted for in the tabulated cash flows.

time (end of year)	cash flow (\$)
1	80,000
2	120,000
3	180,000
4	270,000

The real rate of growth is most nearly ______%. (Fill in the blank.)

238. Equipment selection has been narrowed to two options. The costs, life expectancy, and salvage values are shown. The expected rate of return is 8%.

	equipment	
parameters	1	2
initial cost	\$50,000	\$75,000
annual maintenance	\$15,000	\$10,000
life expectancy (yrs.)	10	15
salvage value	\$5,000	\$12,000

Select two options to identify which equipment is better and the amount the selection saves.

(A) Equipment 1 is a better choice.

(B) Equipment 2 is a better choice.

(C) It saves approximately \$1800.

(D) It saves approximately \$4200.

(E) It saves approximately \$8500.

239. Two alternatives for a construction project are being considered. Alternative A costs \$1M initially and \$100,000 in service and maintenance each year. Alternative B costs \$1.8M and \$65,000 in service and maintenance each year. Assume a service life of 25 years, no salvage value, and no interest. The engineer should recommend that the owner

(A) select alternative A

(B) select alternative B

(C) select either alternative

(D) reject both alternatives and explore a third option

240. An engineering firm presently gets its drawings printed at a local blueprint facility. The firm's owner is considering buying a new printer for printing its drawings in house. The printer costs $40,000, and the annual operational and insurance cost is $3000. The life expectancy of the printer is 5 years with no salvage value. The printer is expected to generate $15,000 per year in savings for the firm. The rate of return on the printer, to the nearest whole percent is, ______. (Fill in the blank.)

241. Oil Company A must decide whether or not to drill for oil at a particular location for which the company has drilling rights. There are two possible outcomes: striking oil and not striking oil. If the decision is made to drill and the company strikes oil, the oil would be worth $13M. The cost of drilling would be $3M. At present, Company A can sell the drilling rights to Company B for $1M, with a contingency to receive another $1M if Company B strikes oil. Based on the geological assessment, the probability of striking the oil at the location is 0.30. The maximum expected payoff to the oil company is most nearly

(A) $900,000

(B) $1,300,000

(C) $3,900,000

(D) $10,000,000

242. A pole is held vertical by two cables, as shown. All connections are pinned, and the tension in cable AB is 100 N.

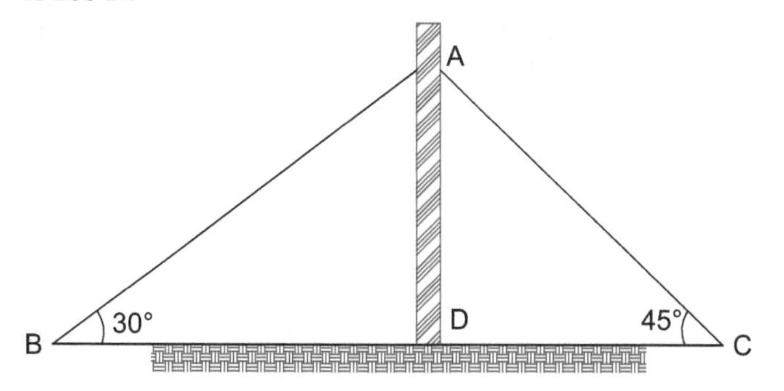

From the options given, select two that denote the type of force in cable AC and its magnitude.

(A) tension

(B) compression

(C) 82 N

(D) 100 N

(E) 122 N

243. A rooftop unit (RTU) hanging from a crane requires a horizontal pull to be placed at its specified location, as shown. The RTU weighs 1200 lbf.

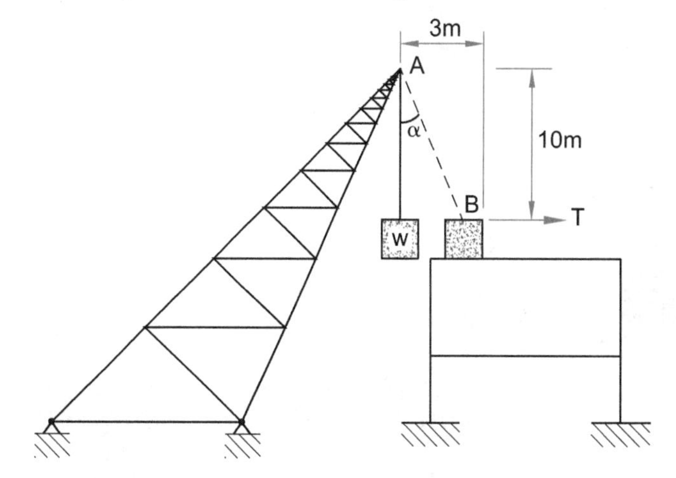

The required pull tension, T, is most nearly

(A) 37 lbf

(B) 360 lbf

(C) 1200 lbf

(D) 11,500 lbf

244. A 10 m long cantilever is carrying a uniformly distributed load of 2 kN/m over its entire span. From the options given, select two that denote the type of moment at support and its magnitude.

(A) clockwise

(B) counterclockwise

(C) 10 kN-m

(D) 100 kN-m

(E) 200 kN-m

245. Consider the rooftop unit shown.

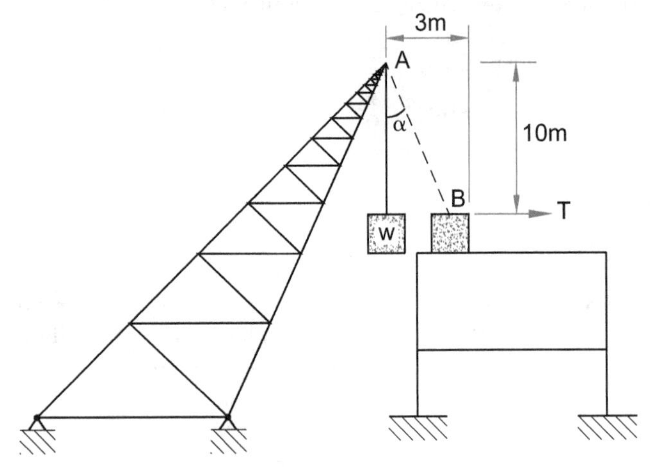

If the unit mass is 1000 kg, the required tension in the pull, T, is most nearly

(A) 3 kN

(B) 300 kN

(C) 1000 kN

(D) 3000 kN

246. An overhang beam carries four point loads, as shown. All loads are in kN, and the distances are in meters.

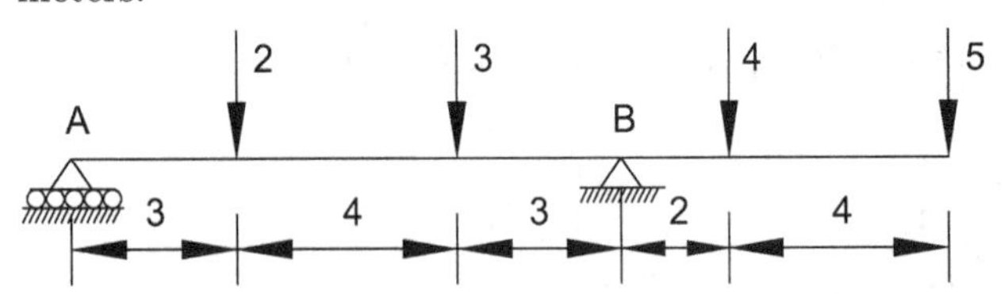

From the options given, choose the two that denote the direction of the reaction at support A and its magnitude.

(A) upward

(B) downward

(C) 1.5 kN

(D) 4.0 kN

(E) 15.0 kN

247. Consider the T-beam section shown.

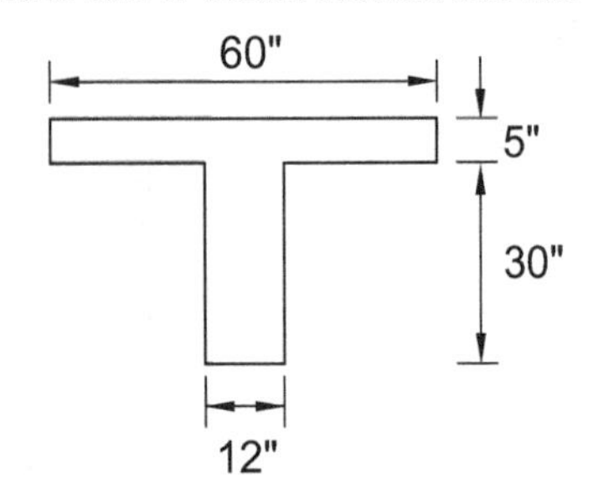

The distance from the bottom of the beam to its centroid is most nearly

(A) 19.8 in

(B) 21.55 in

(C) 22.95 in

(D) 23.45 in

248. Consider the T-beam section shown.

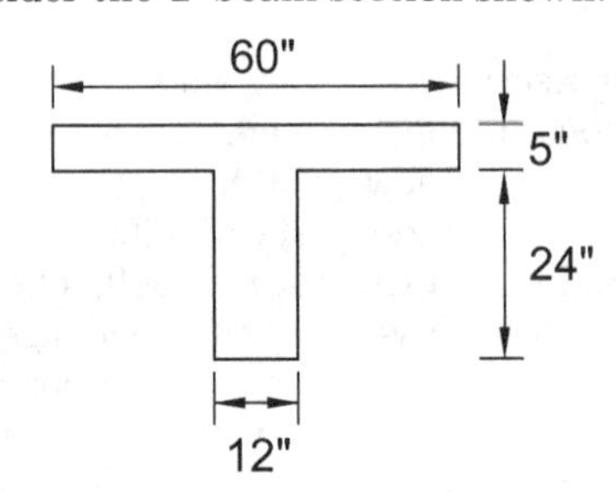

The moment of inertia about its centroidal axis is most nearly

(A) 15,000 in^4

(B) 30,000 in^4

(C) 45,000 in^4

(D) 60,000 in^4

249. A 100 kg block is placed on a sloping plane inclined at 30° to the horizontal. The friction coefficient at the interface between the block and the sloping surface is 0.3. The minimum force required to stop the block from sliding down the slope is most nearly

(A) 240 N

(B) 290 N

(C) 300 N

(D) 980 N

250. A car traveling at a speed of 12 mph on a cliff rolls off and falls on the ground 12 ft below. The distance the car will travel horizontally before hitting the ground is most nearly

(A) 10 ft

(B) 12 ft

(C) 15 ft

(D) 18 ft

251. Two cylinders start from rest and then roll down a ramp. One cylinder is a pipe and is made of a heavier material, and the other is a solid bar made of a lighter material, as shown. Their lengths, masses, coefficients of friction, and outer diameters are identical.

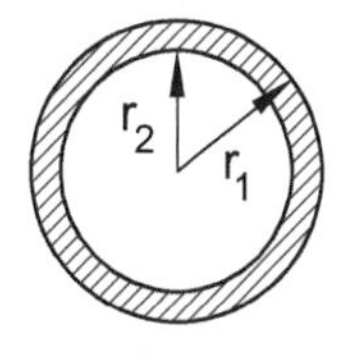

Ring

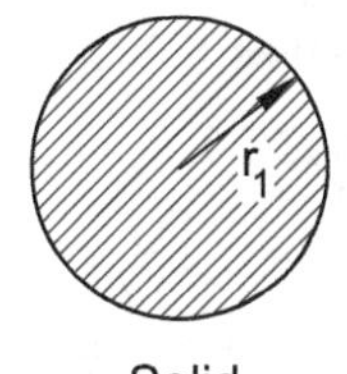

Solid

Which statement regarding the cylinders is correct?

(A) Both cylinders reach the bottom of the ramp at the same time.

(B) The ring reaches the bottom of the ramp first.

(C) The solid cylinder reaches the bottom of the ramp first.

(D) More data is needed to solve the problem.

252. A solid flywheel has a mass of 1000 kg and a radius of gyration of 1 m. It is rotating about its centroidal axis at a speed of 60 rpm. The kinetic energy of rotation of the wheel is most nearly

(A) 1200 N·m

(B) 3600 N·m

(C) 15,000 N·m

(D) 20,000 N·m

253. A flywheel rotates at a speed of 240 rpm. After 20 sec, its speed reduces to 180 rpm. If its deceleration is uniform, the number of revolutions it makes before coming to a full stop is most nearly

(A) 160

(B) 200

(C) 240

(D) 1005

254. A vehicle weighing 30 kN is traveling at 50 km/h. Its kinetic energy is most nearly

(A) 295 kJ

(B) 590 kJ

(C) 1180 kJ

(D) 3800 kJ

255. Consider a two-span beam under a uniformly distributed load (UDL) of 3 N/m.

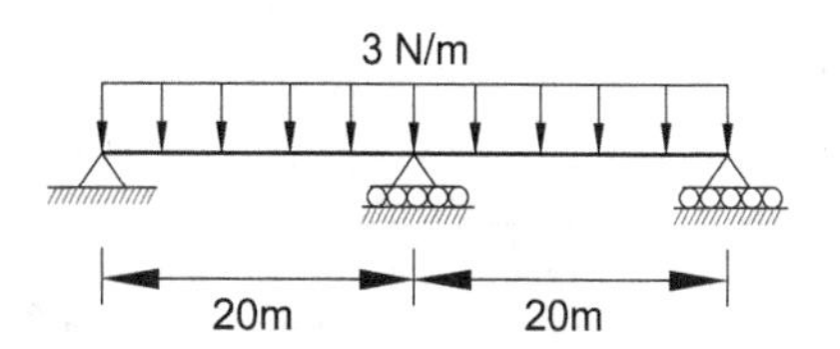

The bending moment of the beam at its central support is most nearly

(A) 0 N·m

(B) 120 N·m

(C) 150 N·m

(D) 600 N·m

256. A beam loading pattern and its shear force diagram are shown.

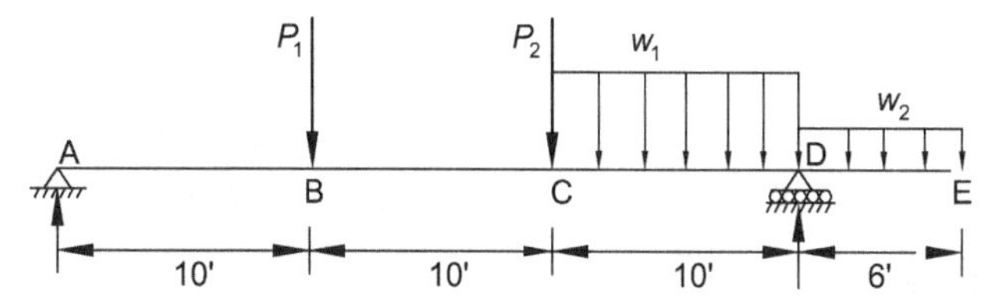

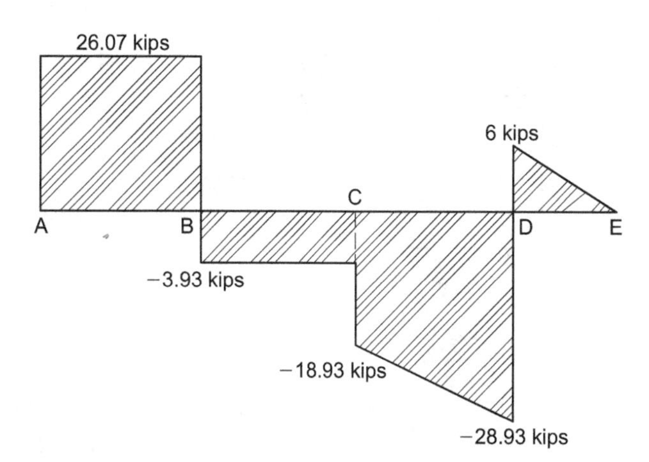

Loads P_1 and P_2 are point loads, and loads w_1 and w_2 are uniformly distributed loads (UDL). The two UDLs w_1 and w_2 are ______ and ______, respectively. (Fill in the blank.)

257. A beam loading pattern and its shear force diagram are shown.

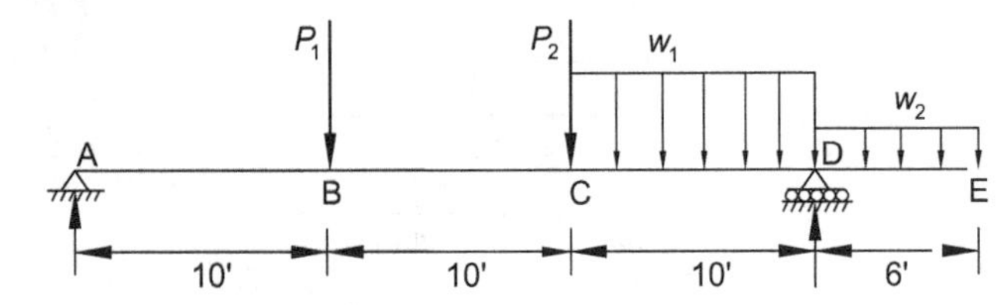

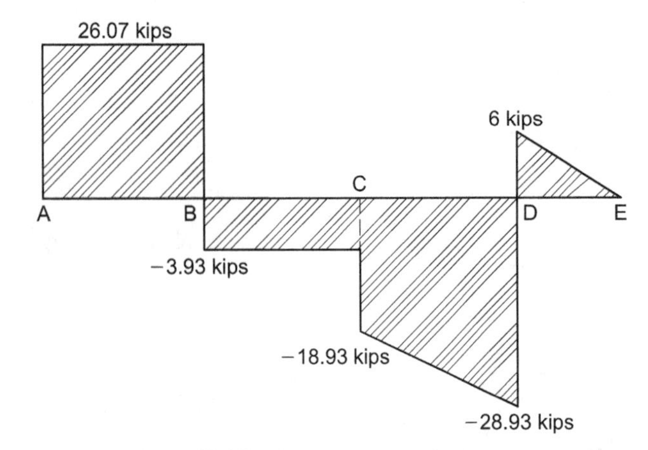

Loads P_1 and P_2 are point loads, and w_1 and w_2 are uniformly distributed loads. The two point loads P_1 and P_2 are ______ and ______, respectively. (Fill in the blank.)

258. A circular shaft is subjected to a torque, T, and bending moment, M, as shown.

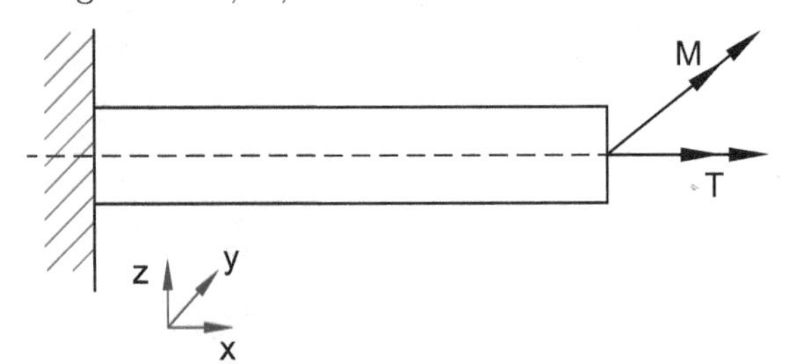

The ratio of the maximum shear stress to the maximum bending stress ratio is

(A) $\dfrac{2T}{M}$

(B) $\dfrac{T}{M}$

(C) $\dfrac{T}{2M}$

(D) $\dfrac{8T}{M}$

259. Consider a 0.5 in thick elastic-plastic steel plate with a yield stress of 50 ksi. The minimum force required to punch a 1 in diameter hole is most nearly

(A) 25 kips

(B) 40 kips

(C) 50 kips

(D) 80 kips

260. Two bars, A and B, of equal lengths are joined and act compositely. Bar A has a higher modulus of thermal expansion than bar B. Point to the bar that will be subjected to compression when the system temperature is raised.

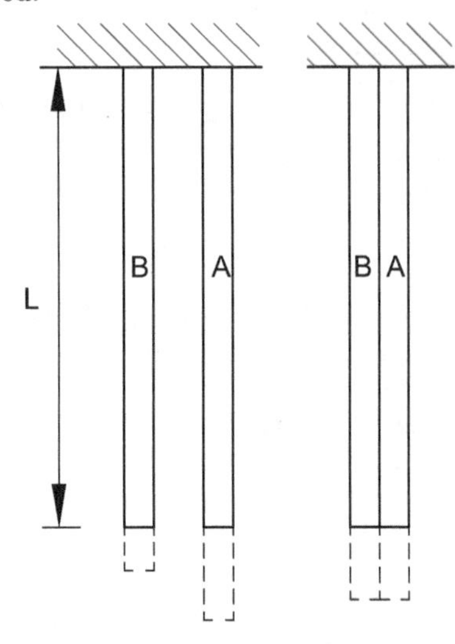

261. The principal stresses on a body are 100 ksi and –100 ksi. The corresponding maximum shear stress is most nearly

(A) –100 ksi

(B) 0 ksi

(C) 100 ksi

(D) 200 ksi

262. A 1 in diameter, 3 ft long bar specimen of a low steel alloy was tested under tension to an elongation of 0.20 in and then unloaded. The steel behavior during the test was elastic-plastic. The yield stress of the steel is 36 ksi. The permanent elongation of the bar is most nearly

(A) 0 in

(B) 0.0012 in

(C) 0.155 in

(D) 0.2 in

263. A Portland cement concrete mix is described as 1:2:4. Which two statements about the mix are INCORRECT?

(A) The ratio of the mix ingredients is by volume.

(B) The mix designation stands for one part water, two parts cement, and four parts aggregates.

(C) Artificial, synthetic, or porous aggregates are allowed in a concrete mix.

(D) The portion of an aggregate that passes 4.75 mm (no. 4) sieve and is retained on the 75 μm (no. 200) sieve is called *fine aggregate.*

(E) The aggregates in the mix weigh six times the Portland cement in the mix.

264. The most common method for protecting steel reinforcement in concrete structures from rusting is

(A) cathodic protection

(B) galvanizing

(C) epoxy coating

(D) membrane protection

265. A concrete mix has a ratio of 1:2:3, and its water/cement ratio is 0.4. Its unit weight is 145 lbf/ft^3. It is known that 611.7 lbf of cement is needed to produce one cubic yard of concrete. The weight of aggregate needed to produce one cubic yard of mix is most nearly

(A) 610 lbf

(B) 1020 lbf

(C) 2050 lbf

(D) 3060 lbf

266. Which of the following is NOT an objective of the performance-based specifications for asphalt concrete mix?

(A) controlling rutting

(B) controlling fatigue cracking

(C) controlling thermal cracking

(D) developing high strength

267. The ability of a material to absorb or store energy without permanent deformation is called

(A) hardness

(B) resilience

(C) ductility

(D) toughness

268. A pipe containing water is fitted with an inverted U-tube with a shutoff valve and connected to two points, as shown. The shutoff valve is left open, and the difference in the water levels between is measured as 15 in.

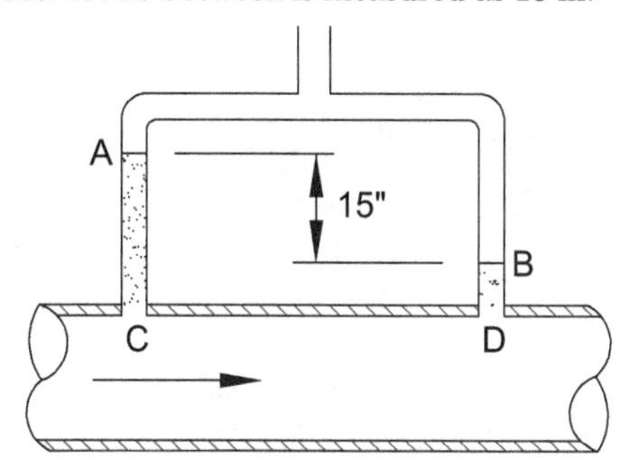

The pressure difference between points A and B is most nearly

(A) 0.54 psi

(B) 2.3 psi

(C) 2.9 psi

(D) 15 psi

269. A 3 m deep tank is filled with oil with a relative density of 0.8. The barometer reading at the surface of the tank is 760 mm Hg. The relative pressure at the bottom of the tank is most nearly

(A) 25 kPa

(B) 50 kPa

(C) 100 kPa

(D) 125 kPa

270. A 10 ft deep tank contains 6 ft of water and 4 ft of oil with a relative density of 0.75. Assume the atmospheric pressure is zero. The pressure at the bottom of the tank is most nearly

(A) 190 psf

(B) 380 psf

(C) 560 psf

(D) 1500 psf

271. A 1 in diameter nozzle delivers a stream of water at a velocity of 190 ft/sec perpendicular to a rigid plate. The force required to keep the plate from moving is most nearly

(A) 280 lbf

(B) 380 lbf

(C) 2.8 kips

(D) 3.8 kips

272. A 2 in diameter nozzle delivers a stream of water at a velocity of 200 ft/sec into an impulse turbine through a straight pipe. The mass flow rate is 8.5 slugs/sec. The maximum power the turbine can generate is most nearly

(A) 20 hp

(B) 80 hp

(C) 150 hp

(D) 300 hp

273. Water at 100°F flows in a piping system. The pressure at which cavitation would result is ______. (Fill in the blank.)

274. Which two statements regarding highway curves are INCORRECT?

(A) Horizontal curves are circular curves.

(B) Horizontal curves are parabolic curves.

(C) Vertical curves are parabolic curves.

(D) Offset curves are laid out so survey stakes are not disturbed by construction activities.

(E) Horizontal curves must maintain a constant elevation along the curve.

275. The coordinates of the vertices of a polygon are (3, 2), (6, 9), (14, 1), and (16, 11). The area of the polygon is most nearly

(A) 84 ft^2

(B) 88 ft^2

(C) 92 ft^2

(D) 150 ft^2

276. Excavated soil from a borrow pit is stockpiled in a conical shape. The stockpile has a diameter of 20 m at its base and a height of 15 m. Its volume is most nearly

(A) 1000 m^3

(B) 1600 m^3

(C) 2500 m^3

(D) 5000 m^3

277. The longitude for Elgin, Illinois is 42.0383°N. On the day of the winter solstice, the sun's rays strike the ground at an angle of about 25° at noon. The sun's rays strike the ground in the Elgin area at a vertical angle of about 72° on the day of the summer solstice. The minimum distance needed between a 60 ft tall building and a photovoltaic solar system so that no part of the system is ever in the building's shadow is most nearly

(A) 30 ft

(B) 60 ft

(C) 130 ft

(D) 180 ft

278. Refer to the vertical curve shown.

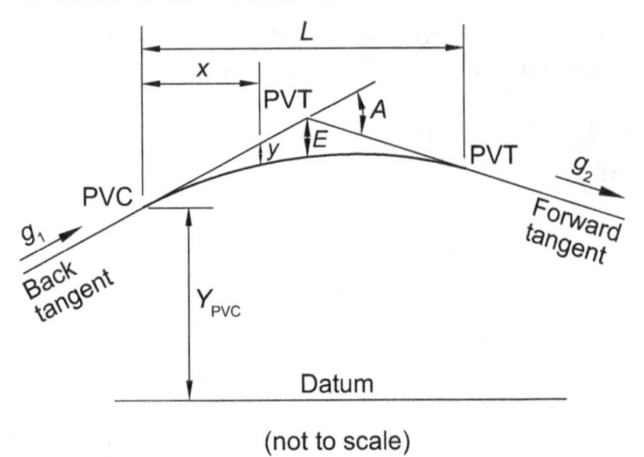

(not to scale)

Which statement regarding the illustration is INCORRECT?

(A) It shows a circular curve starting from the PVC and ending at the PVT.

(B) The tangent offsets vary with the square of the distance from the point of tangency.

(C) The rate of change of the grade is constant throughout the entire length of the curve.

(D) The back tangent and forward tangent grades can both have positive slope.

279. A 1000 ft vertical curve has a back tangent slope of 2% and a forward slope of −4%. The elevation at the point of intersection (PVI) is 2540 ft. The elevation at the point of vertical curvature is most nearly

(A) 1540 ft

(B) 2530 ft

(C) 2540 ft

(D) 2550 ft

280. A 60 ac watershed area consists of four subareas, as shown.

surface type	area (ac)	coefficient, C
buildings	16	0.60
open unpaved space	14	0.20
light industrial	20	0.65
paved roads/ parking lots	10	0.90

The peak discharge from the watershed for a rainfall intensity of 0.1 in/min is most nearly

(A) 206 ac-in/sec

(B) 206 ac-in/min

(C) 206 ac-in/hr

(D) 20.6 ft^3/sec (cfs)

281. A symmetrical open channel is shown.

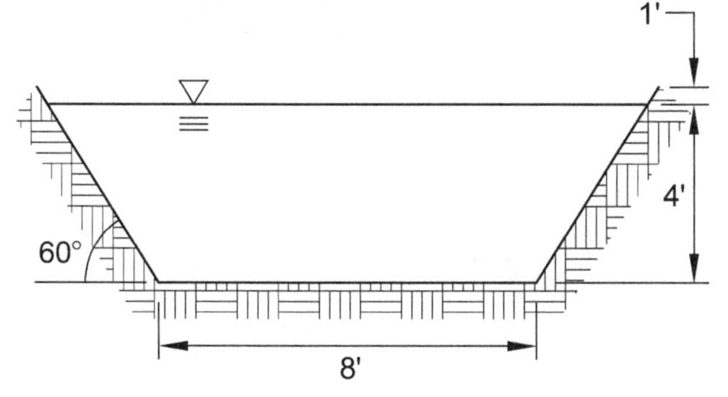

Its hydraulic radius is most nearly

(A) 2.4 ft

(B) 4.8 ft

(C) 13.0 ft

(D) 41.0 ft

282. Out of the following, select the two statements that are correct.

(A) The energy generated by a turbine increases the fluid pressure (head).

(B) The net positive suction head available for a pump decreases as the fluid vapor pressure at the pump inlet increases.

(C) The net positive suction head available for a pump increases as the fluid vapor pressure at the pump inlet increases.

(D) A pump reaches its peak discharge flow rate when it reaches its maximum operating head.

(E) The Ergun equation can be used to estimate pressure loss for both laminar and turbulent flow conditions.

283. A 10 in water main is 3000 ft long and branches into 6 in and 8 in diameter mains, each 1000 ft long, as shown. The total flow is 10 ft^3/sec. All pipelines are horizontal, the head loss due to friction is equal in both pipes, and each pipe has a coefficient of friction of 0.005.

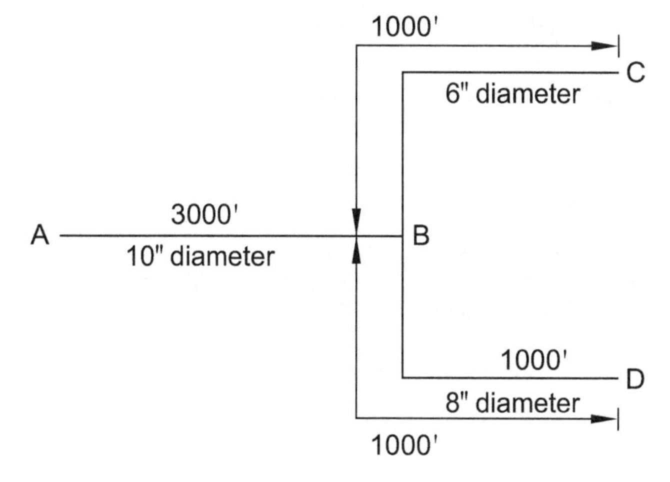

Considering the friction losses only, the flow in the 6 in pipe is most nearly

(A) 1 ft^3/sec

(B) 3 ft^3/sec

(C) 5 ft^3/sec

(D) 7 ft^3/sec

284. A dam is constructed in a 150 ft wide rectangular channel. The depth of the water over the spillway is 3 ft. The rate of water discharge over the spillway is most nearly

(A) 450 ft^3/sec

(B) 1300 ft^3/sec

(C) 2600 ft^3/sec

(D) 13,000 ft^3/sec

285. Based on Darcy's experiments on the flow of water through sand filters, which statement regarding groundwater movement is NOT correct?

(A) The flow rate is proportional to the cross-sectional area.

(B) The flow rate is proportional to the length of the sand filter.

(C) The flow rate is proportional to the pressure head drop.

(D) The flow rate is proportional to the hydraulic conductivity.

286. Select two correct statements from the following.

(A) Polluted water is always contaminated.

(B) Contaminated water is always polluted.

(C) Polluted water always contains pathogenic bacteria.

(D) Polluted water is unfit for drinking and domestic use.

(E) Potable water always tastes good.

287. A wastewater sample is tested and is found to have a 5 day BOD of 240 mg/L. Its 20 day BOD, taken as the ultimate BOD, is 320 mg/L. The BOD rate of logarithmic decay is most nearly

(A) -5.33 days^{-1}

(B) -0.28 days^{-1}

(C) 0.28 days^{-1}

(D) 5.33 days^{-1}

288. A new sanitary sewer line is being designed for a city with a population of 100,000 people. The average flow rate of the sewer line must be 100 gallons per capita per day (GPCD). Five primary clarifiers are planned, each with a 50 ft diameter. The residence time for the wastewater in each clarifier is 2 hr. The required depth of each clarifier must be at least

(A) 11 ft

(B) 12 ft

(C) 25 ft

(D) 50 ft

289. A truck carrying 10,000 kg of HCl (hydrochloric acid) crashes, causing all the acid to leak into a river. A licensed engineer has determined the need to add lime ($Ca(OH)_2$) to the river to neutralize the acid, according to the following reaction.

$$2HCl + Ca(OH)_2 \rightarrow CaCl_2 + 2H_2O$$

The amount of lime needed to neutralize the acid is most nearly

(A) 101 kg

(B) 5010 kg

(C) 10,100 kg

(D) 101,000 kg

290. A town discharges 20,000 m^3/day of wastewater into a stream. The wastewater has a concentration of dissolved oxygen (DO) of 1.5 mg/L. The stream has a flow rate of 1 m^3/s, and its DO is 7.5 mg/L. The DO after mixing is most nearly

(A) 3 mg/L

(B) 4 mg/L

(C) 6 mg/L

(D) 9 mg/L

291. The arch shown is subjected to both uniformly distributed horizontal loads and uniformly distributed vertical loads. The vertical reaction force at support at A is 25 kN, and it is acting upward.

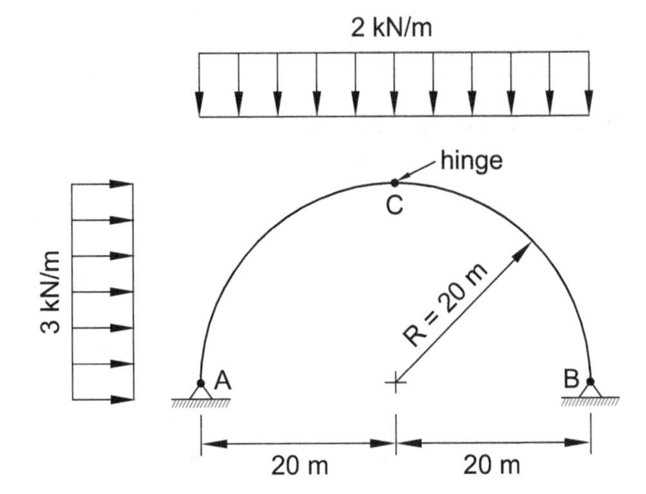

Out of the following options, select the two that describe the horizontal reaction at support A.

(A) 25 kN

(B) 30 kN

(C) 60 kN

(D) toward the center or the arch

(E) away from the center of the arch

292. A single degree of freedom system has a mass of 8000 kg. Its stiffness is 2000 kN/m. Its fundamental period is most nearly

(A) 2 s

(B) 8 s

(C) 12 s

(D) 0.5 Hz

293. A W12 × 50 steel section is used as a column, as shown.

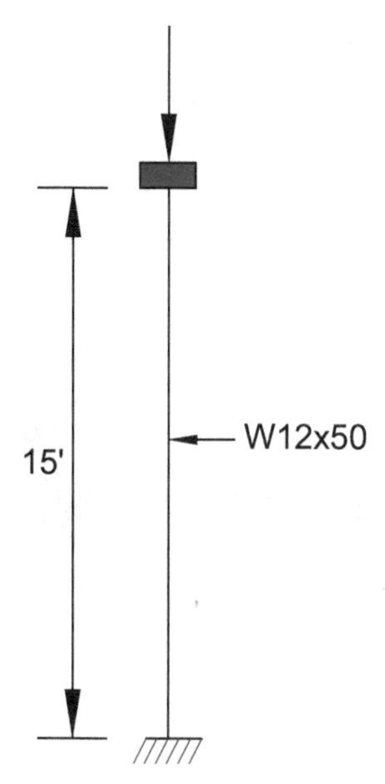

The steel yield stress is 50 ksi. The available strength in axial compression of the column is most nearly:

(A) 90 kips

(B) 270 kips

(C) 360 kips

(D) 660 kips

294. A frame is shown.

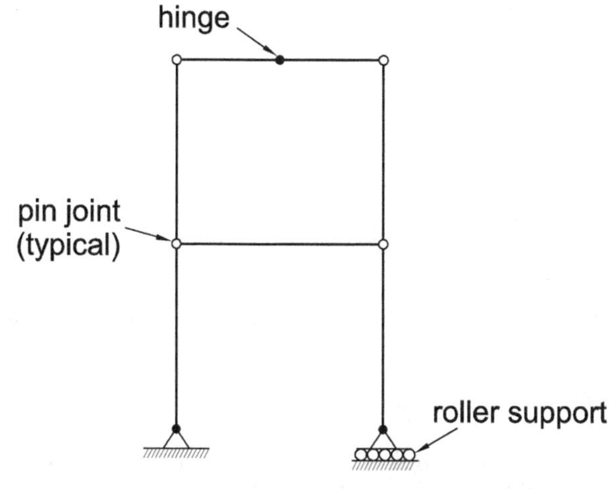

This frame is

(A) unstable

(B) statically determinate

(C) statically indeterminate to the first degree

(D) statically indeterminate to the second degree

295. A W24 × 68 propped cantilever beam spans 25 ft, as shown. The beam is made of structural steel.

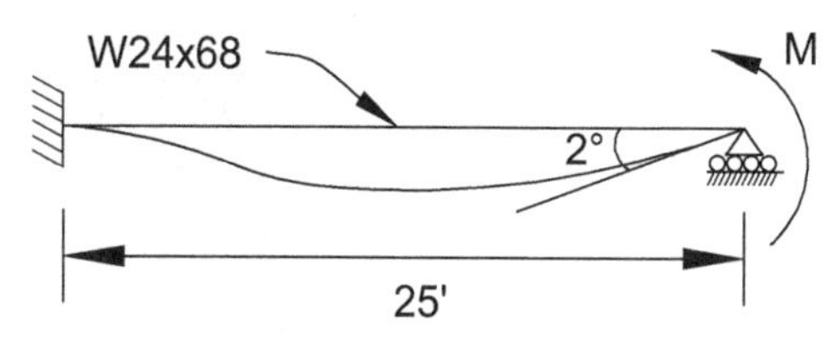

The moment needed to rotate the beam end by 2° is most nearly

(A) 2 ft-kips

(B) 200 ft-kips

(C) 2100 ft-kips

(D) 25,000 ft-kips

296. Consider the fixed-end beam shown.

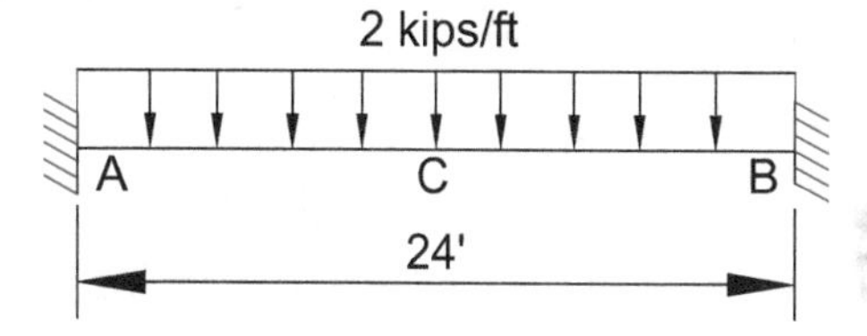

Its midspan bending moment is most nearly

(A) 48 ft-kips

(B) 96 ft-kips

(C) 140 ft-kips

(D) 190 ft-kips

297. A vehicle weighing 72 kips travels at 50 mph on a bridge. The driver applies the brakes and brings the vehicle to a stop in 15 sec. Assume the deceleration is uniform, the bridge is level, and friction is adequate. The longitudinal force exerted on the bridge is most nearly

(A) 0 lbf

(B) 32.2 lbf

(C) 11,000 lbf

(D) 160,000 lbf

298. Which of the following statements regarding the allowable stress design (ASD) and load and resistance factor design (LRFD) methods is NOT correct?

(A) The ASD method uses safety factors.

(B) The LRFD method uses resistance factors.

(C) Both use the same methods of structural analysis.

(D) Both use performance-based design criteria.

299. A steel bar is connected to a gusset plate, as shown.

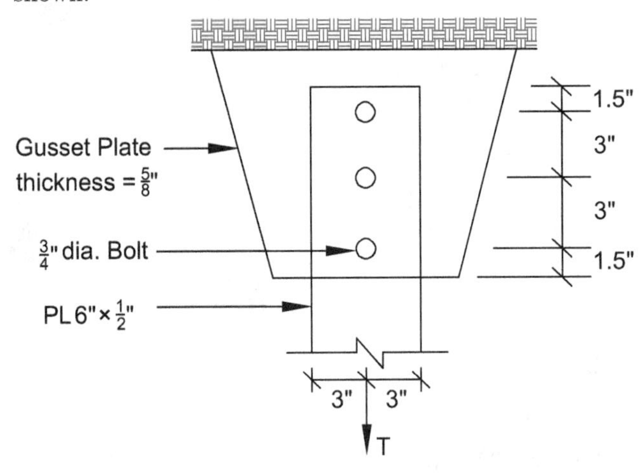

The steel has a yield strength of 50 ksi and an ultimate strength of 65 ksi. The bolts are adequate to carry the connection load. Using load and resistance factor design, the design capacity of the bar in tension is most nearly

(A) 125 kips

(B) 135 kips

(C) 150 kips

(D) 195 kips

300. A 1 in diameter, A325 bolt is used to connect a clevis, as shown. The shear capacity of the bolt is 35.3 kips.

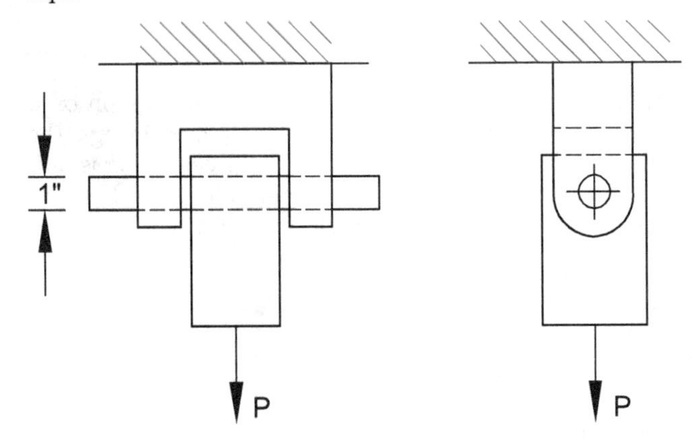

The maximum load the bolt can carry is most nearly

(A) 20 kips

(B) 30 kips

(C) 35 kips

(D) 70 kips

301. The square concrete column shown is subjected to an axial load and a uniaxial bending moment associated with $e/h = 1.2$.

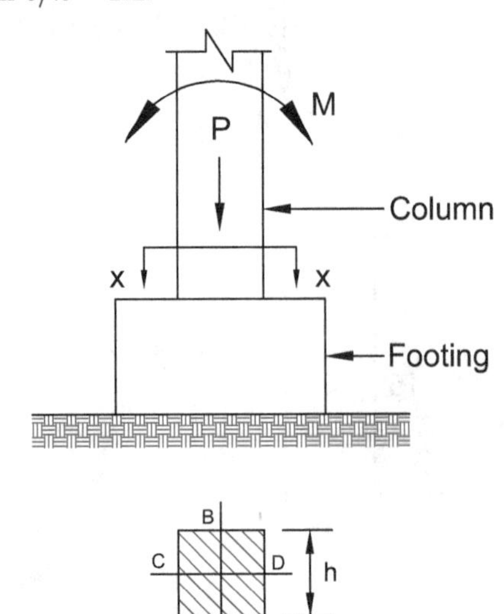

Section x-x

To optimize the column capacity, longitudinal reinforcement should be added to

(A) only one side, A, B, C, or D

(B) two sides, A and B

(C) two sides, C and D

(D) all four sides, equally distributed

302. Which statement about the ACI code resistance factor is NOT correct?

(A) It accounts for the material quality control achievable for each member.

(B) It accounts for the uncertainties in the strength of the member.

(C) The higher the uncertainty, the smaller the value of the factor ϕ.

(D) The higher the uncertainty, the larger the value of the factor.

303. Which statement is correct?

(A) A soil sample that contains particles of similar sizes is called well-graded soil.

(B) A soil sample that contains particles of all sizes is called poorly graded soil.

(C) The symbol *M* in the USCS is used for microbic peat.

(D) A grain of clay cannot be seen with the naked eye.

304. For a given soil sample, the coefficient of uniformity and the coefficient of curvature both equal 3.50. Under USCS, the soil sample is classified as

(A) GC

(B) GP

(C) CH

(D) ML

305. A soil sample weighs 222 g and has a volume of 101 cm^3. The sample is oven dried and its dry weight is 180 g. The dry unit weight of the soil is most nearly

(A) 1.8 g/cm^3

(B) 62 g/cm^3

(C) 110 g/cm^3

(D) 120 g/cm^3

306. A sample of sandy soil was tested, and the minimum and maximum dry unit weights were determined to be 110 lbf/ft^3 and 140 lbf/ft^3, respectively. The soil in the field that the sample was taken from was compacted to 129.4 lbf/ft^3. The soil's relative compaction is most nearly

(A) 90%

(B) 92%

(C) 95%

(D) 100%

307. A footing is placed in sandy soil at a depth of 6 ft below ground level. The soil has a dry density of 120 lbf/ft^3 and a submerged density of 128 lbf/ft^3. The water table is 3.5 ft above the foundation level. The total overburden pressure at the footing depth is most nearly

(A) 220 lbf/ft^2

(B) 650 lbf/ft^2

(C) 750 lbf/ft^2

(D) 770 lbf/ft^2

308. A cantilever sheet piling wall is used to retain soil, as shown.

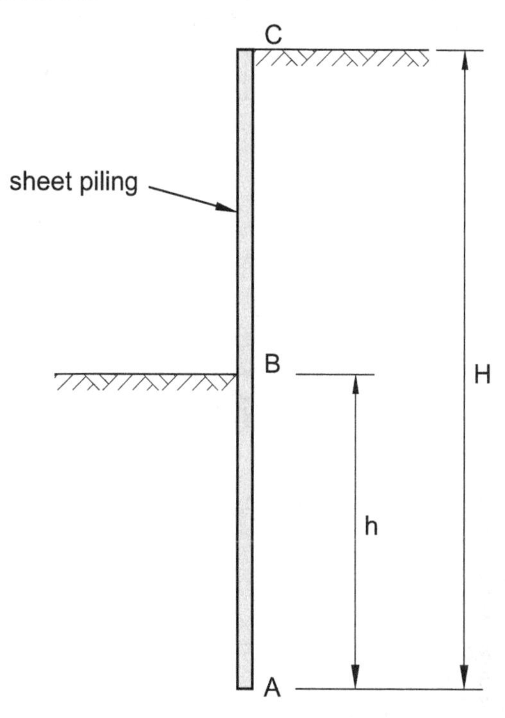

The angle of internal friction of the soil is 30°. The height $H = 30$ ft, and the height $h = 15$ ft. The factor of safety against sliding is most nearly

(A) 1.125

(B) 2

(C) 2.25

(D) 3.33

309. A sand deposit of infinite length has an average slope of 25°. The angle of shear resistance of the sand is 30°. The factor of safety required for the deposit to maintain its side slope is most nearly

(A) 0.81

(B) 1.2

(C) 1.24

(D) 5.0

310. A 3 ft diameter, 30 ft deep caisson is shown. The allowable soil bearing capacity is 10 ksf, and the skin friction at perimeter is 100 psf. The factor of safety is 3.

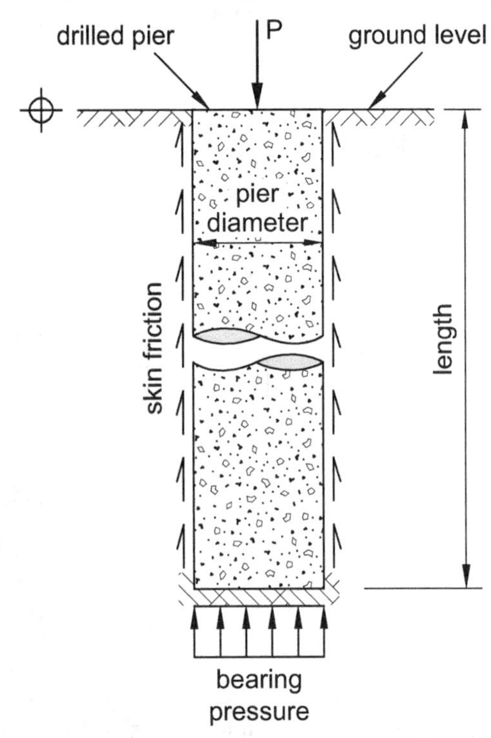

The maximum load-carrying capacity of the caisson is

(A) 33 kips

(B) 100 kips

(C) 295 kips

(D) 300 kips

311. Which of the following statements regarding the consolidation of soils is NOT correct?

(A) A consolidation test performed on an undisturbed sample is also called an odometer test.

(B) Primary consolidation equals an instantaneous void ratio reduction.

(C) Secondary consolidation is a void ratio reduction which takes place over a long period of time.

(D) The longest drainage path a particle can take to escape normal pressure is used to compute the coefficient of consolidation.

312. The basement shown is 12 ft tall. The active and at-rest coefficients of lateral load pressures on the wall are 0.41 and 0.58, respectively. The soil unit weight is 120 pcf.

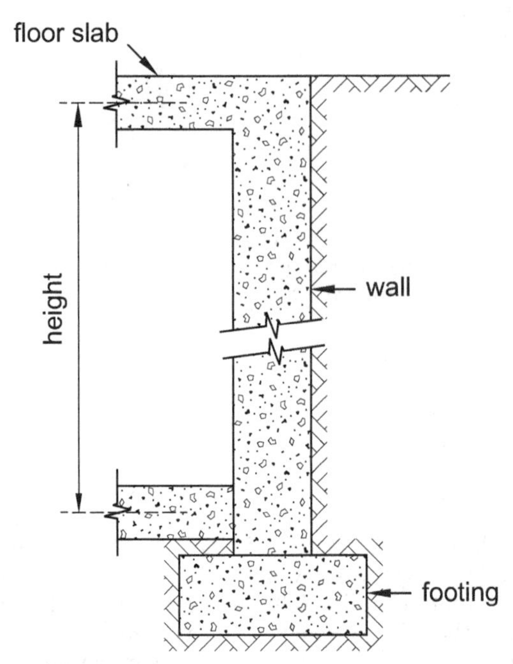

The force the wall exerts on the slab on grade is most nearly

(A) 1700 lbf/ft

(B) 2800 lbf/ft

(C) 3300 lbf/ft

(D) 10,000 lbf/ft

313. To control soil erosion over the sloping area shown, a slope is engineered to create the area marked "X."

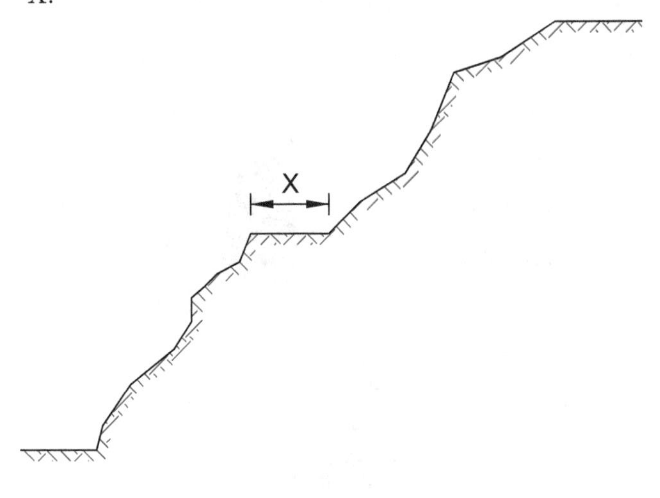

The area is called

(A) bench

(B) sediment trap

(C) sediment basin

(D) hiking trail

314. A car was traveling at a speed of 35 mph. When approaching a stopping line at an intersection, it started decelerating at a point 55 ft away. The deceleration rate was 16 ft/sec^2. Which of the following statements is correct?

(A) The car would be able stop at or before reaching the stopping line.

(B) The car would not be able to stop at the stopping line.

(C) There is insufficient data to determine the stopping distance. The weight of the car is needed.

(D) There is insufficient data to determine the stopping distance. The road friction coefficient and grade incline are needed.

315. A highway has a design speed of 55 mph. It has a 2% grade followed by a −4% grade. The deceleration rate is 11 ft/sec^2, and the reaction time is 2.0 sec. The height of the driver's eyes is 3.5 ft, and the object height is 2.0 ft. The required stopping sight distance (SSD) is 496 ft. The minimum length of curve that will satisfy the SSD is most nearly

(A) 500 ft

(B) 700 ft

(C) 800 ft

(D) 1,000 ft

316. A curve is needed to connect a −5% grade to a −1% grade. The design is 45 mph, and the acceptable deceleration is 10 ft/sec^2. The driver's reaction time is 2.5 sec. The minimum length of the curve to satisfy the riding comfort criteria is most nearly

(A) 75 ft

(B) 170 ft

(C) 300 ft

(D) 400 ft

317. Which of the following statements about AASHTO layer coefficients for pavement design is correct?

(A) The engineer selects the structural number (SN) at will.

(B) The SN is used to protect the surface below.

(C) The SN is used to protect the layer(s) to be installed.

(D) The SN assumes a minimum of three layers.

318. Truck traffic comprises 15% of the total traffic on a highway. For a rolling terrain segment of the highway, the heavy vehicle adjustment factor is most nearly

(A) 0.13

(B) 0.15

(C) 0.77

(D) 0.84

319. A car was traveling at 35 mph on a grade road approaching a signaled intersection. The car was 80 ft away from the intersection when the signal turned yellow. The driver believed that he could not comfortably stop the car at the stopping line and decided to keep moving in order to clear the intersection before the light turned red. The car is 15 ft long, the intersection is 60 ft wide, and the driver reaction time is 1.5 sec. The comfortable deceleration rate is 10 ft/sec^2. The time needed for the car to clear the intersection is most nearly

(A) 2.0 sec

(B) 3.0 sec

(C) 4.5 sec

(D) 5.5 sec

320. A city has determined that a maximum of 8 recorded crashes/million entering vehicles will be tolerated before improvements are made to an intersection. If an intersection has an average daily traffic of 8000 vehicles, the minimum number of crashes that must occur in a year for the city to start improving the intersection is most nearly

(A) 9

(B) 20

(C) 24

(D) 27

321. A speed-density relationship for a roadway was determined based on the data from the traffic study shown.

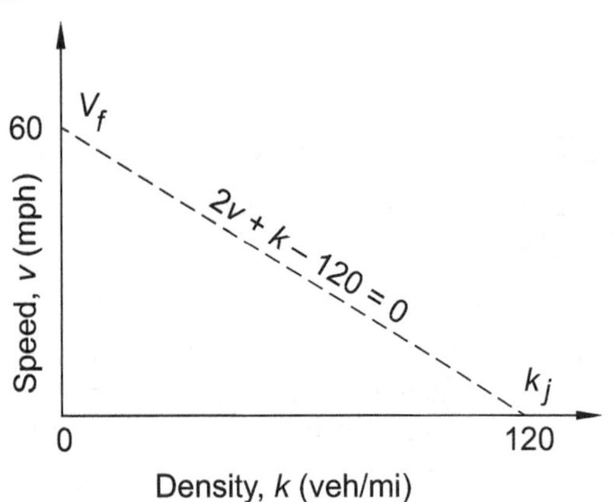

The maximum flow is most nearly

(A) 900 veh/hr

(B) 1800 veh/hr

(C) 3600 veh/hr

(D) 7200 veh/hr

322. A speed-density relationship for a roadway was determined based on the data from the traffic study shown.

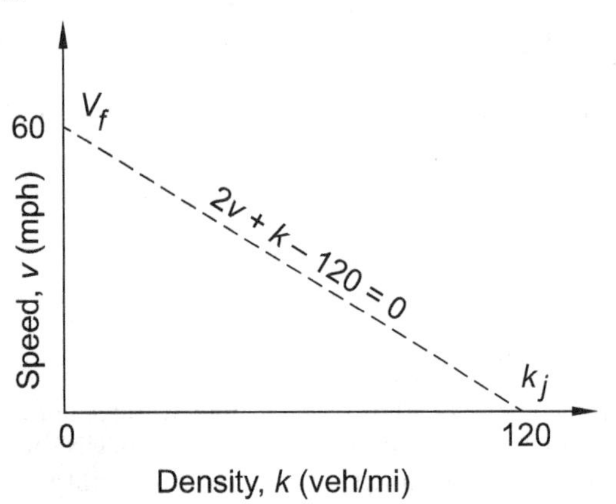

The mean free speed for the roadway is most nearly

(A) 0 mph

(B) 30 mph

(C) 60 mph

(D) 120 mph

323. Which party in a design/construction team is responsible for the design of the concrete formwork?

(A) architect

(B) owner

(C) contractor

(D) ready-mix concrete supplier

324. A contractor is bidding on a project that has a liquidated damages clause in the contract. Liquidated damages are

(A) damages caused by liquids, such as by rainstorms, floods, tsunamis, and oil spills

(B) a penalty imposed on the contractor for every day the contractor is late in completing the project

(C) actual damages that the contractor will pay to the owner if there is a delay in completion of the project

(D) a specified amount of money the contractor will pay to the owner per day if there is a delay in completion of the project

325. Which of the following statements regarding shoring and reshoring during concrete construction is NOT correct?

(A) Shores are designed to carry the weight of the formwork, concrete, and construction loads above the shores.

(B) Before reshoring is installed, original shoring should be removed.

(C) Reshoring is installed to transfer all or part of the load to the hardened concrete floor system.

(D) Reshoring primarily requires bracing and horizontal shoring elements.

326. Fall protection is needed on construction sites when the minimum height above a lower level (excluding scaffolding) is

(A) 2 ft

(B) 4 ft

(C) 6 ft

(D) 8 ft

327. A contractor signed a contract to haul and compact 100,000 yd^3 of fill material in 10 months at a lump sum price of $1.1 million, which includes 10% profit. The contractor allocated uniform resources for the construction period and projected uniform productivity. After two months, the contractor has hauled 15,000 yd^3 and spent $210,000. The contractor's actual cost of the work performed after two months is most nearly

(A) $150,000

(B) $200,000

(C) $210,000

(D) $1,000,000

328. A contractor signed a contract to haul and compact 100,000 yd^3 of fill material in 10 months at a lump sum price of $1.1 million, which includes 10% profit. The contractor allocated uniform resources for the construction period and projected uniform productivity. After two months, the contractor has hauled 15,000 yd^3 and spent $210,000. The contractor's budgeted cost of the work scheduled for the two-month period is most nearly

(A) $150,000

(B) $200,000

(C) $210,000

(D) $1,000,000

329. A 100 ft long, 10 ft tall, 10 in thick wall is to be built in 20 ft long segments. The cost of leasing the forms is $2 per ft^2 for initial erection and $0.50 per ft^2 for subsequent erection. The labor cost of erecting and stripping the forms is $5 per ft^2 and $2 per ft^2, respectively. The total cost of the formwork to build the wall is most nearly

(A) $10,000

(B) $15,600

(C) $16,000

(D) $18,000

330. A preliminary framing plan and a section for a moment frame are shown.

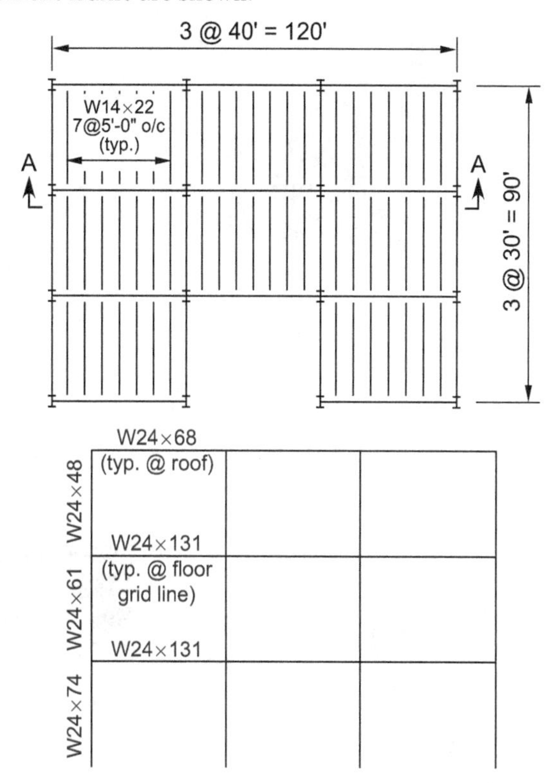

Excluding girders and beams that are part of moment frames, and allowing for 2% overage, the steel tonnage needed for beams at roof level is most nearly

(A) 2.4 tons

(B) 2.7 tons

(C) 19 tons

(D) 56 tons

STOP!

DO NOT CONTINUE!

This concludes the examination. If you finish early, check your work and make sure that you have followed all instructions. After checking your answers, you may turn in your examination booklet and answer sheet and leave the examination room. Once you leave, you will not be permitted to return to work or change your answers.

Exam 4 Instructions

In accordance with the rules established by your state, you may use any approved battery- or solar-powered, silent calculator to work this examination. However, no blank papers, writing tablets, unbound scratch paper, or loose notes are permitted. Sufficient paper will be provided. The *NCEES FE Reference Handbook* is the only reference you are allowed to use during this exam.

You are not permitted to share or exchange materials with other examinees.

You will have six hours in which to work this session of the examination. Your score will be determined by the number of questions that you answer correctly. There is a total of 110 questions. All 110 questions must be worked correctly in order to receive full credit on the exam. There are no optional questions. Each question is worth 1 point. The maximum possible score for this section of the examination is 110 points.

Partial credit is not available. No credit will be given for methodology, assumptions, or work written on scratch paper.

Record all of your answers on the Answer Sheet. Mark your answers with a no. 2 pencil. Answers marked in pen may not be graded correctly. Marks must be dark and must completely fill the bubbles. Record only one answer per question. If you mark more than one answer, you will not receive credit for the question. If you change an answer, be sure the old bubble is erased completely; incomplete erasures may be misinterpreted as answers.

If you finish early, check your work and make sure that you have followed all instructions. After checking your answers, you may submit your answers and leave the examination room. Once you leave, you will not be permitted to return to work or change your answers.

When permission has been given by your proctor, you may begin your examination.

Name: ______________________________________

Last First Middle Initial

Examinee number: ______________

Examination Booklet number: ______________

Fundamentals of Engineering Examination

Exam 4

Exam 4 Answer Sheet

331. ____________

332. ____________

333. ____________

334. ____________

335. ____________

336. ____________

337. ____________

338. ____________

339. ____________

340. (A) (B) (C) (D)

341. (A) (B) (C) (D)

342. (A) (B) (C) (D)

343. (A) (B) (C) (D)

344. (A) (B) (C) (D)

345. (A) (B) (C) (D)

346. ____________

347. (A) (B) (C) (D)

348. ____________

349. ____________

350. ____________

351. ____________

352. (A) (B) (C) (D)

353. (A) (B) (C) (D) (E)

354. (A) (B) (C) (D) (E)

355. (A) (B) (C) (D)

356. (A) (B) (C) (D)

357. (A) (B) (C) (D) (E)

358. (A) (B) (C) (D) (E)

359. ____________

360. (A) (B) (C) (D)

361. (A) (B) (C) (D)

362. (A) (B) (C) (D) (E)

363. (A) (B) (C) (D)

364. (A) (B) (C) (D)

365. (A) (B) (C) (D)

366. (A) (B) (C) (D)

367. (A) (B) (C) (D)

368. (A) (B) (C) (D)

369. (A) (B) (C) (D)

370. (A) (B) (C) (D)

371. (A) (B) (C) (D)

372. (A) (B) (C) (D)

373. (A) (B) (C) (D)

374. (A) (B) (C) (D)

375. (A) (B) (C) (D)

376. (A) (B) (C) (D)

377. (A) (B) (C) (D)

378. (A) (B) (C) (D)

379. (A) (B) (C) (D)

380. (A) (B) (C) (D)

381. (A) (B) (C) (D) (E)

382. (A) (B) (C) (D)

383. ____________

384. (A) (B) (C) (D)

385. (A) (B) (C) (D)

386. ____________

387. ____________

388. (A) (B) (C) (D)

389. (A) (B) (C) (D)

390. (A) (B) (C) (D)

391. (A) (B) (C) (D)

392. ____________

393. ____________

394. (A) (B) (C) (D)

395. (A) (B) (C) (D)

396. (A) (B) (C) (D)

397. (A) (B) (C) (D)

398. (A) (B) (C) (D)

399. (A) (B) (C) (D)

400. (A) (B) (C) (D)

401. (A) (B) (C) (D)

402. ____________

403. (A) (B) (C) (D)

404. (A) (B) (C) (D)

405. ____________

406. (A) (B) (C) (D)

407. (A) (B) (C) (D)

408. (A) (B) (C) (D)

409. (A) (B) (C) (D)

410. (A) (B) (C) (D)

411. (A) (B) (C) (D)

412. (A) (B) (C) (D)

413. (A) (B) (C) (D)

414. (A) (B) (C) (D)

415. (A) (B) (C) (D)

416. (A) (B) (C) (D)

417. (A) (B) (C) (D)

418. (A) (B) (C) (D)

419. ____________

420. (A) (B) (C) (D)

421. (A) (B) (C) (D)

422. (A) (B) (C) (D)

423. (A) (B) (C) (D)

424. (A) (B) (C) (D)

425. (A) (B) (C) (D)

426. (A) (B) (C) (D)

427. (A) (B) (C) (D)

428. (A) (B) (C) (D)

429. (A) (B) (C) (D)

430. (A) (B) (C) (D)

431. (A) (B) (C) (D)

432. (A) (B) (C) (D)

433. (A) (B) (C) (D)

434. (A) (B) (C) (D)

435. (A) (B) (C) (D)

436. (A) (B) (C) (D)

437. (A) (B) (C) (D)

438. (A) (B) (C) (D)

439. ____________

440. ____________

Exam 4

331. The determinant of the following matrix is ______. (Fill in the blank.)

$$\begin{vmatrix} 1 & 2 & -1 \\ 2 & 3 & 2 \\ 1 & -2 & -2 \end{vmatrix}$$

332. In 10 years, a 100 g mass of a radioactive element decreases to 50 g. Radioactive decay is described by the following equation.

$$m = m_0 e^{kt}$$

m is final mass, m_0 is original mass, and t is time elapsed in years.

The constant, k, in the equation with units of yr^{-1} (rounded to three decimal places) is ______. (Fill in the blank.)

333. The sum of the roots of the quadratic equation $ax^2 + bx + c = 0$ is ______. (Fill in the blank.)

334. A quadratic equation has two solutions, $x_1 = 12$ and $x_2 = -2$. The equation is the form

$$Ax^2 + Bx + C = 0$$

Match the following options to the coefficients A, B, and C. (Not all options will be used.)

options	coefficients
−24	$A =$
−10	$B =$
1	$C =$
10	
24	

335. The value of θ (in degrees) to satisfy the equation $2\cos^2\theta + 9\cos\theta - 5 = 0$ is ______. (Fill in the blank.)

336. Two matrices, A and B, are added to form matrix C.

$$A = \begin{vmatrix} 2 & 4 & 5 \\ 3 & 1 & 7 \\ 2 & 6 & 8 \end{vmatrix} \quad B = \begin{vmatrix} 11 & 14 & 12 \\ 21 & 33 & 9 \\ 40 & 9 & 16 \end{vmatrix}$$

The term c_{23} in matrix C is ______. (Fill in the blank.)

337. Consider a parabola with the following equation.

$$y = x - \frac{1}{6}x^2$$

Its radius of curvature at $x = 3$ to the nearest integer is ______. (Fill in the blank.)

338. Out of a pool of eight male and nine female workers, a crew of five workers is to be formed. The probability of getting a crew with at least four men (approximated to two decimal places) is ______. (Fill in the blank.)

339. A data set consists of four points as shown.

x	y
2	9
3	11
5	15
9	22

For the data set, the mean standard error of estimate (MSE), to the nearest two decimal places, is ______. (Fill in the blank.)

340. A relationship between worker-hours and excavation earthwork was developed using regression analysis. For worker-hours ranging from 2 hr to 21 hr, the scatter diagram was fitted to the equation $y = 2 + 5x$, where x is the number of worker-hours required and y is the excavation in cubic yards.

Which of the following statements regarding the fitted line is NOT correct?

(A) The y-intercept means that two worker-hours are needed prior to excavation beginning.

(B) It should take 30 worker-hours to excavate 152 yd^3 of material.

(C) For $x = 3$, the average value of y is 17 yd^3.

(D) The slope of the fitted line means that one worker-hour of work yields 5 yd^3 of excavation.

341. Which of the following statements regarding bribery is INCORRECT?

(A) It helps the free market system by getting people to buy and sell goods and services at a price they agree on in a confidential manner.

(B) It creates inequity in the marketplace.

(C) It destroys the integrity of professional services.

(D) It treats people as commodities whose honor can be bought.

342. An engineer is friends with the mayor of a city on a social media platform. The engineer is awarded a project in the area of his or her expertise by the city. Should the licensee be disciplined as a result of pursuing and accepting such an assignment?

(A) Yes. A licensee should not solicit or accept a professional contract from a governmental body in which he is friends with a principal or an officer of that organization.

(B) Yes, unless he makes full prior disclosure of the potential conflict of interest.

(C) Yes, unless he unfriends the mayor.

(D) No. The licensee may accept the assignment.

343. Which of the following statements is true about managing risk in the engineering profession?

(A) In practicing the profession, the engineer is independent and not obligated to anyone.

(B) In practicing the profession, the engineer can set his or her own standard of care.

(C) Nontechnical ability is needed to succeed in design practice.

(D) All claims and lawsuits are solely caused by the technical incompetency of the design engineer.

344. Which of the following situations constitutes ethical, professional conduct?

(A) fee-splitting with a third party who is not a licensed design professional

(B) practicing beyond the licensee's competency without adequate supervision

(C) willfully making a false report

(D) attending a disciplinary hearing

345. Under what circumstances is a "cost-plus" contract favorable to both an owner and a contractor?

(A) when the cost is of primary importance and the schedule is of secondary importance

(B) when a complete set of project drawings and specifications are available

(C) when the unit quantities are known

(D) when the project is an emergency

346. A city issues a $20,000,000 10 yr bond at an interest rate of 6% per annum to construct a wastewater project. The rate of return of similar quality bonds is 4% per annum. The value of the bond (rounded to nearest million dollars) is likely to be ______. (Fill in the blank.)

347. A lender advertises charging 1% per month as interest. The effective annual interest rate is most nearly

(A) 12.00%

(B) 12.12%

(C) 12.69%

(D) 14.40%

348. A company can manufacture personal protection equipment (PPE) using two alternate methods. The initial costs and per-unit manufacturing costs are as shown.

cost item	alternate 1	alternate 2
set up/initial	$10,000	$150,000
manufacturing	$1.5/unit	$0.6/unit

The demand for the PPEs is anticipated to be 50,000 annually. Neglecting interest, the break-even period (to the nearest whole number of years) is ______. (Fill in the blank.)

349. A contractor buys a soil-boring rig at the cost of $150,000. After six years, its salvage value is estimated to be $10,000. If interest on the loan for the rig is at 8%, the annual cost (to the nearest thousand) is ______. (Fill in the blank.)

350. Your client, Gr8 Oil, must decide whether or not to drill for oil at a particular location for which Gr8 has drilling rights. There are two possible outcomes: striking oil and not striking oil. If the decision is made to drill and Gr8 strikes oil, the oil would be worth $13,000,000. The cost of drilling would be $3,000,000. At present, Gr8 can sell the drilling rights to a wildcatter for $1,000,000, with a contingency to receive another million dollars if the wildcatter strikes the oil. Based on the geological assessment, the probability of striking the oil at the location is 0.30. The payoffs are tabulated for a drill or no-drill decision.

		event	
		oil	no oil
decision	drill	$10,000,000	−$3,000,000
	do not drill	$2,000,000	$1,000,000

If the decision is made to drill and no oil is present, the opportunity loss (rounded to the nearest million dollars) is ______. (Fill in the blank.)

351. Four levy barrier alternatives are being considered for flood protection. The level of protection offered by each alternative and the associated cost varies from other alternatives, as tabulated below.

barrier alternate	maximum flood protection (cfs)	probability of a greater flood occurring in a year	damage cost of the greater flood	cost of barrier
1	2000	0.050	$4,000,000	$1,000,000
2	3000	0.010	$6,000,000	$2,000,000
3	4000	0.005	$8,000,000	$3,000,000
4	5000	0.001	$10,000,000	$4,000,000

Assume that floods occur no more than once a year. Each of the proposed structures has a design life of 40 years. The minimum rate of return is 8%. Considering cost benefits, which of the four alternatives will provide the optimal barrier system? ______. (Fill in the blank.)

352. A 2 in diameter steel shaft is subjected to a torque of 100 lbf-ft. The maximum shear stress in the shaft is most nearly

(A) 400 psi

(B) 600 psi

(C) 800 psi

(D) 1600 psi

353. Consider the set of forces (in kN) shown.

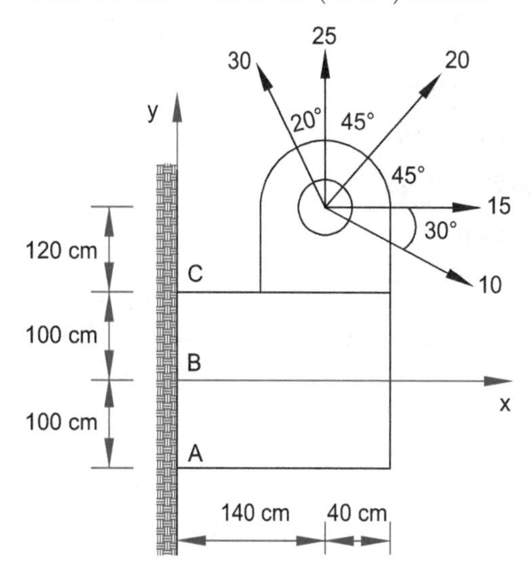

From the following five options, select the two that most nearly represent the magnitude and direction of the resultant force.

(A) 62 kN

(B) 68 kN

(C) 56° from x-axis

(D) 66° from x-axis

(E) 246° from x-axis

354. A connection is being designed to resist the set of forces (in kN) shown.

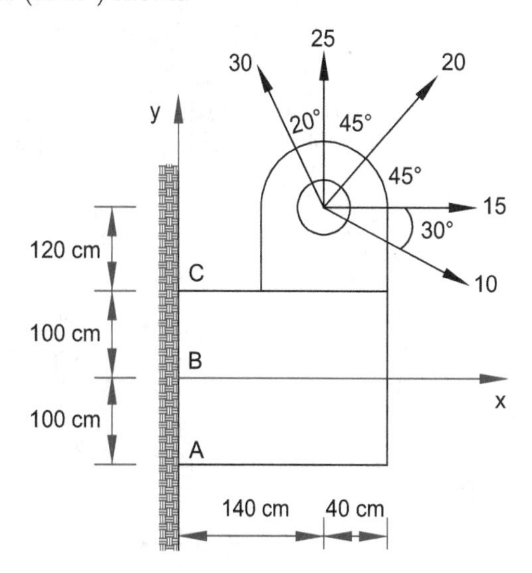

From the five options listed, choose the two that define the resultant moment (kN·m) at the support.

(A) clockwise

(B) counterclockwise

(C) 26.7 kN·m

(D) 66.6 kN·m

(E) 87.3 kN·m

355. Consider the steel girder shown.

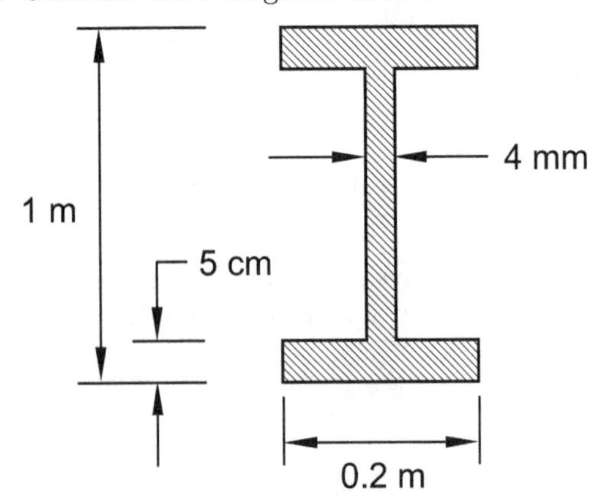

The mass of the girder per meter length is most nearly

(A) 7.8 kg

(B) 184.1 kg

(C) 780 kg

(D) 1805 kg

356. A car wheel strikes a 6 in high curb, as shown. The wheel is carrying 4000 lbf and its diameter is 28 in.

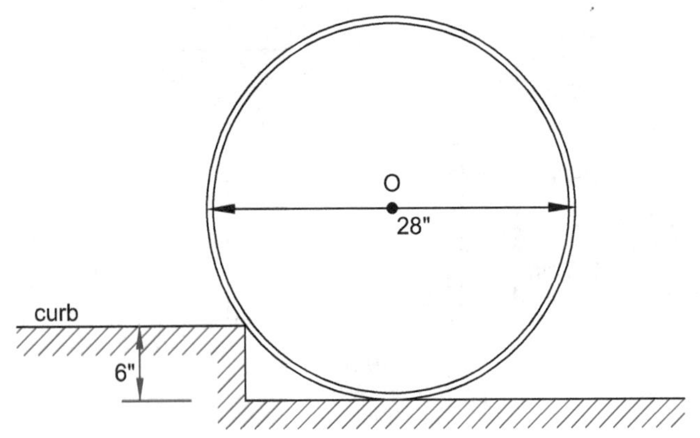

The horizontal force required at the centerline of the wheel for it to roll up onto the curb is most nearly

(A) 2 kips

(B) 4 kips

(C) 6 kips

(D) 8 kips

357. A traffic sign is hung from two points off a fixed pole, as shown.

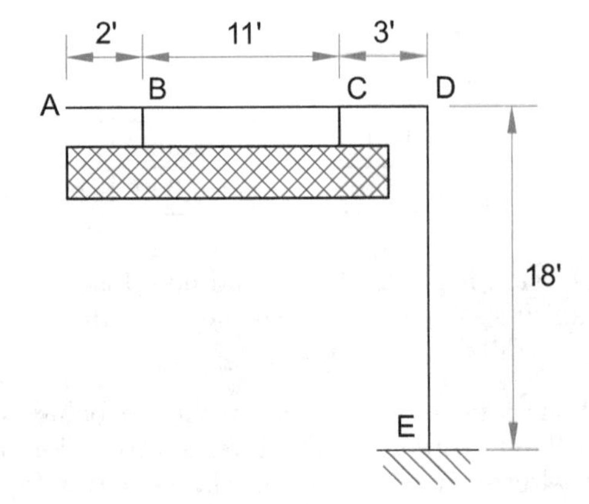

The load at each point is 1 kip. From the options given, choose the correct magnitude and direction of the resultant moment at the base.

(A) clockwise

(B) counterclockwise

(C) 1 ft-kips

(D) 17 ft-kips

(E) 35 ft-kips

358. A traffic sign carries a point load, as shown.

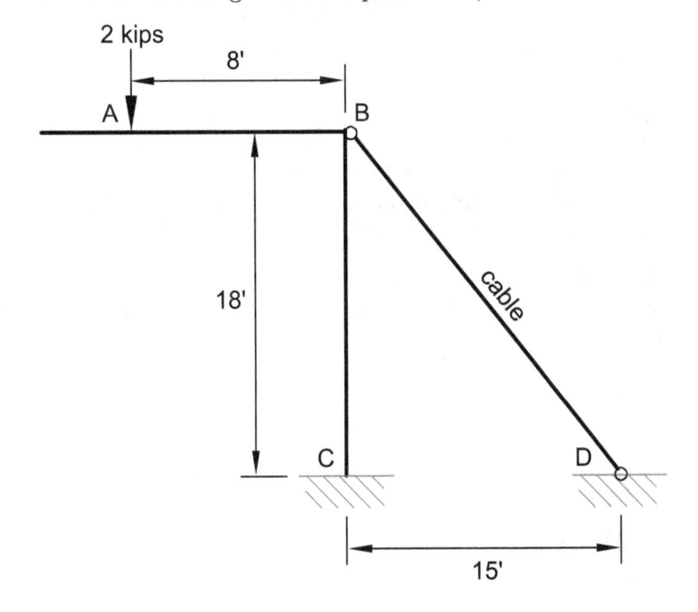

The pole is braced by a cable that has a tensile force of 1100 lbf. From the options below, choose the correct magnitude and direction of the moment at base C.

(A) clockwise

(B) counterclockwise

(C) 3 ft-kips

(D) 8 ft-kips

(E) 16 ft-kips

359. A steel bar is placed on top of an identical bar and the two are welded together, as shown.

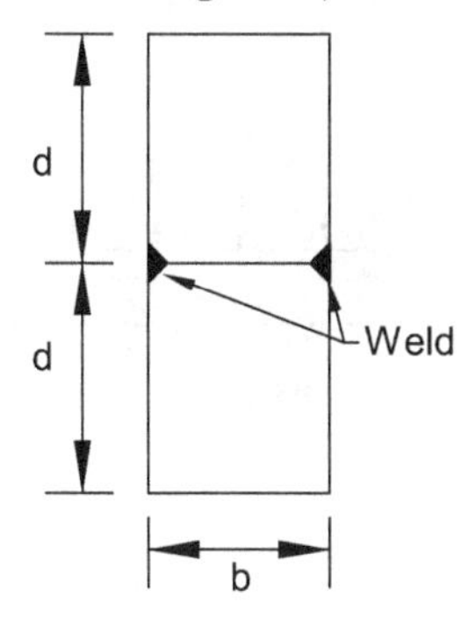

The weld is adequate and no slippage can take place during bending. The ratio of the combined bars' bending stiffness after welding to the bending stiffness of the two bars placed one on top of the other before welding is _____. (Fill in the blank.)

360. A 500 lbf block is placed on a sloping plane with a friction coefficient of 0.5. The block is connected to a cable that is in tension, as shown.

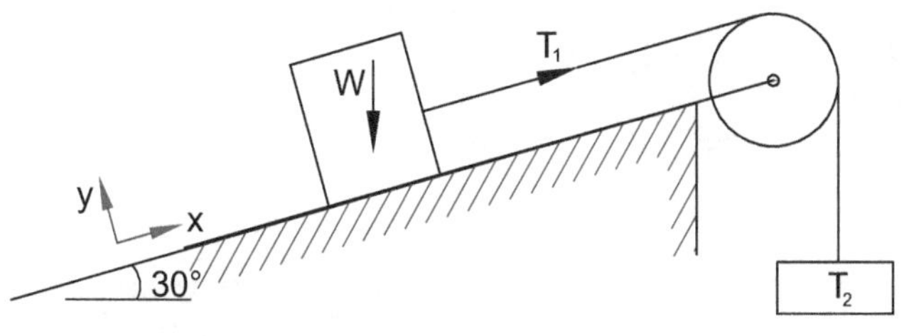

If the friction coefficient between the cable and the pulley is 0.2, the minimum force T_2 required to pull the block up the slope is most nearly

(A) 51 lbf

(B) 101 lbf

(C) 251 lbf

(D) 501 lbf

361. A concrete block weighing 145 lbf is resting on a concrete floor. The bearing surface of the block is an 11 in square. The coefficient of friction is 0.7. The minimum force required to slide the block is most nearly

(A) 1.12 lbf

(B) 86 lbf

(C) 102 lbf

(D) 121 lbf

362. Two concurrent forces are 300 N at 30° and 400 N at −60° from a reference direction. Assume clockwise angles to be positive. Out of the following options, choose the resultant force (in newtons) and its deviation from the reference.

(A) 500 N

(B) $300\sqrt{2}$ N

(C) 0°

(D) −23.1°

(E) 53.1°

363. A hollow cylinder flywheel in a locomotive engine has a mass of 2000 kg and a radius of gyration of 1 m. A constant torque of 2000 N·m is applied to move the wheel. Its angular acceleration is most nearly

(A) 1 rad/s^2

(B) $\pi/2$ rad/s^2

(C) 2π rad/s^2

(D) 9.81 rad/s^2

364. Two springs are used to support a load P of 100 kg, as shown.

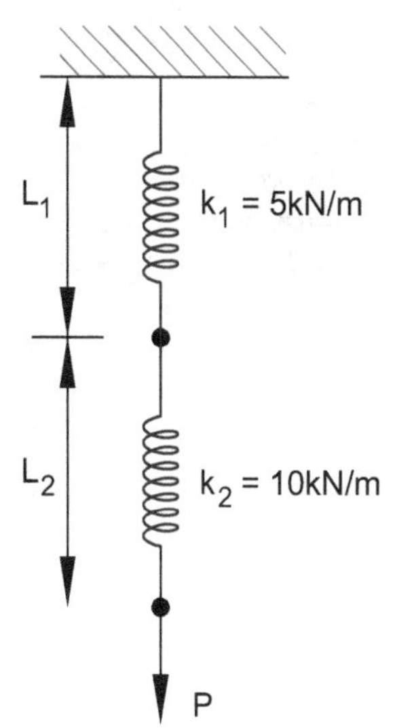

The natural period of vibration of the system is most nearly

(A) 0.92 s

(B) 1.09 s

(C) 6.11 s

(D) 18.1 s

365. An asphalt roller has a total kinetic energy of 70,000 N·m. Assuming the road is flat, the force required to stop the roller within a distance of 5 m is

(A) 7 kN

(B) 14 kN

(C) 28 kN

(D) 137 kN

366. Consider a two-span beam under a uniformly distributed load (UDL) of 3 N/m. Recall that the maximum moment in a propped cantilever under a UDL is $wL^2/8$.

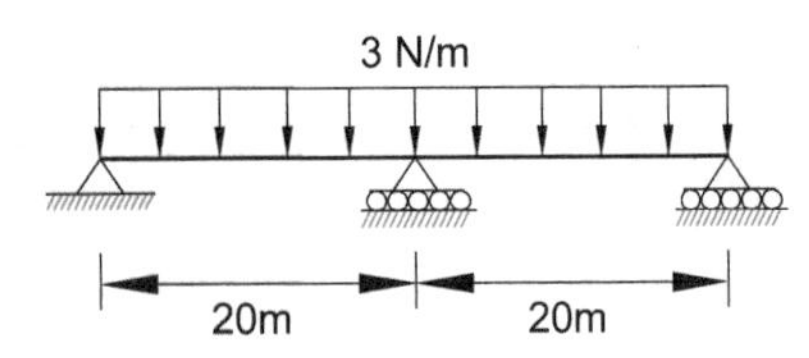

The bending moment of the beam at its central support is

(A) 0 N·m

(B) 120 N·m

(C) 150 N·m

(D) 600 N·m

367. A 1.1 m internal diameter thin-walled cylinder contains a fluid at a gauge pressure of 3 N/mm^2. The permissible circumferential pressure is 60 N/mm^2, and the permissible longitudinal pressure is 45 N/mm^2. The minimum cylinder wall thickness is most nearly

(A) 3 mm

(B) 19 mm

(C) 28 mm

(D) 50 mm

368. A hollow, thin-walled steel pipe has a mean diameter of 30 cm and a wall thickness of 30 mm. If the maximum allowable shear stress in the cross section is 5 MPa, the maximum torque that can be applied to the pipe is most nearly

(A) 2 kN·m

(B) 20 kN·m

(C) 200 kN·m

(D) 2000 kN·m

369. Two steel beams are loaded with a uniformly distributed load, as shown.

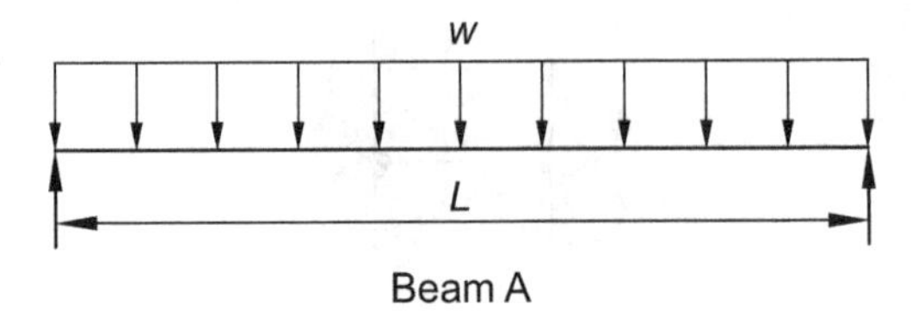

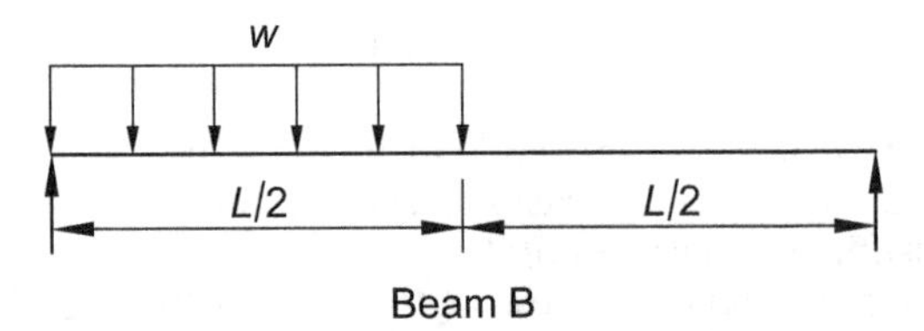

Both beams have identical span lengths and section properties. The ratio of the maximum deflections of beams A and B under the loading is most nearly

(A) 0.5

(B) 1

(C) 1.98

(D) 2

370. Consider the concentric cantilever column shown.

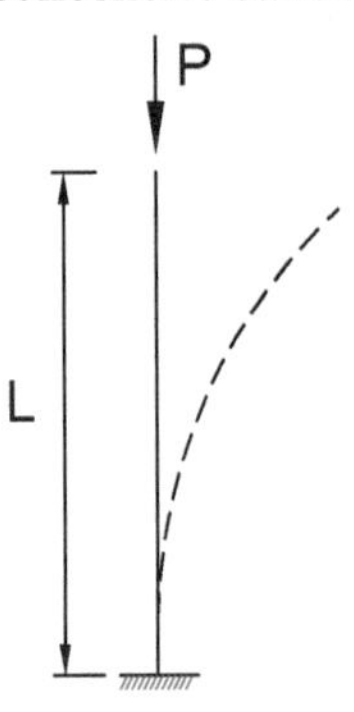

Use the recommended design values for the column effective length factors. The ratio of this column's buckling capacity to that of a pinned-pinned column of an identical size and length is most nearly

(A) 1/4

(B) 1/2

(C) 1

(D) 4

371. A 4 × 8 lumber column has an actual size of 3.5 in × 7.5 in. The bracing conditions are the same for buckling about both the x- and y-axes. The column's controlling radius of gyration is most nearly

(A) 1 in

(B) 1.15 in

(C) 2.2 in

(D) 4 in

372. A section's moment of inertia (MOI) about its x-axis is I_x, and its MOI about its y-axis is I_y. If the section is circular or rectangular, the MOI about its z-axis (along its length) is

(A) $I_x + I_y$

(B) $\frac{I_x + I_y}{2}$

(C) $\sqrt{I_x^2 + I_y^2}$

(D) $\sqrt{I_x I_y}$

373. The primary factor affecting concrete strength is

(A) Portland cement

(B) water

(C) water/cement ratio

(D) fineness modulus of aggregate

374. Two groups of concrete cylinders made from one mix were cured in a laboratory. One group was cured in air and the other in a continuously moist condition. Out of each group, one-half of the cylinders were tested after 28 days and the other half were tested after 180 days of casting. The group that attains the highest concrete strength is

(A) moist-cured for 28 days

(B) air-cured for 28 days

(C) moist-cured for 180 days

(D) air-cured for 180 days

375. A concrete mix has a ratio of 1:2:3, and its water/cement ratio is 0.4. Its unit weight is 145 lbf/ft^3. For the concrete mix, 611.7 lbf of cement is needed to produce one cubic yard of mix. The amount of water needed to produce one cubic yard of mix is most nearly

(A) 30 gal

(B) 60 gal

(C) 120 gal

(D) 240 gal

376. The tensile strength of concrete is given by the following equation, where f_c' is in psi.

$$f_t' = 7.5\sqrt{f_c'}$$

The increase in tensile strength of concrete when its compressive strength is doubled from 3000 psi to 6000 psi is most nearly

(A) 25%

(B) 40%

(C) 100%

(D) 200%

377. The stress-strain diagram of a material is shown.

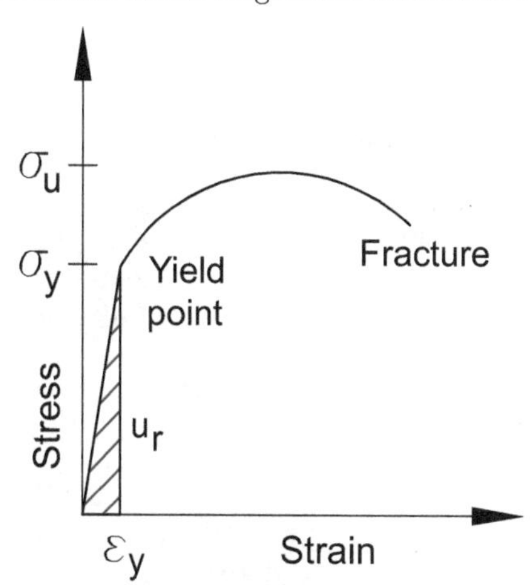

If the yield strength of the material is 60 ksi and the associated strain is 0.002, the modulus of resilience of the material is most nearly

(A) 0.01 ksi

(B) 0.06 ksi

(C) 0.12 ksi

(D) 60 ksi

378. A 12 in diameter pipe carries oil flowing at a rate of 3 MGD. The oil's average velocity in the pipe is most nearly

(A) 1 ft/sec

(B) 6 ft/sec

(C) 24 ft/sec

(D) 72 ft/sec

379. Water drains out of a tank, as shown.

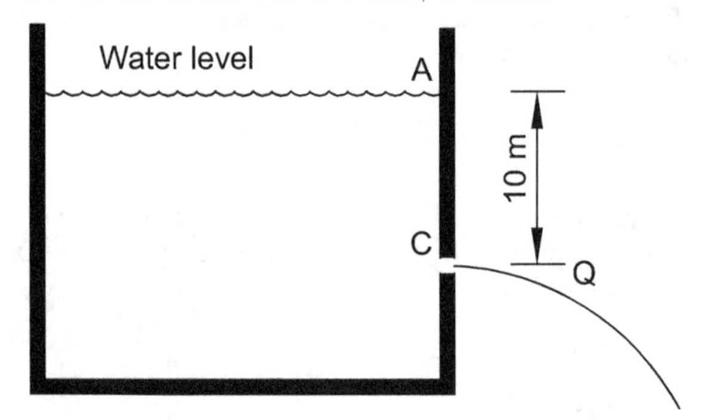

Neglecting all losses, the initial velocity of the discharge at point C is most nearly

(A) 0 m/s

(B) 10 m/s

(C) 14 m/s

(D) additional data is needed

380. Two plates are kept 4 mm apart, and the space between them is filled with an oil with a dynamic viscosity of 0.2 poise. The lower plate is fixed, and the upper plate is movable. For the gap between the plates, assume a linear variation in velocity. If the upper plate moves at a velocity of 1.2 m/s, the shear stress on the lower plate is most nearly

(A) 6 Pa

(B) 60 Pa

(C) 150 Pa

(D) 300 Pa

381. Out of the following options, choose two devices that are used to measure fluid flow in pipes.

(A) orifice

(B) nozzle

(C) Parshall flume

(D) broad-crested weir

(E) sharp-crested weir

382. For the pipe system shown, the total discharge is 10 ft^3/sec.

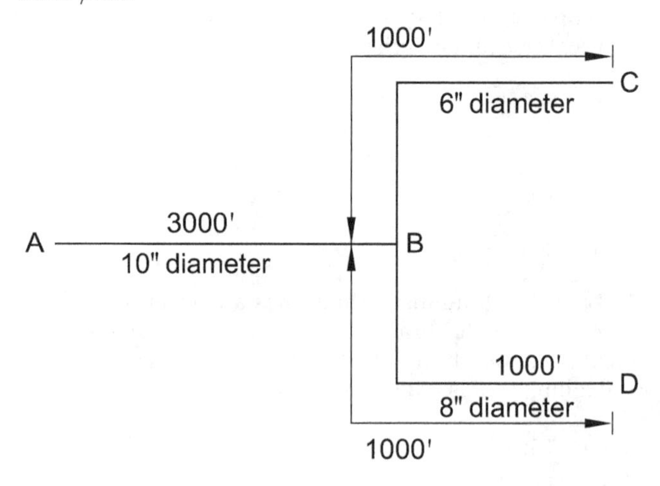

A reaction block stops joint B from moving due to thrust in the direction of flow in the 10 in diameter pipe. The force exerted by water against the reaction block is most nearly

(A) 100 lbf

(B) 350 lbf

(C) 2000 lbf

(D) 12000 lbf

383. A cargo ship is 400 ft long and 60 ft wide. Its total depth is 38 ft, and the waterline depth is 28 ft. Assume that sea water weighs 64 pcf. The total allowable load-carrying capacity of the ship, including its self-weight, in tons, is ______. (Fill in the blank.)

384. A horizontal curve has a radius of 600 ft. The degree of curve using the arc definition is most nearly

(A) 4.78°

(B) 4.78 rad

(C) 9.55°

(D) 9.55 rad

385. A 10° horizontal curve has an intersection angle of 20°. The curve's PC station is located at sta 12 +34.00. The curve's PT station is most nearly

(A) sta 24+68.00

(B) sta 20+34.00

(C) sta 14+34.00

(D) sta 6+17.00

386. A reservoir is being designed to store water. The reservoir base is a 20 ft by 40 ft rectangle. The depth is 10 ft with 1:1 side slope on all sides. The volume of soil that needs to be excavated, to the nearest whole number of cubic yards, is ______. (Fill in the blank.)

387. The longitude for Elgin, Illinois, is 42.0383°N. The earth's radius is 3960 mi at the equator. Assuming earth as a perfect sphere, the distance between Elgin and the equator along the earth's surface, to the nearest whole number of miles, is ______. (Fill in the blank.)

388. A 1000 m long vertical curve has a back tangent slope of −4% and a forward slope of 2%. The elevation at the point of intersection (PVI) is 2540 m above the datum. The elevation at the point of vertical tangency is most nearly

(A) 1540 m

(B) 2520 m

(C) 2530 m

(D) 2550 m

389. A 1000 ft long vertical curve has a back tangent slope of 2% and a forward slope of −4%. The rate of change of grade for the curve is most nearly

(A) −0.6

(B) −0.02

(C) +0.02

(D) +0.06

390. The curve number (CN) in the USDA Natural Resources Conservation Service (NRCS) rainfall-runoff method depends on which of the following factors?

(A) land cover type and soil's hydrologic group

(B) 24 hr rainfall

(C) 100 yr storm

(D) initial abstraction of the drainage area

391. A 24 in diameter sewer pipeline is running half full. The pipe hydraulic radius is most nearly

(A) 5.0 in

(B) 6.0 in

(C) 12 in

(D) 18 in

392. For a pipe connecting two reservoirs as shown, the losses are as follows.

pipe entry loss = 2 ft
pipe exit loss = 3 ft
pipe friction loss = 30 ft
other fitting losses = 5 ft

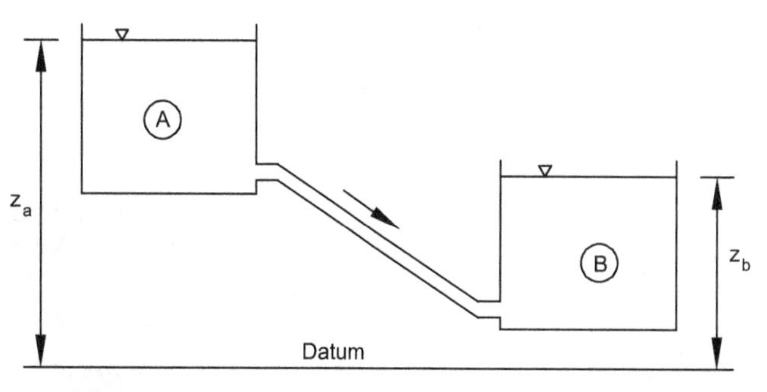

If the elevation of the water table in reservoir A is 100 ft, the elevation of the water table in reservoir B is ______. (Fill in the blank.)

393. A 1 ft diameter, 1200 ft long galvanized steel pipe connects two reservoirs, as shown.

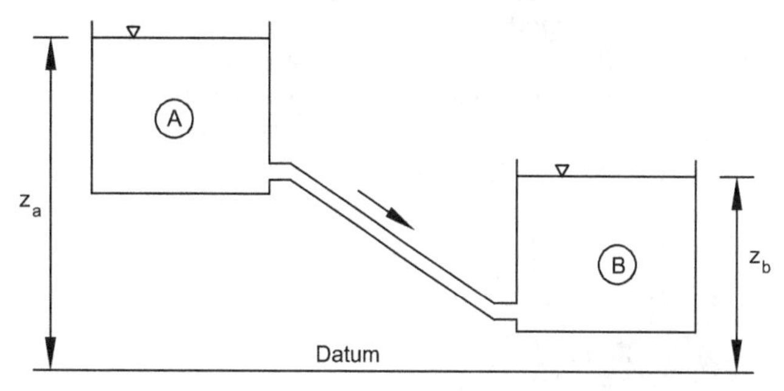

The pipe roughness factor is 0.0005 ft. The pipe discharge rate is 2 cfs at a water temperature of 57°F. Using the Moody diagram in the *NCEES Handbook*, Fluid Mechanics section, draw a line to show the friction factor for the pipe.

394. A 12 in diameter, 1000 ft long new cast-iron pipe connects two reservoirs. The water velocity in the pipe is 11 ft/sec. Neglect all minor losses. Using the Hazen-Williams method, the difference in elevations of the water tables between the reservoirs is most nearly

(A) 25 ft

(B) 31 ft

(C) 43 ft

(D) 65 ft

395. A 100 ft wide and 10 ft deep unconfined aquifer with a hydraulic conductivity of 0.2 ft/hr connects two water channels. The water level is 500 ft in one channel and 450 ft in the other. The clear distance between the channels is 3000 ft. The rate of seepage from one channel to the other is most nearly

(A) 1.2 ft^3/hr

(B) 2.0 ft^3/hr

(C) 3.3 ft^3/hr

(D) 6.7 ft^3/hr

396. An air sample has 100 $\mu g/m^3$ of CO_2 at standard temperature and pressure (STP). Assuming CO_2 is an ideal gas at STP, the concentration of CO_2 is most nearly

(A) 0.051 ppm

(B) 0.51 ppm

(C) 51 ppm

(D) 510 ppm

397. A 20 mL wastewater sample is filtered and placed in an empty crucible dish. The dish with the sample weighs 50.029 g. After evaporation and drying at 103°C, the crucible dish is cooled in a desiccator. The dish weighs 50.002 g. The total solids concentration of the sample is most nearly

(A) 0.00135 mg/L

(B) 0.0270 mg/L

(C) 1.35 mg/L

(D) 1350 mg/L

398. A 10 ft deep, 50 ft diameter circular clarifier handles 2 MGD. Its loading, or overflow, rate is most nearly

(A) 3 ft/hr

(B) 6 ft/hr

(C) 8 ft/hr

(D) 9 ft/hr

399. A water sample contains five cations and five anions as shown.

no.	cation	concentration (mg/L)	anion	concentration (mg/L)
1	Na^+	25	HCO_3^-	75
2	Ca^{2+}	12	NO_3^-	20
3	Mg^{2+}	15	Cl^-	25
4	Fe^{2+}	5	SO_4^{2-}	41
5	K^+	27	CO_3^{2-}	0

The cations contributing to the hardness of the water sample are

(A) 1, 2, 3

(B) 2, 3, 4

(C) 3, 4, 5

(D) 1, 2, 3, 4, 5

400. Three wastewater sewer lines have the flow rates and biochemical oxygen demand (BOD) concentrations shown.

no.	flow (10^6 L/day)	BOD (mg/L)
1	5	300
2	3	350
3	2	500

The sewer lines discharge into a stream that has a BOD of 10 mg/L, and the stream flow is twice that of the influent flow from the sewer lines. The resulting BOD in the stream at the point of discharge is most nearly

(A) 100 mg/L

(B) 125 mg/L

(C) 175 mg/L

(D) 350 mg/L

401. A crane carries a vertical load of 30 kips, as shown.

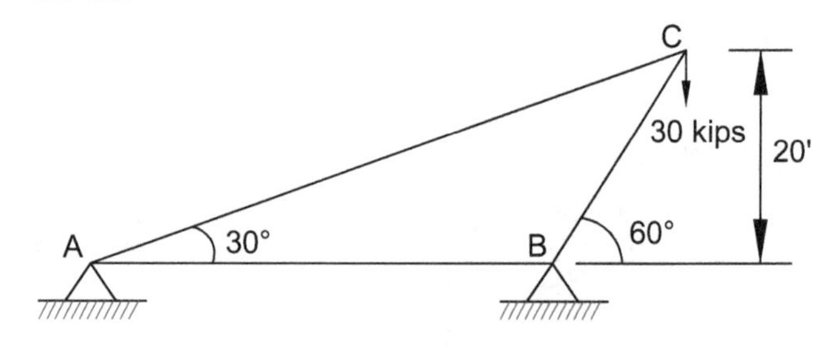

The truss is

(A) unstable

(B) statically determinate

(C) statically indeterminate to the first degree

(D) statically indeterminate to the second degree

402. Consider the crane shown.

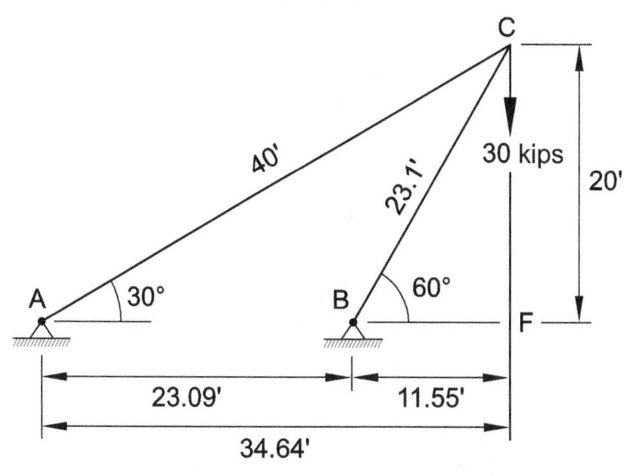

Assuming the negative sign denotes an uplift, the vertical reactions at supports A and B in kips are R_A = ______ and R_B =______, respectively. (Fill in the blank.)

403. A W10 × 45 steel column is 20 ft tall. Its weak axis is braced at its midlength, as shown.

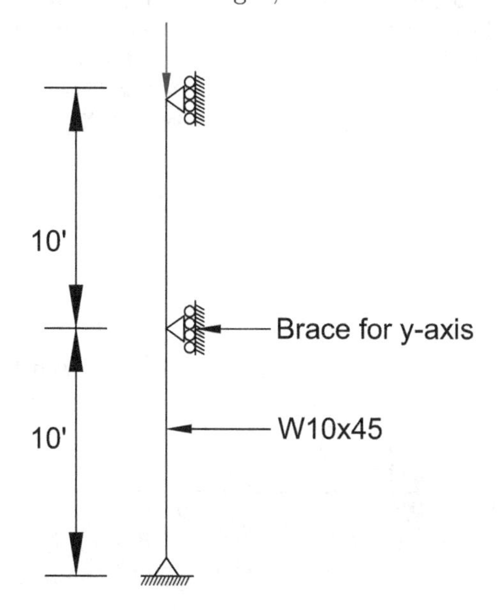

The controlling slenderness ratio of the column is most nearly

(A) 40

(B) 60

(C) 70

(D) 100

404. A truss carries a vertical load, P, of 90 kips.

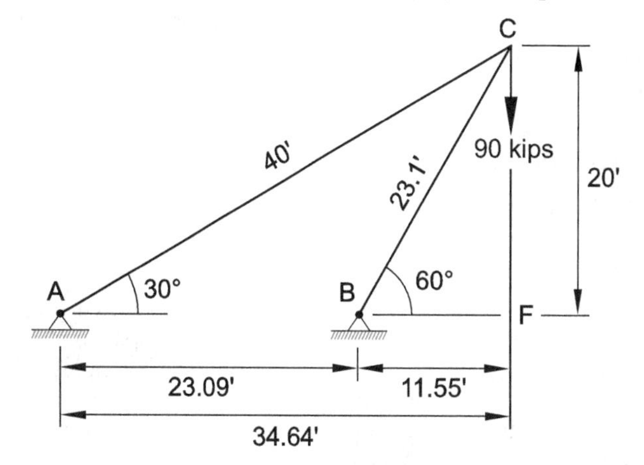

The forces in the truss members AC and BC are 90 kips (tension) and 155.7 kips (compression), respectively. All members have pinned joints and have a standard steel pipe cross-sectional area of 5 in^2. The vertical deflection of the crane in the direction of the applied load is most nearly

(A) 0.2 in

(B) 0.8 in

(C) 1 in

(D) 2 in

405. A cantilever beam carries a uniformly distributed load (UDL) of 1 kip/ft. The cantilever spans 24 ft. It is being propped at its free end to reduce its tip deflection to zero. The vertical force, in kips, needed to prop the cantilever is most nearly ______. (Fill in the blank.)

406. Two aluminum planar parallel rods, each 10 mm in diameter and 5 m long, carry 90 kN load, as shown.

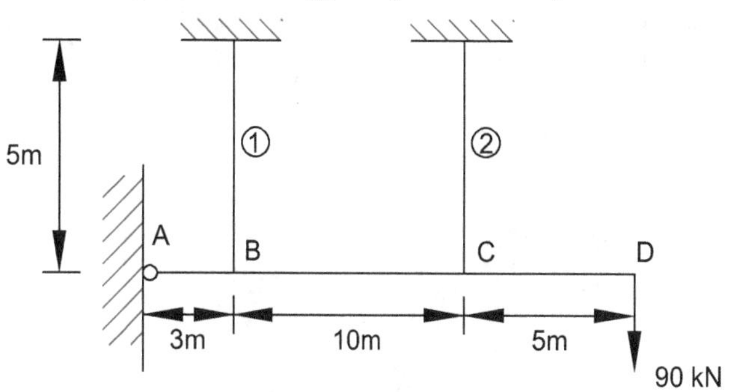

If bar AD is rigid, the axial tensile force in the vertical rod at C is most nearly

(A) −56 kN

(B) 27 kN

(C) 95 kN

(D) 120 kN

407. For a 100 kip rolling load over a 30 ft simply supported beam, the maximum end support reaction is most nearly

(A) 30 kips

(B) 50 kips

(C) 60 kips

(D) 100 kips

408. A steel beam is required to carry a service point load of 50 kips at the midspan. The beam span is 20 ft and its depth cannot exceed 18.5 in. Using the AISC allowable stress design method and a yield strength of 50 ksi, the lightest wide flange section to carry the load is

(A) W18 × 46

(B) W18 × 55

(C) W18 × 65

(D) W21 × 55

409. A beam carries two equal factored loads of 30 kips at its third points along the span. The beam span is 36 ft. The beam is braced every 6 ft along its span. The lightest wide-flange section with the yield strength of 50 ksi that can carry the load is most nearly

(A) W16 × 45

(B) W18 × 46

(C) W21 × 44

(D) W21 × 48

410. A simply supported steel W-section is carrying a uniformly distributed load over its span. If the beam is braced at its supports only, the C_b factor is most nearly

(A) 1.0

(B) 1.12

(C) 1.14

(D) 1.56

411. The rectangular singly reinforced concrete beam shown in the illustration has three #9 grade 60 bars as its bottom reinforcement.

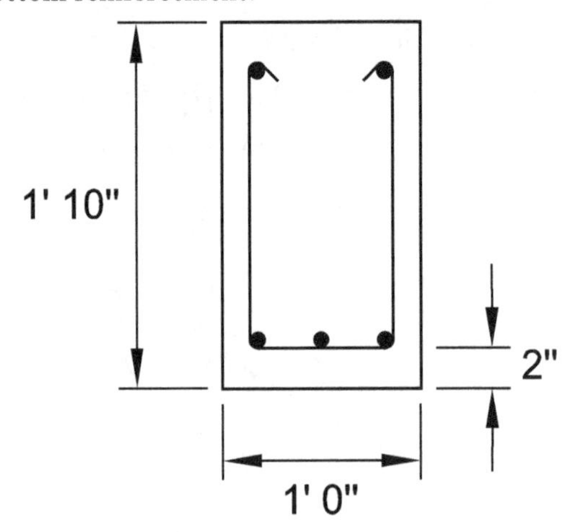

The concrete strength is 5000 psi. The section is tension controlled. The ultimate moment capacity, ϕM_n, of the section is most nearly

(A) 240 ft-kips

(B) 270 ft-kips

(C) 320 ft-kips

(D) 3100 ft-kips

412. A 5000 psi reinforced concrete beam is required to resist a dead load shear force of 10 kips and a live load shear force of 15 kips. The beam is 12 in wide and 23 in deep. The distance from the concrete cover to the center of the bottom steel is 3 in. Use f_y = 60 ksi and f_c' = 5 ksi. The bar size needed for a two-leg stirrup with a spacing of 10 in on the center is most nearly

(A) #3

(B) #4

(C) #5

(D) #6

413. A sieve analysis of a soil sample showed the following results.

sieve no.	retained (%)
4	0
16	20
50	25
80	20
200	25

The percentage of soil particles in the sample that are finer than 75 microns is most nearly

(A) 10%

(B) 30%

(C) 40%

(D) 100%

414. A soil sample weighing 10 lbf has a volume of 0.085 ft^3. The sample is oven-dried, and its weight is 9.25 lbf. The soil's specific gravity is 2.70. The degree of saturation of the soil sample is most nearly

(A) 20%

(B) 40%

(C) 60%

(D) 100%

415. Which of the following statements regarding the permeability through a soil is NOT correct?

(A) The coefficient of permeability can be measured using the constant head test method.

(B) The coefficient of permeability can be measured using the falling head test method.

(C) The coefficient of permeability is a measure of hydraulic conductivity.

(D) The coefficient of permeability is dimensionless.

416. A footing is placed in sandy soil. The following parameters are given.

depth of footing below ground level, z = 6 ft
water table above foundation level, h = 3.5 ft
dry soil density, γ = 120 pcf
submerged density of the soil, γ_{sat} = 128 pcf

The effective overburden pressure at the footing depth is most nearly

(A) 530 psf

(B) 650 psf

(C) 740 psf

(D) 770 psf

417. A cantilever sheet piling wall is used to retain soil, as shown.

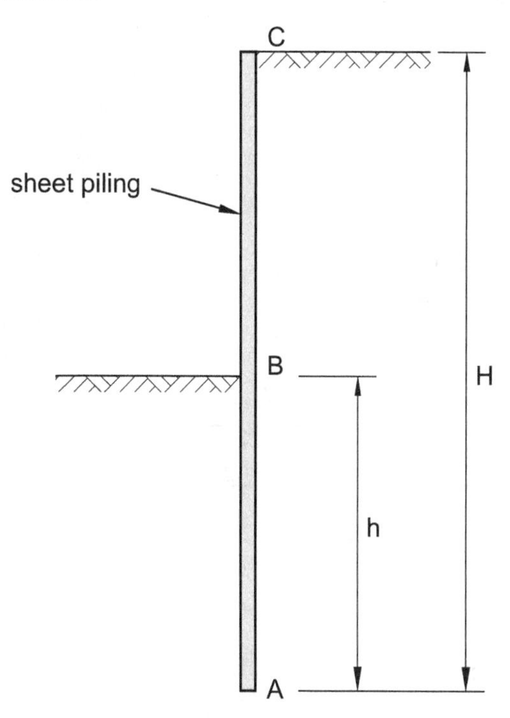

The angle of internal friction of the soil, ϕ, is 30°. Assume the wall rotates about point A. If $H = 30$ ft and $h = 15$ ft, the factor of safety, FS, against rotation is most nearly

(A) 1.1

(B) 2.3

(C) 2.5

(D) 3.6

418. The drained direct shear test was done on two soil samples from a job site. The loading consisted of a constant vertical load and an increasing horizontal shear until the soil sample failed. The test results are

sample	effective vertical stress (psf)	effective shear stress (psf)
1	1000	700
2	4000	2500

Based on the test results, the soil's internal angle of friction is most nearly

(A) 27°

(B) 29°

(C) 31°

(D) 33°

419. A 4 ft wide strip footing is shown. The soil supporting the foundation has a friction angle of 25° and cohesion of 2 ksf. The Terzaghi ultimate bearing equation coefficients are listed below. The soil unit weight is 110 pcf. The water table is 10 ft below the footing base.

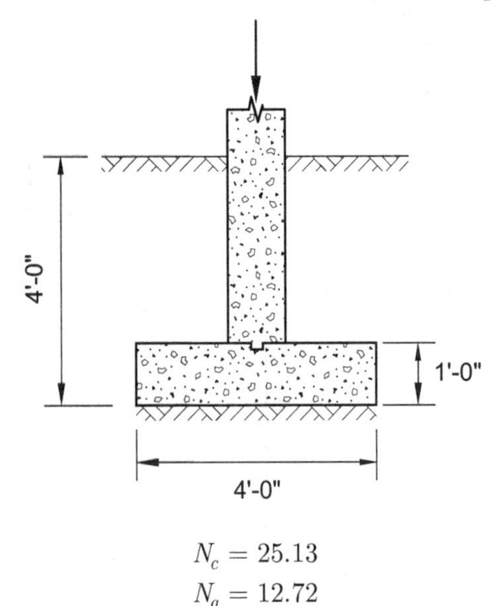

$$N_c = 25.13$$
$$N_q = 12.72$$
$$N_\gamma = 8.34$$

Using a safety factor of 3.0, the allowable soil bearing pressure for footing is ______. (Fill in the blank.)

420. The data for a strip footing is as follows.

width = 5 ft

wall load = 10 kips/ft

wall moment = 10 ft-kips/ft

The soil pressure at the toe is most nearly

(A) −0.4 ksf

(B) 0.9 ksf

(C) 3.4 ksf

(D) 4.4 ksf

421. A basement wall is laterally supported at its top end and the base. The back of the wall is smooth. The backfill is level and overconsolidated. The following properties are given.

- soil density = 120 pcf
- angle of internal friction = 25°
- cohesion coefficient, $c = 0$
- pressure under which the deposit was fully consolidated in the past = 125 psf
- present overburden pressure = 75 psf

The coefficient of lateral load pressure on the wall is most nearly

(A) 0.42

(B) 0.58

(C) 0.72

(D) 0.95

422. A saturated clay stratum of 10 m thickness is bound by sand layers. The coefficient of consolidation of sand is 0.002 cm^2/s. The stratum was subjected to a loading. The minimum days needed for the stratum to undergo 50% of its primary consolidation is most nearly

(A) 1 day

(B) 29 days

(C) 142 days

(D) 285 days

423. Which of the following factors helps decrease the soil losses due to erosion on a terrain due to runoff?

(A) excessive unsheltered length and steepness of terrain

(B) high rainfall erosivity factor

(C) excessive organic contents in soil

(D) wave height and frequency

424. A highway is designed for 55 mph. The grade of the highway is 2%. For a deceleration of 10 ft/sec^2 and a reaction time of 1.5 sec, the stopping sight distance is most nearly

(A) 125 ft

(B) 305 ft

(C) 425 ft

(D) 615 ft

425. An existing 300 ft long vertical curve on a roadway has a 5% grade followed by a −5% grade. Assume the height of the driver's eyes to be 3.5 ft and an object height of 2 ft. The permissible deceleration is 10 ft/sec^2, and the reaction time is 1.5 sec. The stopping sight distance, S, needs to be less than the length of the curve, L. While driving on the worst-case slope, the maximum safe speed on this curve is most nearly

(A) 35 mph

(B) 45 mph

(C) 55 mph

(D) 65 mph

426. A curve is needed to connect a −5% grade to a −1% grade. The design is 45 mph, and the acceptable deceleration is 10 ft/sec^2. The driver's reaction time is 2.5 sec. For the curve, the sight stopping distance is most nearly

(A) 160 ft

(B) 260 ft

(C) 360 ft

(D) 420 ft

427. An existing asphalt pavement has three layers. The layer characteristics are given.

	layer coefficient	layer thickness (in)
asphalt concrete surface	0.4	4
gravel base	0.14	5
sand sub-base	0.11	8.5

Assuming the drainage coefficient as 1.0, the maximum structural number (SN) that the pavement can resist is

(A) 1

(B) 2

(C) 3

(D) 4

428. At an approach to an intersection, 15-minute traffic flow readings were taken during an hour.

15 min interval	volume (veh)
8:00–8:15 a.m.	400
8:15–8:30 a.m.	425
8:30–8:45 a.m.	465
8:45–9:00 a.m.	450

The peak hourly factor equals most nearly

(A) 0.935

(B) 1.07

(C) 435

(D) 1740

429. Which of the following statements is correct about a dilemma zone?

(A) When four motorists are at a four-stop sign intersection and they all are hesitant to move into the intersection, it is called a dilemma zone.

(B) When a motorist is approaching a signalized intersection at a legal speed limit and the green light turns yellow, the motorist is in a dilemma zone.

(C) When a motorist is approaching a signalized intersection at a legal speed limit and the green light turns yellow, the motorist is in a dilemma zone if he or she can stop without encroaching into the intersection but needs to accelerate to clear the intersection before the light turns red.

(D) When a motorist is approaching a signalized intersection at a legal speed limit and the green light turns yellow, the motorist is in a dilemma zone if he or she can neither stop safely before entering the intersection nor continue and clear the intersection without speeding before the light turns red.

430. As the speed of the traffic stream increases, the traffic flow in the lane

(A) increases

(B) decreases

(C) increases initially and then decreases after reaching a maximum speed

(D) decreases initially and then increases after reaching a minimum speed

431. A four-way intersection had 18 accidents in one year. During this year, the average daily traffic entering the intersection from all approaches was 10,000 vehicles per day. The intersection's crash rate per million entering vehicles is most nearly

(A) 0.55

(B) 1.5

(C) 5.0

(D) 18

432. A city implemented countermeasures on an intersection to reduce crashes. The following data was taken.

- ADT (annual average for 3-year study period) = 8,000
- Annual number of crashes (average over 3-year period) = 25
- Crash reduction factor for the countermeasures = 30%
- ADT after countermeasures = 10,000

The number of crashes prevented by the countermeasures is most nearly

(A) 7

(B) 8

(C) 9

(D) 10

433. An owner hires a licensed design professional (LDP) to prepare construction documents for a new office building project. Once the design is complete, the owner plans to have the project bid for construction. The documents the LDP is expected to prepare for and deliver to the owner are

(A) a drawing set and specifications

(B) a drawing set, specifications, and shop drawings

(C) a drawing set, specifications, procurement documents, and contracting documents

(D) a drawing set, specifications, procurement documents, contracting documents, and shop drawings

434. Which of the following is NOT considered a "struck-by" hazard on a construction site?

(A) a vehicle

(B) a falling or flying object

(C) a concrete or masonry wall that collapses

(D) an armed trespasser

435. A contractor budgeted $150,000 to complete part of a project, but spent $210,000 to complete the task. The contractor's cost variance, CV, for the work performed is most nearly

(A) −$60,000

(B) $10,000

(C) $40,000

(D) $60,000

436. For earthwork, a balanced region is defined as the distance between two stations in which

(A) the cut and fill materials have equal masses

(B) the cut and fill materials have equal volumes

(C) the proposed profile matches the existing profile so that no cut or fill is needed

(D) the cut and fill volume or mass may differ, but there is no hauling cost for the earthwork

437. The total cost of formwork in the United States can be as much as ______ of the completed concrete structure in place. (Select the highest correct option.)

(A) 10%

(B) 20%

(C) 40%

(D) 60%

438. An arrow diagram created using the critical path method, CPM, is shown.

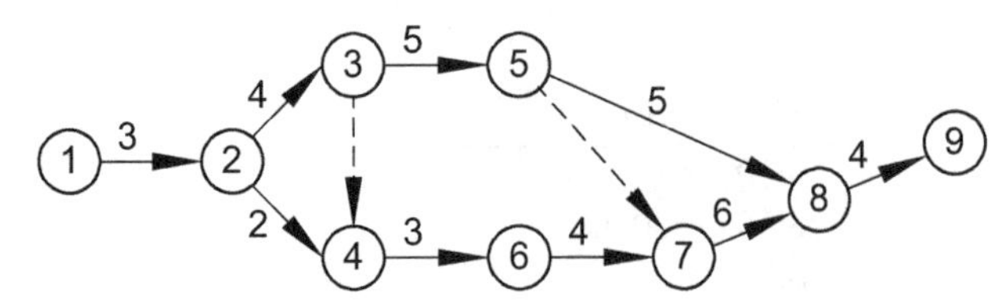

The dotted lines represent dummy activities, which consume no time or resource. The minimum number of days to complete the project is

(A) 21

(B) 22

(C) 23

(D) 24

439. A basement wall section is shown. The wall is designed to be laterally supported at its top end and at the base.

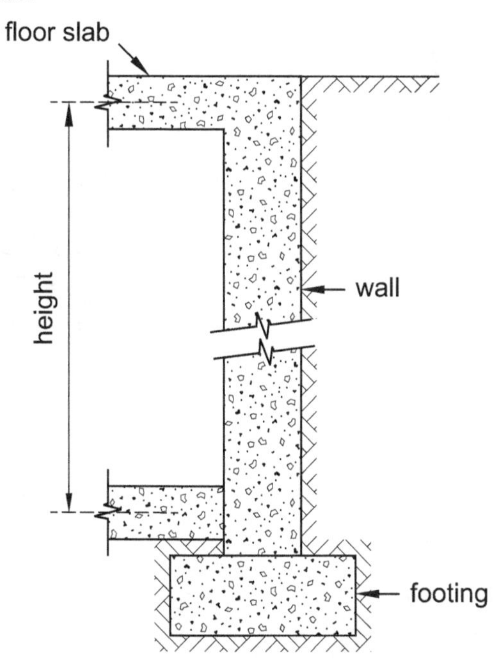

Out of the several construction activities listed below, rearrange the activities to be carried out in the proper sequence to construct the wall.

(A) Excavate and pour the foundation.

(B) Build the wall.

(C) Backfill with engineered fill.

(D) Pour the slab on grade and floor slab.

correct sequence	
1.	
2.	
3.	
4.	

440. A construction project is divided into four activities. The workforce resources required vary on a weekly basis, as tabulated below.

activity	weekly workers needed	week no.	
		start	end
A	6	9	15
B	4	11	20
C	3	15	22
D	7	13	24

During which week or weeks is the workforce demand at the maximum? ______. (Fill in the blank.)

STOP!

DO NOT CONTINUE!

This concludes the examination. If you finish early, check your work and make sure that you have followed all instructions. After checking your answers, you may turn in your examination booklet and answer sheet and leave the examination room. Once you leave, you will not be permitted to return to work or change your answers.

Exam 5 Instructions

In accordance with the rules established by your state, you may use any approved battery- or solar-powered, silent calculator to work this examination. However, no blank papers, writing tablets, unbound scratch paper, or loose notes are permitted. Sufficient paper will be provided. The *NCEES FE Reference Handbook* is the only reference you are allowed to use during this exam.

You are not permitted to share or exchange materials with other examinees.

You will have six hours in which to work this session of the examination. Your score will be determined by the number of questions that you answer correctly. There is a total of 110 questions. All 110 questions must be worked correctly in order to receive full credit on the exam. There are no optional questions. Each question is worth 1 point. The maximum possible score for this section of the examination is 110 points.

Partial credit is not available. No credit will be given for methodology, assumptions, or work written on scratch paper.

Record all of your answers on the Answer Sheet. Mark your answers with a no. 2 pencil. Answers marked in pen may not be graded correctly. Marks must be dark and must completely fill the bubbles. Record only one answer per question. If you mark more than one answer, you will not receive credit for the question. If you change an answer, be sure the old bubble is erased completely; incomplete erasures may be misinterpreted as answers.

If you finish early, check your work and make sure that you have followed all instructions. After checking your answers, you may submit your answers and leave the examination room. Once you leave, you will not be permitted to return to work or change your answers.

When permission has been given by your proctor, you may begin your examination.

Name: ______________________________________

Last First Middle Initial

Examinee number: ______________

Examination Booklet number: ______________

Fundamentals of Engineering Examination

Exam 5

Exam 5 Answer Sheet

441. (A) (B) (C) (D)
442. ____________
443. (A) (B) (C) (D)
444. (A) (B) (C) (D)
445. (A) (B) (C) (D)
446. ____________
447. ____________
448. (A) (B) (C) (D)
449. (A) (B) (C) (D)
450. ____________
451. (A) (B) (C) (D)
452. (A) (B) (C) (D)
453. ____________
454. (A) (B) (C) (D)
455. (A) (B) (C) (D)
456. (A) (B) (C) (D)
457. (A) (B) (C) (D)
458. ____________
459. (A) (B) (C) (D)
460. ____________
461. ____________
462. (A) (B) (C) (D)
463. (A) (B) (C) (D) (E)
464. (A) (B) (C) (D)
465. (A) (B) (C) (D) (E)
466. ____________
467. (A) (B) (C) (D)
468. ____________
469. ____________
470. (A) (B) (C) (D)
471. (A) (B) (C) (D)
472. (A) (B) (C) (D)
473. (A) (B) (C) (D)
474. (A) (B) (C) (D)
475. (A) (B) (C) (D)
476. (A) (B) (C) (D)
477. ____________
478. ____________
479. (A) (B) (C) (D)
480. (A) (B) (C) (D)
481. (A) (B) (C) (D)
482. ____________
483. (A) (B) (C) (D)
484. (A) (B) (C) (D)
485. (A) (B) (C) (D)
486. (A) (B) (C) (D) (E)
487. (A) (B) (C) (D)
488. (A) (B) (C) (D)
489. (A) (B) (C) (D)
490. (A) (B) (C) (D)
491. ____________
492. ____________
493. ____________
494. (A) (B) (C) (D)
495. ____________
496. ____________
497. (A) (B) (C) (D)
498. (A) (B) (C) (D)
499. (A) (B) (C) (D)
500. (A) (B) (C) (D)
501. (A) (B) (C) (D)
502. (A) (B) (C) (D)
503. (A) (B) (C) (D)
504. (A) (B) (C) (D)
505. (A) (B) (C) (D)
506. (A) (B) (C) (D)
507. (A) (B) (C) (D)
508. ____________
509. (A) (B) (C) (D)
510. ____________
511. (A) (B) (C) (D)
512. (A) (B) (C) (D)
513. (A) (B) (C) (D)
514. (A) (B) (C) (D)
515. (A) (B) (C) (D)
516. (A) (B) (C) (D)
517. (A) (B) (C) (D)
518. (A) (B) (C) (D)
519. (A) (B) (C) (D)
520. (A) (B) (C) (D)
521. (A) (B) (C) (D)
522. (A) (B) (C) (D)
523. (A) (B) (C) (D)
524. (A) (B) (C) (D)
525. (A) (B) (C) (D)
526. (A) (B) (C) (D)
527. (A) (B) (C) (D)
528. (A) (B) (C) (D)
529. (A) (B) (C) (D)
530. (A) (B) (C) (D)
531. (A) (B) (C) (D)
532. (A) (B) (C) (D)
533. (A) (B) (C) (D)
534. (A) (B) (C) (D)
535. (A) (B) (C) (D)
536. (A) (B) (C) (D)
537. (A) (B) (C) (D)
538. (A) (B) (C) (D)
539. (A) (B) (C) (D)
540. (A) (B) (C) (D)
541. (A) (B) (C) (D)
542. (A) (B) (C) (D)
543. ____________
544. ____________
545. ____________
546. (A) (B) (C) (D)
547. ____________
548. ____________
549. (A) (B) (C) (D)
550. ____________

Exam 5

441. For the quadratic equation $ax^2 + bx + c = 0$, the product of its roots is

(A) bc/a

(B) $b/(4a)$

(C) b^2c

(D) c/a

442. A precast concrete plant casts one piece each day during the first week. During the second week, it casts two pieces each day. The third week, it casts three pieces each day and so on. Assuming five working days per week, the total number of pieces the plant will have cast after the tenth week is ______. (Fill in the blank.)

443. Two matrices are given.

$$A = \begin{vmatrix} 1 & 2 \\ 5 & 4 \\ 3 & 7 \end{vmatrix}$$

$$B = \begin{vmatrix} 2 & 8 & 9 \\ 10 & 1 & 6 \end{vmatrix}$$

Matrices A and B are multiplied to obtain a product matrix C. Which statement about the matrix multiplication is true?

(A) The product matrix C will have 2 rows and 3 columns.

(B) The product matrix C will have 2 rows and 2 columns.

(C) The product matrix C will have 3 rows and 3 columns.

(D) The multiplication is not possible since the matrices A and B do not have the same number of rows and columns.

444. For the given polar equation, the rectangular form of the equation is plotted.

$$r = 4 \cos \theta$$

Which statement regarding the plot is correct?

(A) Its center is located at (0, 0), and the radius of the circle is 2.

(B) Its center is located at (2, 0), and the radius of the circle is 2.

(C) Its center is located at (0, 2), and the radius of the circle is 2.

(D) Its center is located at (4, 0), and the radius of the circle is 4.

445. A circular disk is shown.

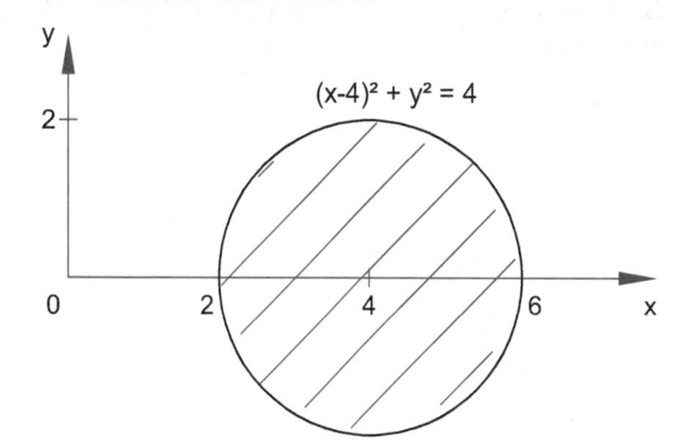

The disk is revolved about the y-axis to form a donut, as shown.

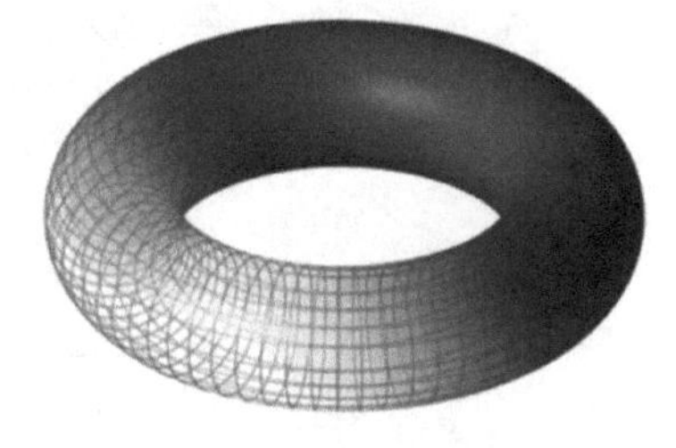

The donut volume is

(A) $18\pi^2$

(B) $24\pi^2$

(C) $32\pi^2$

(D) $36\pi^2$

446. The value of the integral $\int_{-2}^{1}(3x^2+2x-9)\,dx$ is ______. (Fill in the blank.)

447. Three vectors denoting a set of three forces are

$$\mathbf{F_1} = 4\mathbf{i} + 7\mathbf{j} + 2\mathbf{k}$$
$$\mathbf{F_2} = 5\mathbf{i} + 8\mathbf{j} + 3\mathbf{k}$$
$$\mathbf{F_3} = 6\mathbf{i} + 9\mathbf{j} - 4\mathbf{k}$$

The magnitude of the resultant force is ______. (Fill in the blank.)

448. The probability of tossing a die twice and getting a 5 and a 6 is

(A) 1/36

(B) 1/6

(C) 2/6

(D) 5/6

449. Five coins are tossed. The probability that three coins show heads and two show tails is

(A) 1/32

(B) 3/32

(C) 5/32

(D) 5/16

450. A factory employs 100 workers. Their wages vary according to the tasks they perform, as shown.

number of workers	hourly wage
10	$10.00
10	$30.00
15	$25.00
20	$12.00
20	$15.00
25	$20.00

The median hourly wage (in $/hr) is ______. (Fill in the blank.)

451. A licensee receives gifts from the licensee's work-related suppliers and distributors. For which of the following gifts should the licensee NOT inform his/her employer?

(A) raffle prize at a conference

(B) cash rebates

(C) tickets to supporting events the supplier does not attend

(D) a sit-down dinner without alcohol

452. Which statement regarding the low-bid competitive bidding process is NOT correct?

(A) It is a procurement method in which bids from competing contractors, suppliers, or vendors are invited by openly advertising the scope, specifications, and terms and conditions of the proposed contract, as well as the criteria by which the bids will be evaluated.

(B) It is used to obtain goods and services at the lowest prices by stimulating competition and preventing favoritism.

(C) The bids are always opened in full view of all who may wish to witness the bid opening.

(D) A major drawback of this method is the possibility of awarding a construction contract to a contractor who submits, either accidentally or deliberately, an unrealistically low bid price, which in turn promotes disputes, increased costs, and schedule delays.

453. A licensed engineer inspected a lakeside property under a contract with the client. The engineer found no structural issues at the property. The client was unhappy with the engineer's work product and complained to the State Board of Professional Engineers. The Board determined that the engineer's conclusion was arrived at without sufficient research or investigation. The Board further determined that the engineer failed to act with reasonable care and competence and apply the technical knowledge and skill that are ordinarily applied by professional engineers of good standing, practicing in the jurisdiction. The engineer agreed with the Board's determination. The ______ section of the *NCEES Rules of Professional Conduct* gives the Board the power to discipline the engineer. (Use the format in section 123.45.X6.) (Fill in the blank.)

454. An owner has a budget of $2,000,000 to build an office building. The owner signs a contract with a contractor on a cost-plus fee basis. The contractor's fee is 5% of the cost approved by the owner's engineer. At completion, the engineer approved $2,556,900 as the contractor's cost. The owner is obligated to pay the contractor most nearly

(A) $2,065,000

(B) $2,125,000

(C) $2,557,000

(D) $2,685,000

455. An owner has a budget of $100,000 to route and seal concrete floor cracks in a parking garage. It is estimated that the cracks total 25,000 linear ft and that the unit cost is $4/ft. The owner signs a unit price contract with the contractor at a unit repair cost of $5/ft. The price includes 10% net profit for the contractor. At completion, it is determined that the cracks total 30,000 linear ft. The owner's obligation to pay the contractor for completing the project is most nearly

(A) $100,000

(B) $110,000

(C) $150,000

(D) $165,000

456. An engineer is planning to start his own firm. He expects his firm to have net operating losses of $120,000 per year for the first two years, followed by net profit of $240,000 per year for the next eight years. The minimum attractive rate of return (MARR) is 12%. The net present worth of the expected cash flow is most nearly

(A) $747,000

(B) $951,000

(C) $1,150,000

(D) $1,680,000

457. An engineer is planning to start his own firm. He expects his firm to have net operating losses of $120,000 per year for the first two years, followed by a net profit of $240,000 per year for the next eight years. He plans to borrow $10,000 at the end of each month for the next 24 months to meet his operating losses. The interest rate is 0.5% monthly. If the debt is to be paid in monthly installments as soon as possible, the minimum number of years it would take for the engineer to be debt-free would be most nearly

(A) 2

(B) 3

(C) 4

(D) 5

458. A company manufactures a product and faces incremental marketing cost.

description	cost
initial setup	$100,000
labor	$1.5/unit
materials	$0.5/unit
marketing	$0.3/unit for the first 100,000 units
	$0.6/unit for units exceeding 100,000
sale price	$2.95 per unit

The break-even point to be profitable in manufacturing the product is ______. (Fill in the blank.)

459. A repair project will take 1 year to complete. It will cost $100,000 now and $50,000 at the beginning of the second year. The restoration would extend the life of the facility by 3 years after completion. Considering a 10% annual rate of return, the yearly minimum income the project should produce to make the project feasible is

(A) $64,000

(B) $65,000

(C) $66,000

(D) $67,000

460. A city is considering widening a roadway to accommodate bike lanes. The project costs and its return are

land	$1,000,000
construction	$11,000,000
annual maintenance	$500,000
annual benefit	$1,300,000

If the project life is 25 years and the rate of interest is 6%, the benefit-to-cost ratio, rounded to one decimal place, is ______. (Fill in the blank.)

461. A ready-mix company delivers concrete at the rate of $110 per truck. The gross profit per truck is $20. Historically, 1% of all concrete produced is rejected, not meeting the specifications. It costs the company $500 to properly dispose of it. The profit per truck that the company should expect is ______. (Fill in the blank.)

462. A mechanical shaft is subjected to four point loads (kN), as shown.

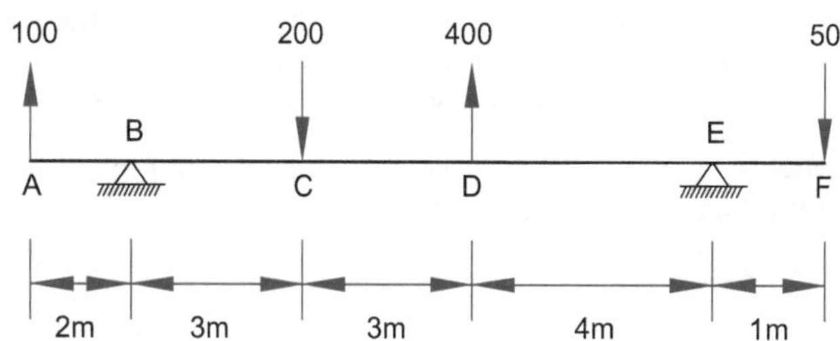

Assume an upward force as positive. The equivalent applied loading at point A is most nearly

(A) 250 kN, 1550 kN·m (ccw)

(B) 350 kN, 1550 kN·m (cw)

(C) 750 kN, 4850 kN·m (cw)

(D) 1200 kN, 0 kN·m

463. A mechanical shaft supports four belts and pulleys. It is subjected to a set of four-point loads, as shown.

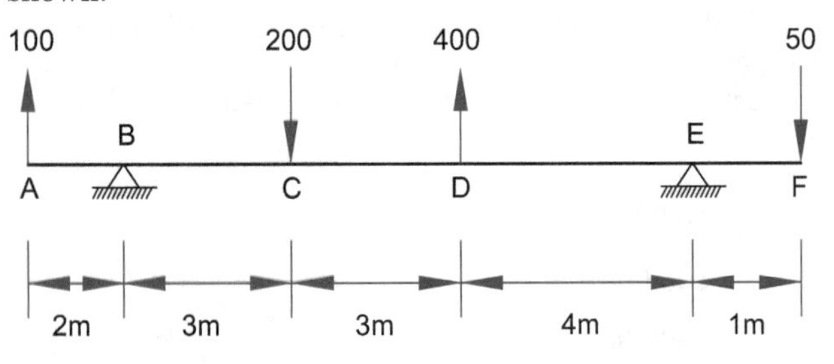

Assume upward force as positive. Out of the options listed, choose the two that describe the resultant force and its location from A.

(A) 250 kN

(B) 750 kN

(C) 0 m

(D) 3.6 m

(E) 6.2 m

464. Which of the following statements is INCORRECT?

(A) A force is a vector quantity.

(B) A system of two forces that are equal in magnitude, opposite in direction, and parallel to each other is called a couple.

(C) The polar moment of inertia of an area about a point is equal to the sum of the moments of inertia of the area about two perpendicular axes in the area that pass through the centroid.

(D) The polar moment of inertia of an area about a point is equal to the sum of the moments of inertia of the area about any two perpendicular axes in the area and passing through the same point.

465. A 500 N mass is hung from point O supported by cables OA and OB, as shown.

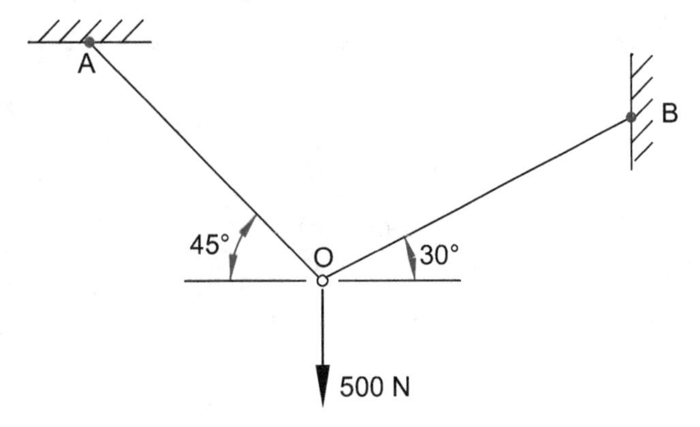

Assume a tension force as positive. Forces F_{OA} and F_{OB} respectively pertain to which of the following options?

(A) −448 N

(B) −366 N

(C) 0.0 N

(D) 366 N

(E) 448 N

466. The truss tower shown is built with 9 joints and 17 members. All joints are pinned, and the bars comprising the X-bracings are not connected at their respective cross-points.

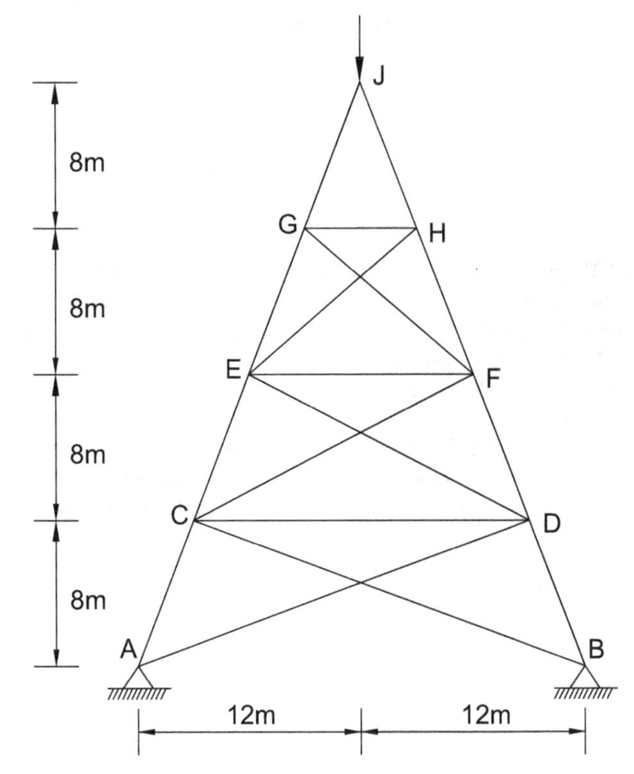

The total degree of indeterminacy of the tower is ______. (Fill in the blank.)

467. A baseball with a mass of 149 g mass travels at 30 m/s and is caught by a player and brought to rest in 0.1 s. The average force applied to the ball is most nearly

(A) 9.8 N

(B) 45 N

(C) 150 N

(D) 200 N

468. A truss tower is built as shown.

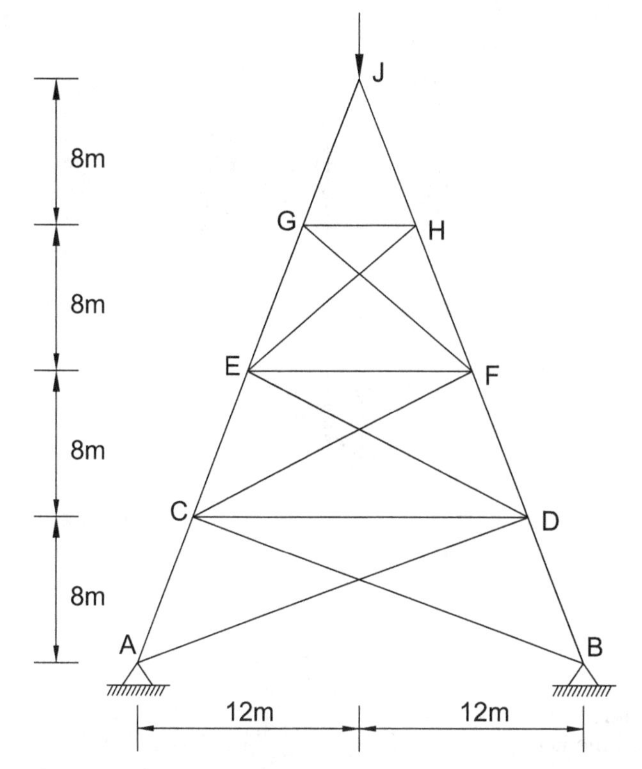

It has an antenna weighing 500 kN at joint J. The vertical reaction at support A is ______. (Fill in the blank.)

469. A traffic sign is hung from two points off a fixed pole, as shown.

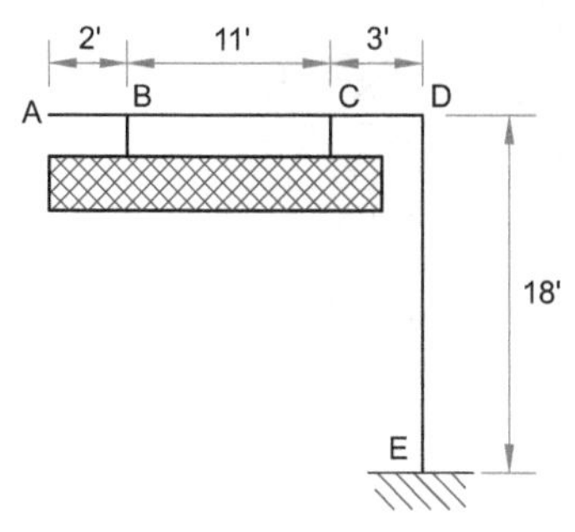

The load at points B and C is 1 kip. Assuming counterclockwise moment as positive, what is the resulting moment (ft-kips) at the base? ______. (Fill in the blank.)

470. Consider the ladder shown.

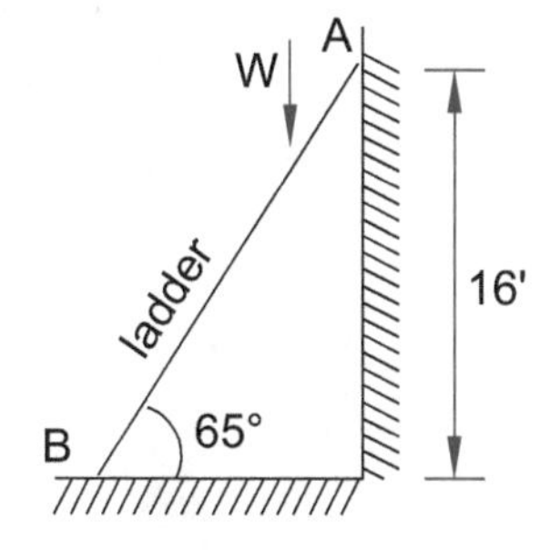

Assume zero friction between the ladder and wall. The minimum coefficient of friction needed at the floor for a 200 lbf person to safely use the ladder is most nearly

(A) 0.2

(B) 0.3

(C) 0.4

(D) 0.5

471. A wheel rotates at an angular velocity of 900 rpm. A brake is applied at an angular rate of 10 rad/sec^2. The number of revolutions the wheel makes before it stops is most nearly

(A) 70

(B) 150

(C) 300

(D) 450

472. A 200 mm diameter steel cantilever shaft carries a 1000 kg disk on its cantilevered end. The radius of gyration of the disk is 1 m. The shaft is 1 m long. The torsional frequency of the shaft is most nearly

(A) 6 Hz

(B) 18 Hz

(C) 110 Hz

(D) 350 Hz

473. An asphalt pavement roller has a mass of 10,000 kg. Its wheels have a radius of gyration of 0.5 m and a mass of 5000 kg each. The diameter of each wheel is 1.1 m. If the roller moves at a speed of 10 km/h, its kinetic energy is most nearly

(A) 15,000 N·m

(B) 30,000 N·m

(C) 40,000 N·m

(D) 70,000 N·m

474. A wheel comes off a vehicle traveling at 100 km/hr. Assume that the tire mass is well-distributed and neglect friction losses. If the road slope is 5% uphill, the horizontal distance the tire will roll before stopping is most nearly

(A) 80 m

(B) 680 m

(C) 1180 m

(D) 2300 m

475. A 200 mm diameter steel shaft is loaded as shown.

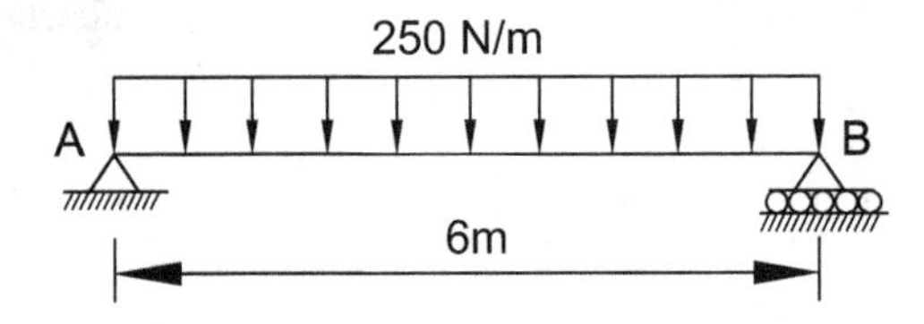

Neglecting self-weight, the shaft's maximum bending stress is most nearly

(A) 0.0014 MPa

(B) 1.4 MPa

(C) 14 MPa

(D) 140 MPa

476. Bar ABCD is subjected to axial loads (kN) along its length, as shown.

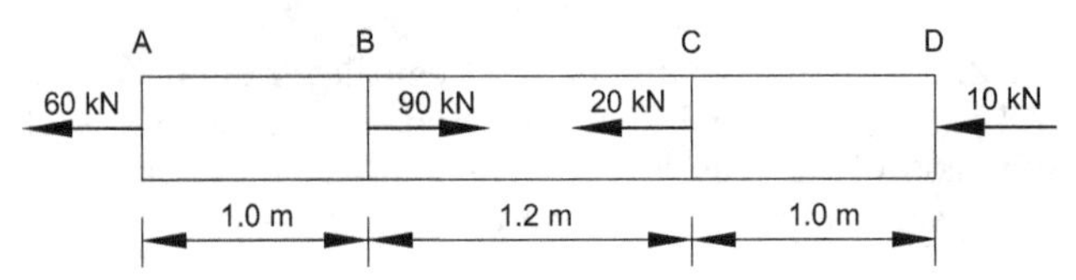

The force in bar segment BC is most nearly

(A) 10 kN

(B) 20 kN

(C) 30 kN

(D) 90 kN

477. Bar ABCD is subjected to axial loads (kN) along its length, as shown.

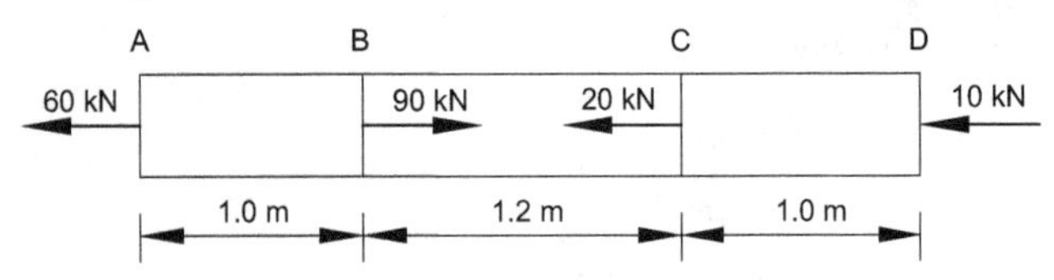

If the product of the bar's cross-sectional area, A, and its modulus of elasticity, E, is 1000 kN, the elongation of the bar (in mm) is ______. (Fill in the blank.)

478. A stepped fixed-free column carries two 15,000 lbf concentric loads, as shown. The cross-sectional area of segment AB is 100 in^2 and the cross-sectional area of segment BC is 150 in^2.

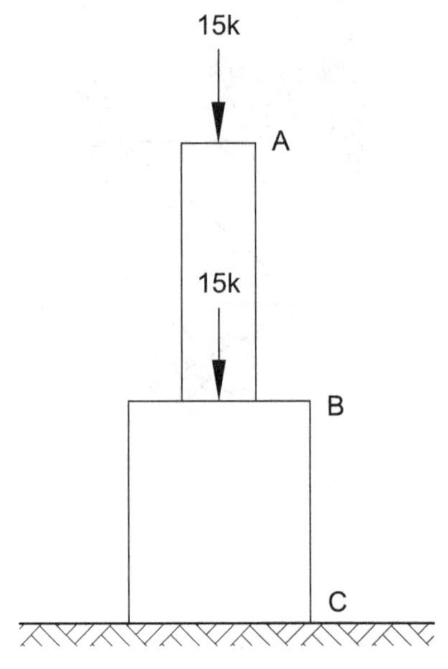

The maximum normal stress (psi) in the column at plane B in segment BC is ______. (Fill in the blank.)

479. Two 30 ft long W10 × 50 sections, each with a cross-sectional area of 14.7 in^2, form a truss and carry 50 kips, as shown.

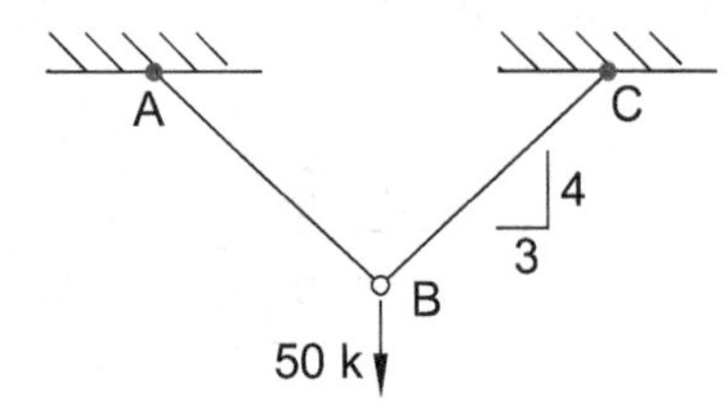

The vertical deflection of joint B is most nearly

(A) 0.01 in

(B) 0.013 in

(C) 0.026 in

(D) 0.033 in

480. A 12 in outer diameter steel tube has a wall thickness of 1 in. The tube is 10 ft long and is subjected to a torque of 100 ft-kips at its ends. The angle of twist between its ends is most nearly

(A) 0.01°

(B) 0.1°

(C) 0.7°

(D) 10°

481. A 50 kN tensile load is gradually applied to a 5 cm diameter, 4 m long steel bar. The energy absorbed in the bar is most nearly

(A) 0.0025 J

(B) 0.012 J

(C) 12 J

(D) 48 J

482. A uniformly loaded fixed-end beam is shown.

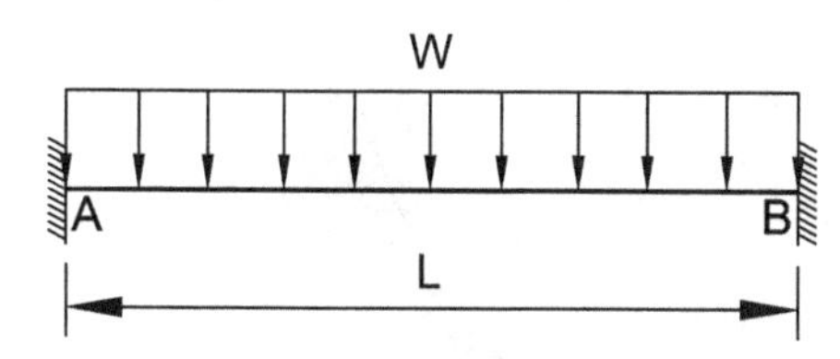

For a uniformly distributed load of w, the beam's moment at supports and midspan is expressed as

$$M = \frac{wL^2}{x}$$

The beam is loaded beyond its elastic limit, to its nominal moment capacity. At that level, the value of x in the expression is ______. (Fill in the blank.)

483. Concrete mix strength increases the most when

(A) both the water/cement ratio and the entrained air decrease

(B) both the water/cement ratio and the entrained air increase

(C) the water/cement ratio increases and the entrained air decreases

(D) the water/cement ratio decreases and the entrained air increases

484. The maximum feasible slump a concrete mix can have is most nearly

(A) 0 in

(B) 6 in

(C) 11 in

(D) 17 in

485. An asphalt concrete mix has a total volume of 1 ft^3. By volume, the mix has voids of 0.04 ft^3, a binder of 0.12 ft^3, and aggregate of 0.84 ft^3. The percentage of voids filled by the asphalt binder is most nearly

(A) 25%

(B) 50%

(C) 75%

(D) 100%

486. An open parking garage is being planned in Chicago, IL, where temperatures frequently fall below freezing during the winter. Of the following options, which two are best practices to design a durable structure?

(A) Concrete should be dense so that it has no entrained air or voids.

(B) Concrete air voids should not exceed 3%.

(C) Concrete should have entrained air not less than 6%.

(D) Air voids should be clustered.

(E) Air voids should be well distributed.

487. Which of the following statements regarding the uniaxial stress-strain relationship of metals is NOT correct?

(A) The slope of the linear portion of the curve equals the modulus of elasticity.

(B) The yield strength at 0.2 percent strain offset is called the yield stress.

(C) The engineering stress is defined as the load divided by the initial cross-sectional area.

(D) The true stress is defined as the average of the yield stress and the ultimate stress.

488. Oil flows through a pipe at a velocity of 10 m/s. Its specific gravity is 2.0, and its kinematic viscosity is 0.0003 m^2/s. For a 1 in diameter pipe, the Reynolds number of the oil is most nearly

(A) 420

(B) 850

(C) 1700

(D) 33,000

489. A turbine is operating under a constant head of 10 m. The water flows through a 0.15 m diameter exit pipe at a velocity of 5 m/s.

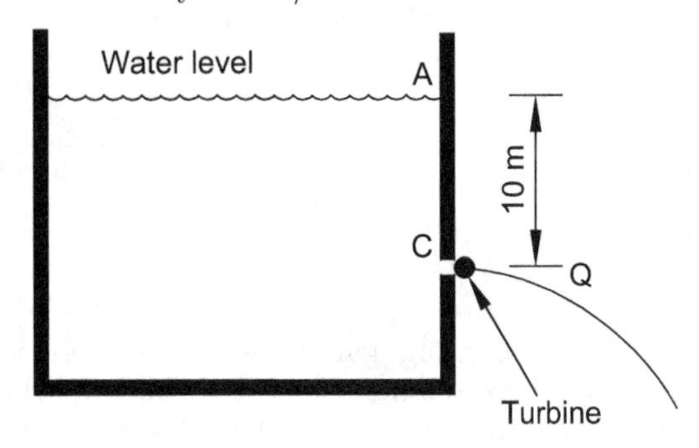

Neglecting all losses, the turbine's maximum power output is most nearly

(A) 1 hp

(B) 10 hp

(C) 100 hp

(D) 250 hp

490. A nozzle of 1 in diameter delivers a stream of water at a velocity of 300 ft/sec impinging on a vane that moves at a velocity of 100 ft/sec and deflects the jet 60°.

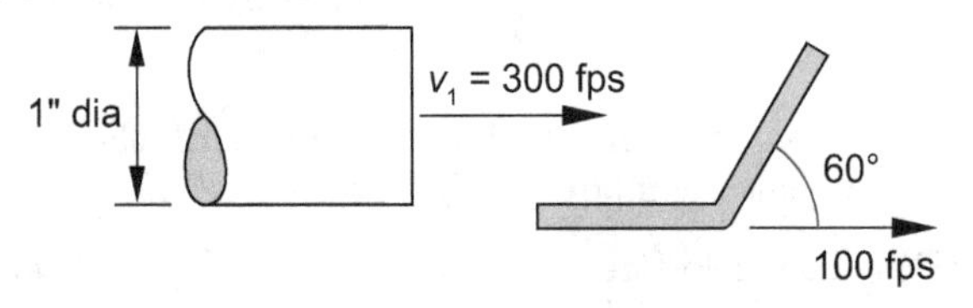

The horsepower of the turbine is most nearly

(A) 1.6 hp

(B) 3.2 hp

(C) 32 hp

(D) 58 hp

491. A torque of 2.0 N·m is needed to rotate a cylinder at 1000 rad/s.

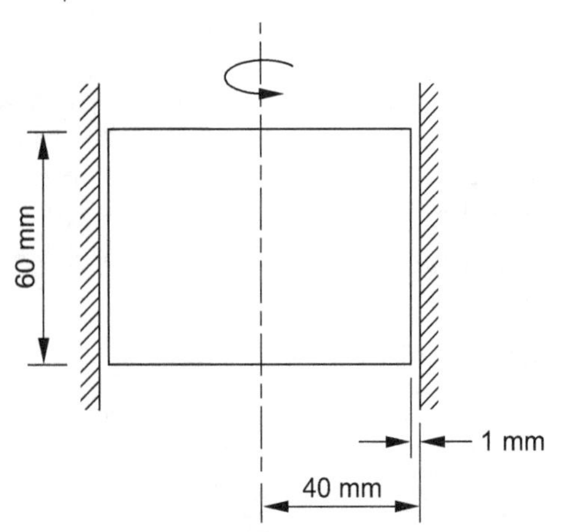

Assuming thin Newtonian film and linear velocity profile, the viscosity at the interface (in N·m/s^2 rounded to 2 decimal places) is ______. (Fill in the blank.)

492. The profile of a concrete dam is parabolic.

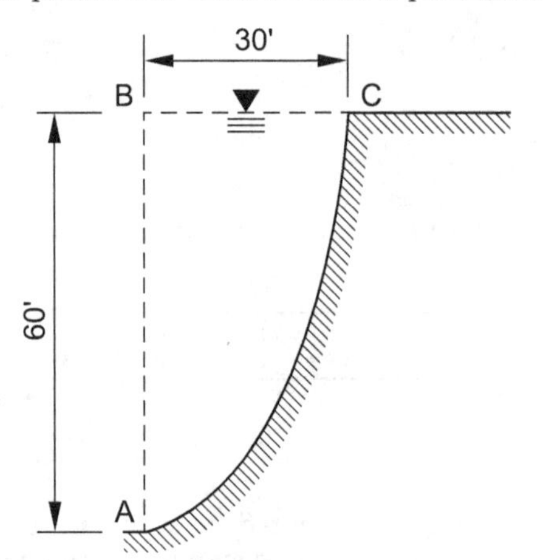

The dam retains water to a depth of 60 ft. What are the magnitude of the resultant thrust per foot run of the dam and its inclination to the horizontal in degrees? Select your response from the options given below.

thrust per foot of run: ______

inclination angle: ______

options
112 kips
135 kips
34°
56°

493. A billboard is 2 m wide and 1 m tall, mounted on top of a 5 m long pole. The pole diameter is 20 cm. The billboard is subjected to 65 km/h wind. The kinematic viscosity of air is 1.47 × 10^{-5} m^2/s.

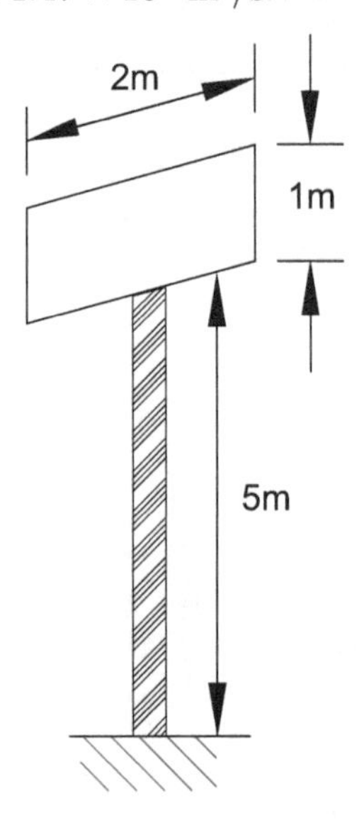

Point on the graph Drag Coefficient for Spheres, Disks, and Cylinders given in the Fluid Mechanics section of the *NCEES Handbook* to determine the drag coefficient (rounded to the nearest digit) on the pole.

494. A surveyor measures a 120 m long traverse segment AB. The surveyor then measures angle BAC to be 55° and angle ABC to be 45° to a benchmark C. The distance AC is most nearly

(A) 86.15 m

(B) 99.81 m

(C) 144.3 m

(D) 167.2 m

495. A parcel is described in a public survey land section in a township as

$$\text{S}\ \frac{1}{2}\ \text{of the NE}\ \frac{1}{4}\ \text{of the NW}\ \frac{1}{4}\ \text{of the SW}\ \frac{1}{4}\ \text{of Section 11}\ldots$$

The area of the parcel in acres is ______. (Fill in the blank.)

496. A road is being planned to connect 1°10′20″ north and 5°20′23″ south along a meridian. Assume the earth is a perfect sphere and its mean radius is 20,906,000 ft. The length of road, in miles, is ______. (Fill in the blank.)

497. A horizontal curve has a radius of 350 ft. Its intersection angle is 66°. Its PC is at station 10+67.50. The point of tangent (PT) station of the curve is most nearly

(A) sta 14+70.67

(B) sta 76+67.50

(C) sta 55+32.50

(D) sta 360+67.50

498. A 1000 ft vertical curve has a back tangent slope of −2% and a forward slope of 4%. At the point of intersection (PVI), the elevation is 2540 ft and is located at station 12+50. The curve elevation at station 10+00 is most nearly

(A) 2501 ft

(B) 2513 ft

(C) 2526 ft

(D) 2547 ft

499. A 500 ft long vertical curve has a back tangent slope of −4% and a forward tangent slope of 3%. The PVI station is 21+50 at an elevation of 51.50 ft above the datum. The curve's PVC and PVT stations and elevations are most nearly

	PVC		PVT	
	station	elevation	station	elevation
A.	19+00.00	61.50	24+00.00	59.00
B.	24+00.00	61.50	19+00.00	59.00
C.	19+00.00	41.50	24+00.00	44.00
D.	19+00.00	42.75	24+00.00	42.75

500. The NRCS rainfall-runoff method assumes that the initial abstraction (depression storage, evaporation, and interception losses) is most nearly

(A) 0% of the maximum basin storage or retention capacity

(B) 10% of the maximum basin storage or retention capacity

(C) 20% of the maximum basin storage or retention capacity

(D) 80% of the maximum basin storage or retention capacity

501. A 36 in diameter sewer pipe is running three-quarters full in terms of depth. Its hydraulic radius is most nearly

(A) 9.0 in

(B) 11 in

(C) 15 in

(D) 18 in

502. A water tank has a capacity of 200,000 gal and is full. The demand for water is estimated to be 800 gpm. The inflow into the tank is 300 gpm. The time needed to empty 95% of the tank is most nearly

(A) 4 hr

(B) 5 hr

(C) 6 hr

(D) 10 hr

503. A pipe connects two reservoirs, as shown.

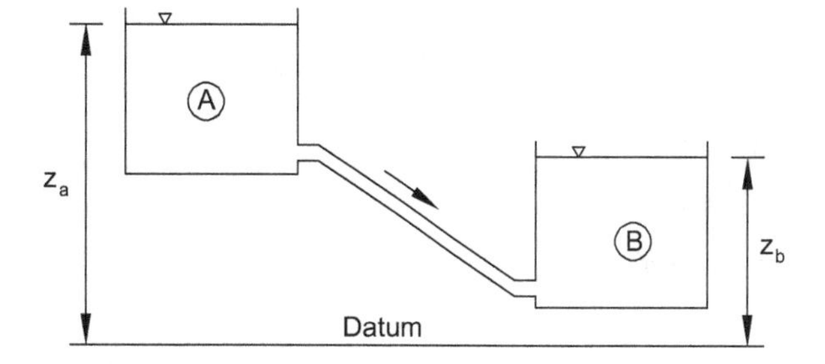

The losses are given.

pipe entry loss = 1.5 ft

pipe exit loss = 4 ft

pipe friction loss = 36 ft

other fitting losses = 7 ft

The elevation of the water table in reservoir A is 180 ft. The elevation of reservoir B is most nearly

(A) 110 ft

(B) 130 ft

(C) 150 ft

(D) 170 ft

504. A 12 in diameter pipe is used to carry water at a flow rate of 3 ft/s from reservoir A to reservoir B, as shown.

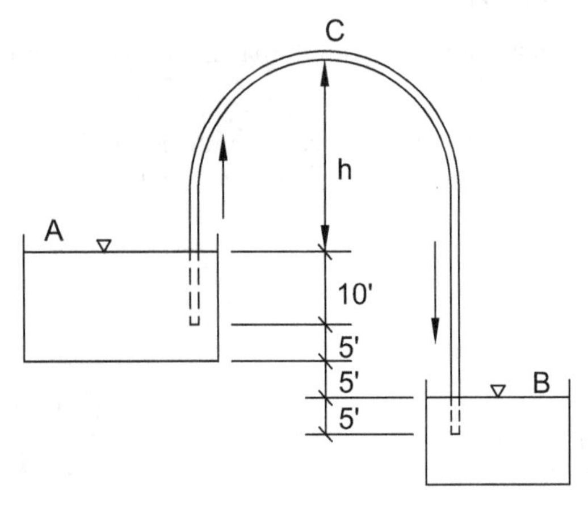

Separation of dissolved gases occurs at an absolute pressure of 8 ft of water height, and the change in the water level in each tank is negligible. Assume full pipe flow, and use the factors listed.

entry loss = 1.0 ft

exit loss = 0.1 ft

pipe friction loss between points A and C = 15 ft

barometric pressure = 30 in of mercury (or 34 ft of water)

The maximum height that can be used for siphoning is most nearly

(A) 10 ft

(B) 12 ft

(C) 15 ft

(D) 30 ft

505. A 12 in water main is 4500 ft long and branches into a 6 in and a 10 in diameter main, each 2000 ft long, as shown.

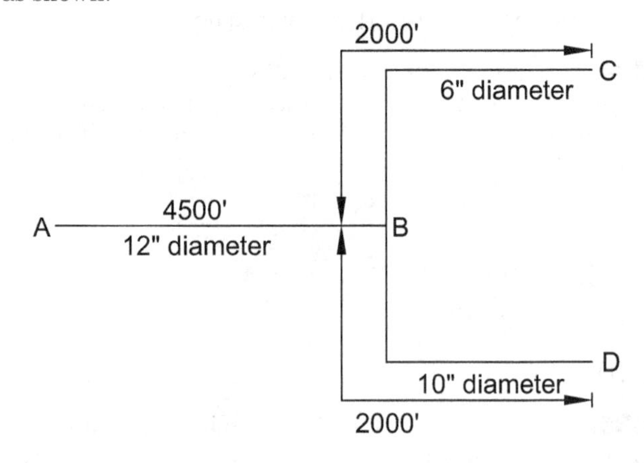

The total flow is 18 ft^3/sec. Assume all pipelines are horizontal, the head loss due to friction is equal in both pipes, and each pipe has a coefficient of friction of 0.005. Assuming friction losses only, the flow in the 6 in pipe is most nearly

(A) 2 ft^3/sec

(B) 4 ft^3/sec

(C) 6 ft^3/sec

(D) 8 ft^3/sec

506. Which statement regarding the hydraulic radius is NOT correct?

(A) $R_H = \frac{\text{cross-sectional area}}{\text{wetted perimeter}}$

(B) It is also called hydraulic mean depth.

(C) It can be used for both circular and noncircular shapes.

(D) $D_H = 2$ (hydraulic radius, R_H)

507. The total noise pollution from several sound pressure levels is computed using the logarithmic equation shown.

$$\text{SPL}_{\text{total}} = 10\log_{10}\sum 10^{\text{SPL}/10}$$

The noise pollution that results from three sound levels of 60 dB, 70 dB, and 80 dB is most nearly

(A) 70 dB

(B) 80 dB

(C) 105 dB

(D) 210 dB

508. A 3 ft diameter pan was used to measure evaporation. The observations were taken from 7 a.m. to 7 p.m. to obtain the data shown.

precipitation during the period = 4 in

drop in water level during the period = 2 in

Assuming the loss of water from the pan was due to evaporation, the evaporation rate (in/hr/ft^2) to the second decimal place is ______. (Fill in the blank.)

509. A river bend in alluvial soil is characterized by

(A) scouring on concave side and silting on convex side

(B) scouring on convex side and silting on concave side

(C) scouring or silting on either concave or convex side

(D) river width narrowing at the bend

510. A dam is being planned to resist a specified flood and to have a design life of 100 years. Assume that the flood events are independent events. In order to be confident with 90% probability that the design flood would not occur during the design life of the dam, the design flood recurrence interval (in years) should be ______. (Fill in the blank.)

511. A crane is shown.

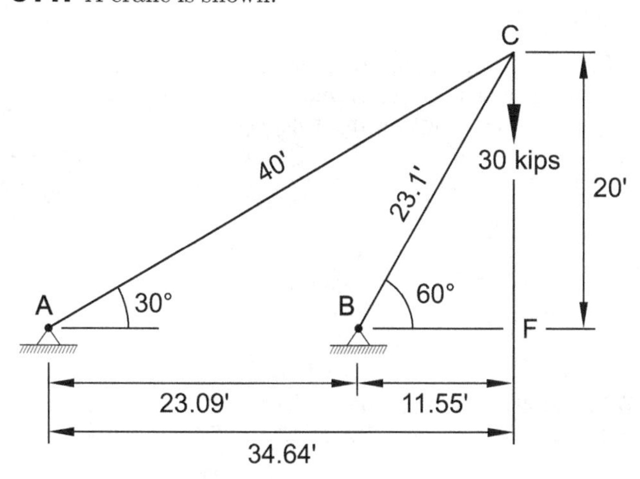

The respective member forces in member AC and member BC are most nearly

(A) 15 kips (compression), 45 kips (tension)

(B) 15 kips (tension), 45 kips (compression)

(C) 30 kips (compression), 52 kips (tension)

(D) 30 kips (tension), 52 kips (compression)

512. A rectangular wood beam is required to support a uniformly distributed load (UDL) of 480 lbf/ft over its span of 15 ft. The UDL includes the beam's self-weight. The modulus of elasticity of the wood material is 2×10^6 psi. The maximum allowable short-term deflection is 1 in. If the beam is 3.5 in wide, its minimum depth for design is

(A) 7.5 in

(B) 9.5 in

(C) 11.5 in

(D) 13.5 in

513. A W10 × 49 steel column carries 100 kips of service dead load. Its controlling slenderness ratio is 120, and the yield stress is 50 ksi. The service live load capacity of the column is most nearly

(A) 65 kips

(B) 100 kips

(C) 150 kips

(D) 225 kips

514. A steel shaft has a moment of inertia of 0.00055 m^4 about its centroid. The shaft is simply supported at its ends, which are 12 m apart. It carries a uniformly distributed load and has a maximum deflection of 12 mm. The shaft is propped at its midspan so that its midspan deflection is reduced to 2 mm. The force needed for propping is most nearly

(A) 25 kN

(B) 30 kN

(C) 35 kN

(D) 40 kN

515. A set of six wheel loads is moving across a 60 ft span girder from A to B. The relative positions of the wheels are shown.

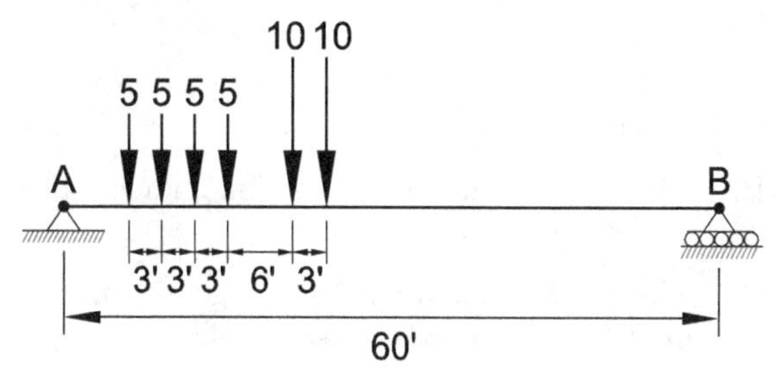

The resultant of the loads is 40 kips, and it is located at a distance of 4.5 ft from the 10 kip load and 3 ft from the 5 kip load. The maximum bending moment in the girder occurs when the minimum distance between support A and the 10 kip wheel load on the right is most nearly

(A) 27 ft

(B) 30 ft

(C) 33 ft

(D) 35 ft

516. A set of six-wheel loads is moving across a 60 ft span girder from point A to point B, as shown.

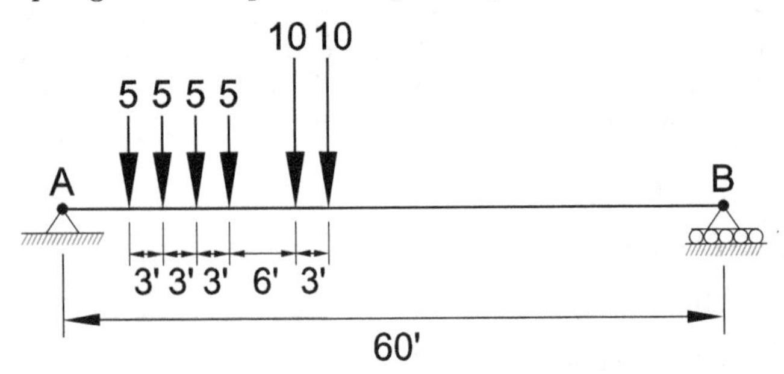

The maximum bending moment in the girder occurs when the wheel load on the right is 35.25 ft from point A, as shown.

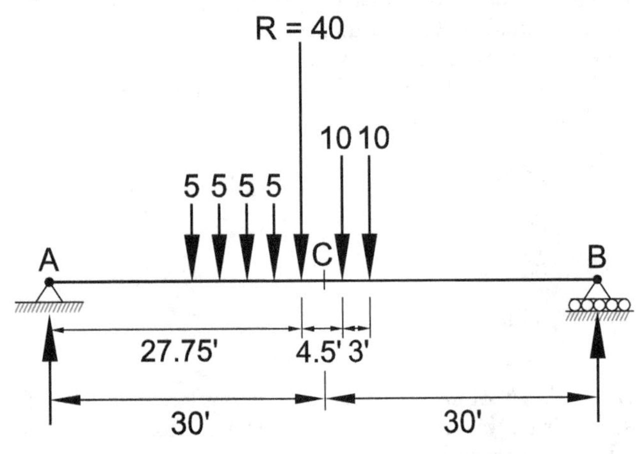

Which of the statements is correct?

(A) The absolute maximum bending moment occurs under the load resultant R.

(B) The absolute maximum bending moment occurs at the midspan.

(C) The absolute maximum bending moment occurs under the 10 kip wheel load on the right.

(D) The absolute maximum bending moment occurs under the 10 kip wheel load on the left.

517. The plan view of a three-story reinforced building is shown.

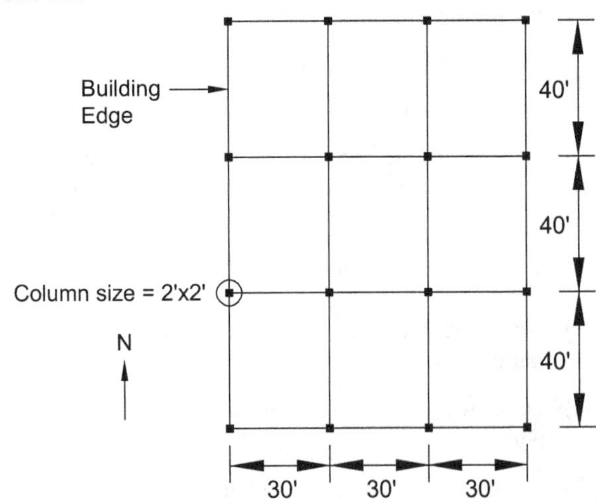

All columns are 2 ft squares and are centered on the grid-lines. The concrete floors and roof are one-way slabs supported on the beams. The floor system weighs 120 psf, including the dead load of the columns. The service dead load on an interior column at its base is most nearly

(A) 200 kips

(B) 400 kips

(C) 600 kips

(D) 750 kips

518. A simply supported W24 × 68 beam has an unbraced length of 20 ft, and its C_b factor is 1.67. The beam design strength is most nearly

(A) 366 ft-kips

(B) 611 ft-kips

(C) 650 ft-kips

(D) 678 ft-kips

519. A straight steel bar is fastened to a gusset plate with three $^3/_4$ in diameter ASTM A-307 bolts, as shown.

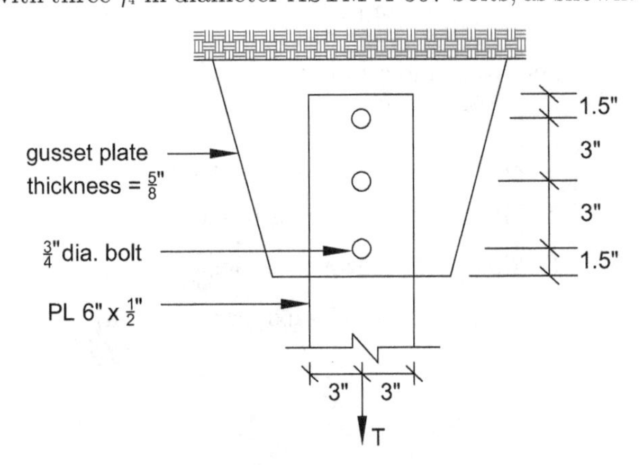

The bolt capacity in single shear is 7.95 kips. Assume that plate capacity is 126.3 kips. Using standard size holes, the design shear strength of the connection is most nearly

(A) 15.9 kips

(B) 23.9 kips

(C) 47.7 kips

(D) 76.3 kips

520. A singly reinforced concrete beam with 5 ksi concrete and grade 60 steel is used. The beam is 14 in wide and 32 in deep. The concrete cover to the center of the bottom steel is 3 in. The maximum steel area permitted by the ACI code is

$$A_{s,\max} = \frac{3d_t}{7}\left(\frac{0.85f'_c\beta_1 b}{f_y}\right)$$

The maximum longitudinal steel area permitted in the beam is most nearly

(A) seven #11 bars

(B) nine #10 bars

(C) ten #9 bars

(D) nine #9 bars

521. A singly reinforced concrete beam is 14 in wide and 32 in deep. The concrete strength is 6 ksi. The beam has three #9 grade 60 steel bars as bottom reinforcement. The concrete cover to the center of the steel is 3 in. The steel strain under the ultimate condition is most nearly

(A) 0.003

(B) 0.030

(C) 0.023

(D) 60.00

522. A reinforced beam is specified to be installed using 6000 psi concrete. However, it is instead placed using 3000 psi concrete. The concrete density remains unchanged at 150 lbf/ft^3. The ratio of the deflection of the as-built beam to the deflection of the designed beam is most nearly

(A) 0.5

(B) 1.0

(C) 1.5

(D) 2.0

523. AASHTO defines sand as particles that pass through sieve no. 10 and are retained on sieve no. 200. A sieve analysis was performed on a soil sample, and results are tabulated in the following table.

sieve no.	retained (%)
4	0
10	15
50	22
80	25
200	10

According to AASHTO, the percentage of sand in the sample is most nearly

(A) 0%

(B) 15%

(C) 57%

(D) 72%

524. Which of the following statements is INCORRECT?

(A) If 50% or more of the soil passes through sieve no. 200, it is called fine-grained soil.

(B) The liquid limit (LL) represents the moisture content at which a soil sample becomes liquid.

(C) The plasticity index (PI) equals the liquid limit minus the plastic limit.

(D) The higher the group index (GI) of a soil, the more suitable the soil is for use as a subgrade material.

525. A soil sample weighs 10 lbf, and its volume is 0.085 ft^3. The sample is oven dried, and its weight is then 9.25 lbf. The soil's specific gravity is 2.70. Its porosity and void ratio are respectively most nearly

(A) 0.35, 0.55

(B) 0.35, 0.95

(C) 0.55, 0.35

(D) 2.85, 1.82

526. A soil sample has a water content of 18% and a void ratio of 0.55. The soil's specific gravity is 2.70. Its porosity and degree of saturation are respectively most nearly

(A) 0.25, 88 %

(B) 0.25, 95%

(C) 0.35, 88%

(D) 0.35, 95%

527. A soil sample is tested for permeability using the constant head method. The sample is 4 cm in diameter and 100 cm in height. One liter of water was collected in 15 minutes under a head of 2 meters. The coefficient of permeability is most nearly

(A) 0.04 cm/s

(B) 0.09 cm/s

(C) 0.44 cm/s

(D) 4.40 cm/s

528. A silty soil sample is tested for permeability at a temperature of 20°C using the constant head method. The sample is 4 cm in diameter and 100 cm in height. One liter of water was collected in 15 minutes under a constant head of 2 m. The average diameter of the soil grains is 0.35 mm. Based on the Reynolds number, the flow type is

(A) laminar

(B) transitional

(C) partially turbulent

(D) fully turbulent

529. The data for a strip footing is

width = 6 ft

wall load = 10 kips/ft

wall moment = 10 ft-kips/ft

The load eccentricity is most nearly

(A) 1.0 in

(B) 3.0 in

(C) 12 in

(D) 36 in

530. A soil sample from a proposed highway site was tested in a laboratory. The sample has the following properties.

soil passing sieve no. 200 = 55%

liquid limit (LL) = 46%

plasticity index (PI) = 20%

Using AASHTO criteria for the combined thickness of surface, base, and sub-base for heavy traffic, the GI-subgrade thickness relationship can be estimated using the equation shown.

$$t_{\text{total}} = 8.5 + \text{GI}$$

For the proposed highway, the combined thickness of surface, base, and sub-base for heavy traffic is most nearly

(A) 15 in

(B) 17 in

(C) 20 in

(D) 24 in

531. Two soil samples from a site were subjected to drained direct shear tests. The loading consisted of a constant vertical load and an increasing horizontal shear until the soil sample failed. The test results are shown.

sample	effective vertical stress (psf)	effective shear stress (psf)
1	1000	700
2	4000	2500

The test results showed that the angle of internal friction is 31°. Based on the results, the soil's cohesion is most nearly

(A) 0 psf

(B) 100 psf

(C) 200 psf

(D) 300 psf

532. A 3 ft long mechanical anchor was installed to resist uplift. The soil weighs 120 pcf. The failure surface slopes 2 (vertical) to 1 (horizontal), and the soil cohesion is 70 psf. Assuming that the anchor will not fracture, the uplift capacity of the anchor is most nearly

(A) 1.54 kips

(B) 1.84 kips

(C) 1.95 kips

(D) 2.09 kips

533. A rectangular shallow footing supports two columns, as shown.

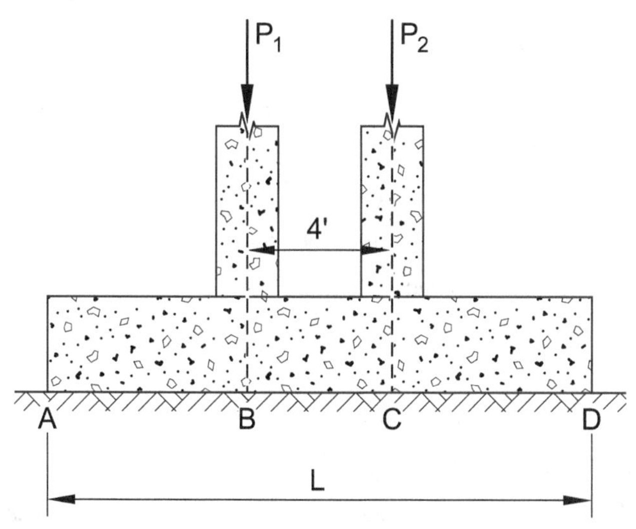

column 1 load = 200 kips

column 2 load = 400 kips

allowable soil pressure = 6 ksf

The footing is 7 ft. wide. The minimum length of the combined footing is most nearly

(A) 12 ft

(B) 14 ft

(C) 15 ft

(D) 18 ft

534. A car was traveling at a speed of 50 mi/hr when the driver unexpectedly saw a stop sign and applied the brakes. The driver's perception-reaction time is 2.0 sec. The distance the car travelled before the driver applied the brakes was most nearly

(A) 0 ft

(B) 25 ft

(C) 100 ft

(D) 150 ft

535. Consider an existing horizontal roadway curve with a 200 ft radius. The minimum length of a spiral curve necessary for a smooth transition from a tangent alignment to the horizontal curve for a design speed of 35 mph is most nearly

(A) 200 ft

(B) 350 ft

(C) 525 ft

(D) 675 ft

536. A curve is needed to connect a −5% grade to a −1% grade. The design is 45 mph, and the acceptable deceleration is 10 ft/sec^2. The driver's reaction time is 2.5 sec. The stopping sight distance for the curve is 424 ft. The minimum curve length based on the standard headlight criterion is most nearly

(A) 130 ft

(B) 200 ft

(C) 300 ft

(D) 400 ft

537. A community consists of 5000 people in which 40% of the households have no automobiles. The rest have only one automobile per household. The surveys conducted on similar communities show that zero-auto households and one-auto households make 0.51 and 0.25 daily transit trips, respectively. The number of daily transit trips for the population is most nearly

(A) 350 trips/day

(B) 750 trips/day

(C) 1000 trips/day

(D) 1770 trips/day

538. Which of the following statements about basic free-flow speed (BFFS) is correct?

(A) BFFS pertains to a section of the freeway that is outside of the influence of exit ramps and weaving areas.

(B) BFFS includes the influence of heavy traffic such as trucks and RVs.

(C) BFFS is based on lane widths up to 11 ft 11 in.

(D) BFFS is always less than the free-flow speed (FFS).

539. A car was traveling at a posted speed of 40 mph near a signaled intersection. As the car approached the intersection, the signal turned yellow, and the driver believed that he could not comfortably stop the car at the stopping line and decided to keep moving in order to clear the intersection before the light turned red. However, the yellow signal turned red while he was in still the intersection. He was cited for crossing the red light. He pleaded not guilty. Which of the following would be a valid argument for the nonguilty plea?

(A) The yellow light interval was longer than the time required.

(B) The yellow light interval was shorter than the time required.

(C) The driver entered the intersection before the light turned red.

(D) The driver was driving an electric car and thus has immunity.

540. A two-lane road is needed for truck traffic only to carry gravel from a stone pit. All trucks have three axles. The axle loads are as follows.

	front axle (single)	middle axles (tandem)	rear axle (single)
fully loaded truck	18,000 lbf	30,000 lbf each	20,000 lbf
empty truck	10,000 lbf	12,000 lbf	8,000 lbf

It is estimated that five trucks will make 20 trips daily, 5 days per week, and 50 weeks per year for 10 years. The load factor for passenger cars and other vehicles using the road is zero, and there is no growth foreseen in the truck traffic. The design ESAL for the roadway traffic from the stone pit is most nearly

(A) 10,000 ESAL

(B) 100,000 ESAL

(C) 1,000,000 ESAL

(D) 2,000,000 ESAL

541. Two types of vehicles use a roadway. The first type is comprised of light vehicles, such as passenger cars and light pickup trucks, that have two single axles. One axle carries 2000 lbf, and the other carries 4000 lbf. The second vehicle type is comprised of HS-20 trucks that have three axles. The front axle has a load of 8000 lbf, and the other two carry 32,000 lbf each. The number of light vehicles needed to cause a level of damage or wear and tear to the roadway equal to that caused by a HS-20 truck is most nearly

(A) 1200 vehicles

(B) 1800 vehicles

(C) 8000 vehicles

(D) 10,000 vehicles

542. A freeway has the listed characteristics.

basic free-flow speed = 70 mph

number of lanes in each direction = 4

lane width = 11 ft

right shoulder width = 3 ft

exit ramp density = 1 per mile

Applying the adjustment factors, its free-flow speed is most nearly

(A) 60 mph

(B) 64 mph

(C) 70 mph

(D) 80 mph

543. An arrow diagram created using the critical path method (CPM) is shown.

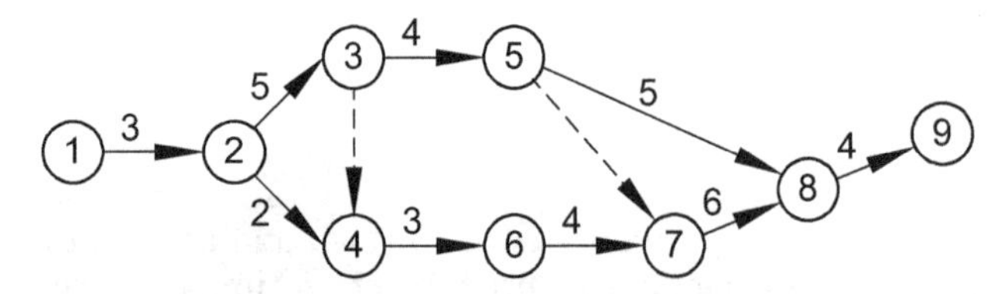

The dotted lines represent dummy activities, which consume no time or resource. The earliest start time for activity 4-6 is ______. (Fill in the blank.)

544. A cross section with varying height is shown.

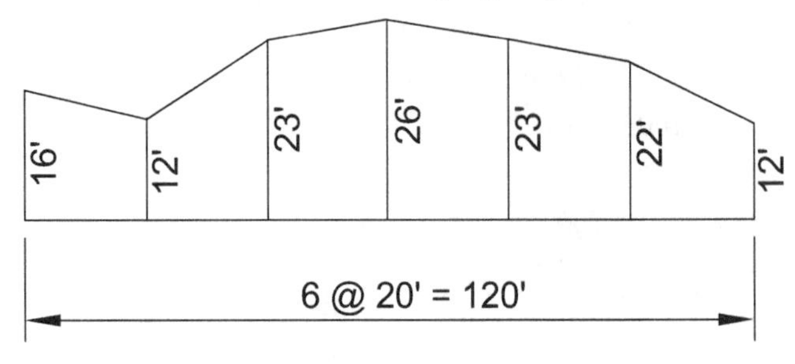

The area in square feet is ______. (Fill in the blank.)

545. A mass concrete subcontractor produces concrete on site and is reimbursed monthly for the materials used on a construction project at a rate of $100 per cubic yard. A total of 4998 sacks of cement were used in a month. The mix proportion for the concrete is 1:2:4 with a water/cement ratio of 0.50. Each sack of cement weighs 94 lbf, and the concrete quantity is estimated to the closest full cubic yard. The amount the contractor earned is ______. (Fill in the blank.)

546. A fence is installed on a construction site as a part of temporary erosion control. Where should it be installed?

(A) along the boundary, enclosing the property

(B) along the boundary line, facing the main street

(C) along a contour line

(D) perpendicular to contour lines

547. A contour map with gridlines for an area is shown.

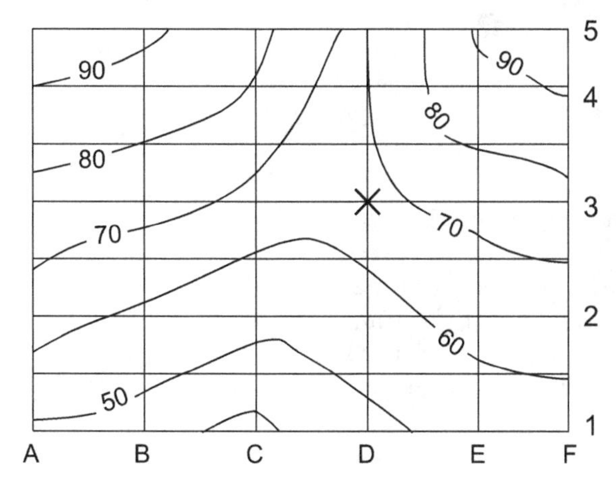

Assume that the area soil is saturated or otherwise will not absorb any precipitation. A drop of rain starts from grid point D-3 and begins to flow as part of the runoff. Point to the nearest grid point on the map shown.

548. An earned-value analysis of a construction project is plotted as shown.

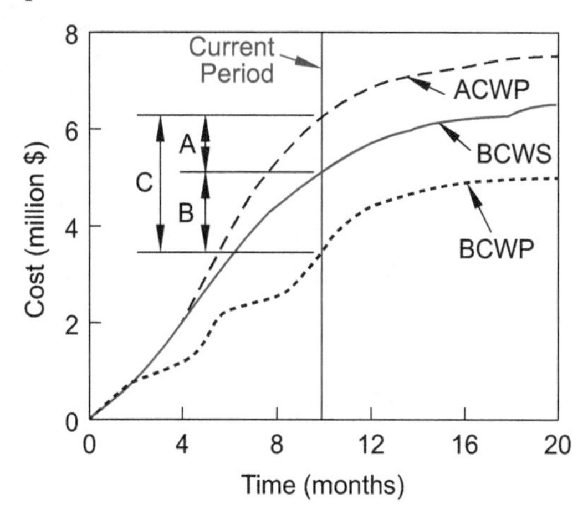

Three variances for the current period are noted as A, B, and C. Point to the cost variance noted therein.

549. A construction manager is assigned to improve the job site safety. Which of the following is a correct statement?

(A) The experience modification rating (EMR) offers a reflection of a company's safety program in relation to other similar employers. The higher the EMR, the better the safety program.

(B) The backbone of a safety program is to have a firm handshake understanding with the construction superintendent.

(C) It is more cost-effective to pay medical bills related to small medical injuries than to report the injuries to the workers' compensation carrier.

(D) Safety is a component of any project.

550. A 3 yd^3 loader will be used to load trucks from a quarry stockpile of processed aggregate having a maximum size of 1 in. The haul distance is negligible, and cycle time is 30 sec. The aggregate has loose unit weight of 3000 lbf per cubic yard, and its swell/shrinkage factor is 0.8. Loader efficiency is 50 minutes per hour. The loader production, in tons per hour, is ______. (Fill in the blank.)

STOP!

DO NOT CONTINUE!

This concludes the examination. If you finish early, check your work and make sure that you have followed all instructions. After checking your answers, you may turn in your examination booklet and answer sheet and leave the examination room. Once you leave, you will not be permitted to return to work or change your answers.

Exam 1 Answer Key

1.	A
2.	C
3.	20
4.	360
5.	B
6.	4/3
7.	B
8.	1000
9.	C
10.	C
11.	A, B, C
12.	D
13.	B, E
14.	E
15.	C
16.	D
17.	B
18.	D
19.	C
20.	A
21.	C
22.	D
23.	C
24.	D
25.	See Sol. 25.
26.	C
27.	B
28.	C
29.	D
30.	B
31.	B
32.	D
33.	B
34.	C
35.	A
36.	C
37.	B
38.	B
39.	A
40.	B
41.	A
42.	C
43.	C
44.	B
45.	B
46.	D
47.	See Sol. 47.
48.	D
49.	C
50.	B
51.	B
52.	B
53.	B
54.	C
55.	B
56.	D
57.	B
58.	D
59.	A
60.	B
61.	D
62.	A
63.	C
64.	C
65.	B
66.	A
67.	D
68.	B
69.	D
70.	A
71.	B
72.	B
73.	C
74.	D

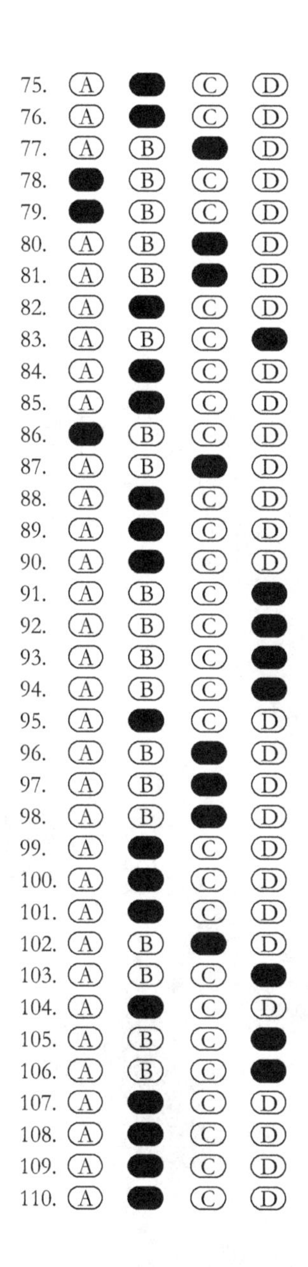

75.	B
76.	B
77.	C
78.	A
79.	A
80.	C
81.	C
82.	B
83.	D
84.	B
85.	B
86.	A
87.	C
88.	B
89.	B
90.	B
91.	D
92.	D
93.	D
94.	D
95.	B
96.	C
97.	C
98.	C
99.	B
100.	B
101.	B
102.	C
103.	D
104.	B
105.	D
106.	D
107.	B
108.	B
109.	B
110.	B

Exam 2 Answer Key

111. 153°C
112. 1
113. (A) (B) ● (D)
114. 16 m
115. 3/4
116. (A) ● (C) (D)
117. 48
118. 3960
119. (A) ● (C) (D)
120. (A) (B) ● (D)
121. ● ● ● ● (E)
122. (A) (B) (C) ●
123. (A) ● (C) (D) ●
124. (A) (B) ● (D)
125. (A) ● (C) (D)
126. (A) (B) ● (D)
127. (A) (B) (C) ●
128. (A) ● (C) (D)
129. ● (B) (C) (D)
130. (A) ● (C) (D)
131. $2,000,000
132. See Sol. 132.
133. (A) ● (C) (D)
134. See Sol. 134.
135. (A) (B) ● (D)
136. See Sol. 136.
137. 16
138. (A) (B) ● (D)
139. ● (B) (C) (D)
140. (A) (B) ● (D)
141. (A) ● (C) ● (E)
142. (A) ● (C) (D)
143. (A) (B) ● (D)
144. (A) (B) (C) ●
145. See Sol. 145.
146. (A) ● (C) (D)
147. (A) (B) (C) ●
148. (A) ● (C) (D)
149. (A) (B) ● (D)
150. ● (B) (C) (D)
151. (A) ● (C) (D)
152. (A) (B) (C) ●
153. ● (B) (C) (D)
154. (A) (B) ● (D) ●
155. (A) (B) ● (D)
156. (A) (B) ● (D)
157. (A) (B) ● (D)
158. (A) (B) (C) ●
159. (A) (B) (C) ●
160. (A) ● (C) (D)
161. ● (B) (C) (D)
162. ● (B) (C) (D)
163. (A) ● (C) (D)
164. 37.3 ft
165. ● (B) (C) (D)
166. ● (B) (C) (D)
167. (A) ● (C) (D)
168. (A) ● (C) (D)
169. (A) ● (C) (D)
170. ● (B) (C) (D)
171. (A) (B) (C) ●
172. (A) (B) (C) ●
173. (A) (B) ● (D)
174. (A) ● (C) (D)
175. (A) ● (C) (D)
176. (A) (B) (C) ●
177. (A) ● (C) (D)
178. (A) ● (C) (D)
179. (A) (B) (C) ●
180. (A) (B) ● (D)
181. ● (B) (C) (D)
182. (A) (B) (C) ●
183. ● (B) (C) (D)
184. (A) ● (C) (D)
185. (A) ● (C) (D)
186. (A) (B) (C) ●
187. (A) (B) (C) ●
188. (A) ● (C) (D)
189. (A) (B) (C) ●
190. (A) (B) ● (D)
191. ● (B) (C) (D)
192. See Sol. 192.
193. ● (B) (C) (D)
194. ● (B) (C) (D)
195. ● (B) (C) (D)
196. (A) (B) ● (D)
197. (A) ● (C) (D)
198. (A) (B) (C) ●
199. (A) ● (C) (D)
200. (A) ● (C) (D)
201. ● (B) (C) (D)
202. (A) (B) ● (D)
203. (A) ● (C) (D)
204. (A) (B) ● (D)
205. (A) ● (C) (D)
206. (A) (B) (C) ●
207. ● (B) (C) (D)
208. (A) (B) ● (D)
209. (A) (B) ● (D)
210. (A) ● (C) (D)
211. (A) (B) ● (D)
212. 5.48
213. (A) (B) (C) ●
214. (A) (B) ● (D)
215. (A) (B) ● (D)
216. (A) (B) (C) ●
217. $150,000
218. (A) (B) ● (D)
219. (A) (B) (C) ●
220. (A) ● (C) (D)

Exam 3 Answer Key

221. (A) (B) (C) ●
222. 1.26
223. 25π
224. 79.02°
225. 163 yd³
226. –1/3
227. 1/720
228. $115,000,000
229. 22
230. 5.46, 1.85
231. ● (B) (C) (D) ●
232. (A) (B) (C) ●
233. (A) (B) ● (D)
234. (A) (B) (C) ●
235. ● (B) (C) (D)
236. $7150
237. 36
238. ● (B) (C) (D) ●
239. (A) ● (C) (D)
240. 15%
241. (A) ● (C) (D)
242. ● (B) (C) (D) ●
243. (A) ● (C) (D)
244. ● (B) (C) ● (E)
245. ● (B) (C) (D)
246. (A) ● ● (D) (E)
247. (A) (B) ● (D)
248. (A) (B) ● (D)
249. ● (B) (C) (D)
250. (A) (B) ● (D)
251. (A) (B) ● (D)
252. (A) (B) (C) ●
253. ● (B) (C) (D)
254. ● (B) (C) (D)
255. (A) (B) ● (D)
256. 1 kip/ft, 1 kip/ft
257. 30 kips, 15 kips
258. (A) (B) ● (D)
259. (A) (B) (C) ●
260. See Sol. 260.
261. (A) (B) ● (D)
262. (A) (B) ● (D)
263. ● ● (C) (D) (E)
264. (A) (B) ● (D)
265. (A) (B) (C) ●
266. (A) (B) (C) ●
267. (A) ● (C) (D)
268. ● (B) (C) (D)
269. (A) (B) (C) ●
270. (A) (B) ● (D)
271. (A) ● (C) (D)
272. (A) (B) (C) ●
273. 0.95 psi
274. (A) ● (C) (D) ●
275. (A) ● (C) (D)
276. (A) ● (C) (D)
277. (A) (B) ● (D)
278. ● (B) (C) (D)
279. (A) ● (C) (D)
280. (A) (B) ● (D)
281. ● (B) (C) (D)
282. (A) ● (C) (D) ●
283. (A) ● (C) (D)
284. (A) (B) ● (D)
285. (A) ● (C) (D)
286. (A) ● (C) ● (E)
287. (A) (B) ● (D)
288. (A) ● (C) (D)
289. (A) (B) ● (D)
290. (A) (B) ● (D)
291. ● (B) (C) (D) ●
292. (A) (B) ● (D)
293. (A) ● (C) (D)
294. ● (B) (C) (D)
295. (A) (B) ● (D)
296. ● (B) (C) (D)
297. (A) (B) ● (D)
298. (A) (B) (C) ●
299. ● (B) (C) (D)
300. (A) (B) (C) ●
301. (A) (B) ● (D)
302. (A) (B) (C) ●
303. (A) (B) (C) ●
304. (A) ● (C) (D)
305. ● (B) (C) (D)
306. (A) ● (C) (D)
307. (A) (B) ● (D)
308. (A) (B) ● (D)
309. (A) (B) ● (D)
310. (A) (B) ● (D)
311. (A) ● (C) (D)
312. (A) (B) ● (D)
313. ● (B) (C) (D)
314. (A) ● (C) (D)
315. (A) ● (C) (D)
316. (A) ● (C) (D)
317. (A) ● (C) (D)
318. (A) (B) ● (D)
319. (A) ● (C) (D)
320. (A) (B) ● (D)
321. (A) ● (C) (D)
322. (A) (B) ● (D)
323. (A) (B) ● (D)
324. (A) (B) (C) ●
325. (A) (B) (C) ●
326. (A) (B) ● (D)
327. (A) (B) ● (D)
328. (A) ● (C) (D)
329. (A) (B) ● (D)
330. (A) (B) ● (D)

Exam 4 Answer Key

331. 17
332. -0.069
333. $-b/a$
334. 1, −10, −24
335. 60°
336. 16
337. 3
338. 0.15
339. 0.06
340. (A) ● (C) (D)
341. ● (B) (C) (D)
342. (A) (B) (C) ●
343. (A) (B) ● (D)
344. (A) (B) (C) ●
345. (A) (B) (C) ●
346. $23 million
347. (A) (B) ● (D)
348. 3 years
349. $31,000
350. $4,000,000
351. alternative 2
352. (A) (B) ● (D)
353. (A) ● (C) ● (E)
354. (A) ● ● (D) (E)
355. (A) ● (C) (D)
356. (A) (B) ● (D)
357. (A) ● (C) ● (E)
358. (A) ● ● (D) (E)
359. 4
360. ● (B) (C) (D)
361. (A) (B) ● (D)
362. ● (B) (C) ● (E)
363. ● (B) (C) (D)
364. (A) ● (C) (D)
365. (A) ● (C) (D)
366. (A) (B) ● (D)
367. (A) (B) ● (D)
368. (A) ● (C) (D)
369. (A) (B) ● (D)
370. ● (B) (C) (D)
371. ● (B) (C) (D)
372. ● (B) (C) (D)
373. (A) (B) ● (D)
374. (A) (B) ● (D)
375. ● (B) (C) (D)
376. (A) ● (C) (D)
377. (A) ● (C) (D)
378. (A) ● (C) (D)
379. (A) (B) ● (D)
380. ● (B) (C) (D)
381. ● ● (C) (D) (E)
382. (A) ● (C) (D)
383. 21,504 tons
384. (A) (B) ● (D)
385. (A) (B) ● (D)
386. 568 yd^3
387. 2905 mi
388. (A) (B) (C) ●
389. ● (B) (C) (D)
390. ● (B) (C) (D)
391. (A) ● (C) (D)
392. 60 ft
393. See Sol. 393.
394. (A) ● (C) (D)
395. (A) (B) ● (D)
396. ● (B) (C) (D)
397. (A) (B) (C) ●
398. (A) ● (C) (D)
399. (A) ● (C) (D)
400. (A) ● (C) (D)
401. (A) ● (C) (D)
402. −15 kips, 45 kips
403. (A) ● (C) (D)
404. (A) ● (C) (D)
405. 9 kips
406. (A) (B) (C) ●
407. (A) (B) (C) ●
408. (A) ● (C) (D)
409. (A) (B) (C) ●
410. (A) (B) ● (D)
411. ● (B) (C) (D)
412. ● (B) (C) (D)
413. ● (B) (C) (D)
414. (A) ● (C) (D)
415. (A) (B) (C) ●
416. ● (B) (C) (D)
417. ● (B) (C) (D)
418. (A) (B) ● (D)
419. 4 ksf
420. (A) (B) (C) ●
421. (A) (B) ● (D)
422. (A) (B) (C) ●
423. (A) (B) ● (D)
424. (A) (B) ● (D)
425. ● (B) (C) (D)
426. (A) (B) (C) ●
427. (A) (B) ● (D)
428. ● (B) (C) (D)
429. (A) (B) (C) ●
430. (A) (B) ● (D)
431. (A) (B) ● (D)
432. (A) (B) ● (D)
433. (A) (B) ● (D)
434. (A) (B) (C) ●
435. ● (B) (C) (D)
436. ● (B) (C) (D)
437. (A) (B) (C) ●
438. (A) (B) (C) ●
439. See Sol. 439.
440. week 15

Exam 5 Answer Key

441. (A) (B) (C) ●
442. 275 pieces
443. (A) (B) ● (D)
444. (A) ● (C) (D)
445. (A) (B) ● (D)
446. −21
447. 28.3
448. ● (B) (C) (D)
449. (A) (B) (C) ●
450. $17.50
451. ● (B) (C) (D)
452. (A) (B) ● (D)
453. Sec. 150.10.A(2)
454. (A) (B) (C) ●
455. (A) (B) ● (D)
456. ● (B) (C) (D)
457. (A) (B) ● (D)
458. 200,000 units
459. (A) ● (C) (D)
460. 0.9
461. $14.80
462. ● (B) (C) (D)
463. ● (B) (C) (D) ●
464. (A) (B) ● (D)
465. (A) (B) (C) ● ●
466. 3
467. (A) ● (C) (D)
468. 250 kN
469. 17 ft-kips
470. (A) (B) (C) ●
471. ● (B) (C) (D)
472. (A) ● (C) (D)
473. (A) (B) (C) ●
474. (A) (B) ● (D)
475. (A) ● (C) (D)
476. (A) (B) ● (D)
477. 2 mm
478. 200 psi
479. (A) (B) (C) ●
480. (A) (B) ● (D)
481. (A) (B) ● (D)
482. 16
483. ● (B) (C) (D)
484. (A) (B) ● (D)
485. (A) (B) ● (D)
486. (A) (B) ● (D) ●
487. (A) (B) (C) ●
488. (A) ● (C) (D)
489. (A) ● (C) (D)
490. (A) (B) (C) ●
491. 0.083 N·s/m^2
492. 135 kips, 34°
493. 1.0
494. ● (B) (C) (D)
495. 5 acres
496. 450 miles
497. ● (B) (C) (D)
498. (A) (B) (C) ●
499. ● (B) (C) (D)
500. (A) (B) ● (D)
501. (A) ● (C) (D)
502. (A) (B) ● (D)
503. (A) ● (C) (D)
504. ● (B) (C) (D)
505. (A) ● (C) (D)
506. (A) (B) (C) ●
507. (A) ● (C) (D)
508. 0.0707 in/hr/ft^2
509. ● (B) (C) (D)
510. 1 in 950 years
511. (A) (B) (C) ●
512. (A) (B) ● (D)
513. ● (B) (C) (D)
514. (A) ● (C) (D)
515. (A) (B) (C) ●
516. (A) (B) (C) ●
517. (A) ● (C) (D)
518. (A) ● (C) (D)
519. (A) ● (C) (D)
520. (A) (B) ● (D)
521. (A) (B) ● (D)
522. (A) (B) ● (D)
523. (A) (B) ● (D)
524. (A) (B) (C) ●
525. ● (B) (C) (D)
526. (A) (B) ● (D)
527. ● (B) (C) (D)
528. ● (B) (C) (D)
529. (A) (B) ● (D)
530. (A) ● (C) (D)
531. (A) ● (C) (D)
532. (A) (B) ● (D)
533. (A) (B) ● (D)
534. (A) (B) (C) ●
535. (A) (B) (C) ●
536. (A) (B) (C) ●
537. (A) (B) (C) ●
538. ● (B) (C) (D)
539. (A) ● (C) (D)
540. (A) (B) ● (D)
541. (A) (B) ● (D)
542. (A) ● (C) (D)
543. 8 days
544. 2400 ft^2
545. $90,000
546. (A) (B) ● (D)
547. See Sol. 547.
548. See Sol. 548.
549. (A) (B) (C) ●
550. 450 tons per hour

Solutions

Exam 1

1. See the Mathematics section of the *NCEES Handbook*. The general equation of a straight line is

$$y - y_1 = m(x - x_1)$$

The slope, m, of a line passing through points (x_1, y_1) and (x_2, y_2) is

$$m = \frac{y_2 - y_1}{x_2 - x_1}$$

For the two given points, the slope is

$$\begin{aligned} m &= \frac{y_2 - y_1}{x_2 - x_1} = \frac{0-2}{5-1} = \frac{-2}{4} \\ &= -\frac{1}{2} \end{aligned}$$

It can be represented graphically as shown.

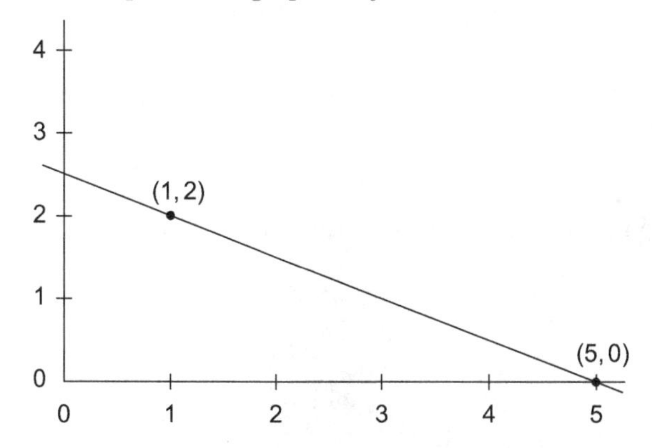

Use the slope and one of the original points to calculate the standard form of the line.

$$\begin{aligned} y - y_1 &= m(x - x_1) \\ y &= mx - mx_1 + y_1 \\ y - mx &= -mx_1 + y_1 \\ y - \left(-\frac{1}{2}\right)(x) &= -\left(-\frac{1}{2}\right)(1) + 2 \\ y + \frac{x}{2} &= \frac{5}{2} \\ 2y + x &= 5 \end{aligned}$$

The standard form of the line passing through the points (1, 2) and (5, 0) is

$$2y + x = 5$$

The answer is (A).

2. See the Mathematics section of the *NCEES Handbook*. Simplify the relation using the identity

$$\tan\theta = \frac{\sin\theta}{\cos\theta}$$

$$\tan\theta + \cot\theta = \frac{\sin\theta}{\cos\theta} + \frac{\cos\theta}{\sin\theta} = \frac{\sin^2\theta + \cos^2\theta}{\sin\theta\cos\theta}$$

Use the following identities to simplify further.

$$\begin{aligned} \sin^2\theta + \cos^2\theta &= 1 \\ \sin 2\theta &= 2\sin\theta\cos\theta \end{aligned}$$

The relation $\tan\theta + \cot\theta$ can be written as

$$\tan\theta + \cot\theta = \frac{2}{\sin 2\theta} = 2\csc 2\theta$$

The answer is (C).

3. See the Mathematics section of the *NCEES Handbook*. The parallelogram can be drawn as

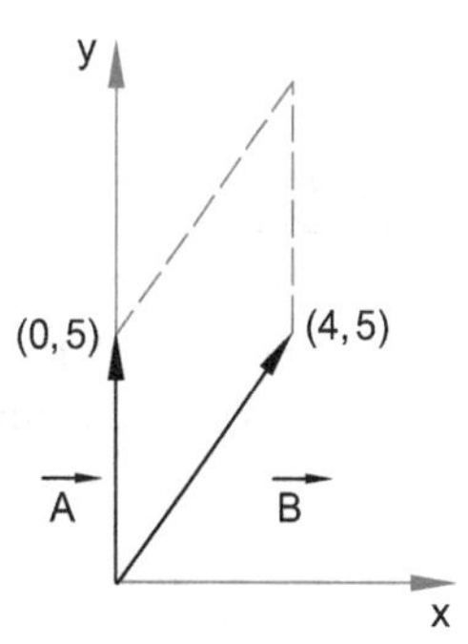

The two vectors are

$$\begin{aligned} \vec{A} &= 0i + 5j \\ \vec{B} &= 4i + 5j \end{aligned}$$

The rule: The magnitude of determinant is the area of parallelogram, with vectors arranged in columns. For two given vectors, the area of the parallelogram is

$$\text{area} = \begin{vmatrix} 0 & 4 \\ 5 & 5 \end{vmatrix} = |(0)(5) - (4)(5)| = |0 - 20| = 20$$

The answer is 20.

4. See the Mathematics section of the *NCEES Handbook*. To determine if the given sequence of wall construction forms arithmetic progression series, subtract each number from the following number. If the differences are equal, the series is arithmetic. In this case, each row uses one less brick than the row below it. Therefore, it represents an arithmetic progression series. The bottom row is the first term, a, and there are 16 terms, n.

The common difference is

$$d = n_{i\text{th row}} - n_{(i-1)\text{th row}} = -1$$

The sum of n terms, s, is

$$\begin{aligned} s &= \frac{n\big(2a + (n-1)d\big)}{2} \\ &= \frac{16\big(2(30) + (16-1)(-1)\big)}{2} \\ &= 360 \end{aligned}$$

The answer is 360.

5. For the triangle, sides b and c and angle A are known. The triangle can be represented graphically as

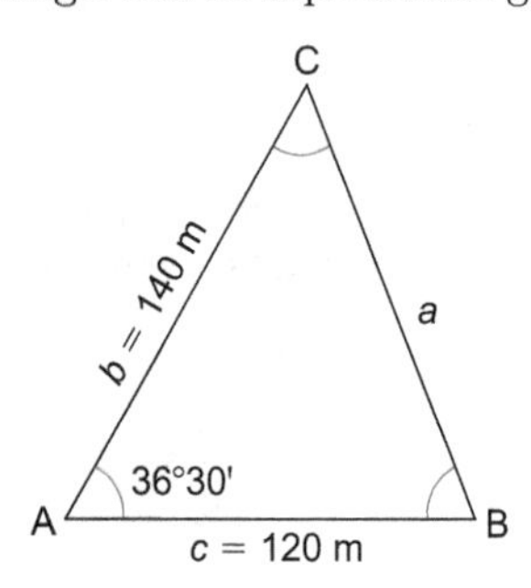

The law of sines applies if three sides of a triangle are given. Use the law of cosines to calculate the length of side a.

$$\begin{aligned} a^2 &= b^2 + c^2 - 2bc(\cos \text{A}) \\ &= (140\ \text{m})^2 + (120\ \text{m})^2 - 2(140\ \text{m})(120\ \text{m})\cos(36.5°) \\ &= 19{,}600\ \text{m}^2 + 14{,}400\ \text{m}^2 - 27{,}010\ \text{m}^2 \\ &= 6990\ \text{m}^2 \\ a &= 83.61\ \text{m} \end{aligned}$$

The perimeter is

$$\begin{aligned} P &= a + b + c = 83.61\ \text{m} + 140\ \text{m} + 120\ \text{m} \\ &= 343.61\ \text{m} \quad (344\ \text{m}) \end{aligned}$$

The answer is (B).

6. Use the factor method to simplify.

$$\begin{aligned} \lim_{x\to 1} \frac{x^2 + 2x - 3}{x^2 + x - 2} &= \lim_{x\to 1} \frac{(x-1)(x+3)}{(x-1)(x+2)} \\ &= \lim_{x\to 1} \frac{(x+3)}{(x+2)} \\ &= \frac{1+3}{1+2} = \frac{4}{3} \end{aligned}$$

The answer is 4/3.

7. The general equation of an ellipse is

$$\frac{x^2}{a^2} + \frac{y^2}{b^2} = 1$$

Comparing this with the given ellipse equation, $a = 4$ and $b = 3$.

The area of the ellipse is

$$A_{\text{ellipse}} = \pi ab = \pi(4)(3) = 12\pi$$

The general equation of a circle is

$$x^2 + y^2 = r^2$$

Comparing the above equation with the given,

$$\begin{aligned} x^2 + y^2 &= 16 \\ r^2 &= 16 \\ r &= 4 \end{aligned}$$

The net area is

$$A_{\text{net}} = A_{\text{circle}} - A_{\text{ellipse}} = 16\pi - 12\pi = 4\pi$$

The answer is (B).

8. In this problem, the order matters. For example, if the required combination to the safe is 123, then 321 or 213, etc. would not work. Therefore, this is a permutation problem. There are ten numbers (0 through 9) to select from for each required number. The number of different permutations is

$$
\begin{aligned}
n &= 10 \\
r &= 1 \\
nPr &= \frac{n!}{(n-r)!} = (10)(P)(1) = \frac{10!}{(10-1)!} = 10
\end{aligned}
$$

The process is repeated three times (one for each entry number). The total number of permutations is

$$P_{\text{total}} = (10)(10)(10) = 1000$$

The answer is 1000.

9. Determine the expected cost based on each report.

$$
\begin{aligned}
\text{report A:}\quad \text{cost}_A &= (500{,}000\ \text{ft}^2) \\
&\quad \times\left[0.25\left(\frac{\$100}{\text{ft}^2}\right) + 0.75\left(\frac{\$300}{\text{ft}^2}\right)\right] \\
&= \$125\ \text{million} \\
\text{report B:}\quad \text{cost}_B &= (500{,}000\ \text{ft}^2) \\
&\quad \times\left[0.55\left(\frac{\$100}{\text{ft}^2}\right) + 0.45\left(\frac{\$300}{\text{ft}^2}\right)\right] \\
&= \$95\ \text{million}
\end{aligned}
$$

Use the "arithmetic mean of a set of values" equation from the Engineering Probability and Statistics section of the *NCEES Handbook.*

$$\overline{X} = \left(\frac{1}{n}\right)\sum_{i=1}^{n} X_i$$

$$
\begin{aligned}
\text{expected cost:}\quad \overline{X} &= \frac{\text{cost}_A + \text{cost}_B}{2} \\
&= \frac{\$125\ \text{million} + \$95\ \text{million}}{2} \\
&= \$110\ \text{million}
\end{aligned}
$$

The answer is (C).

10. A function is said to be normally distributed if its density function is given by an expression of the form

$$f(x) = \frac{1}{\sigma\sqrt{2\pi}} e^{-\frac{1}{2}\left(\frac{x-\mu}{\sigma}\right)^2}$$

The given function is in the form of the normal distribution with $\mu = 15$ and $\sigma = 1$.

Integrate the function to obtain the fraction.

$$
\begin{aligned}
X &= \int_{14}^{16} \frac{1}{\sqrt{2\pi}} e^{-\frac{1}{2}(x-15)^2} \\
Z &= \frac{x-\mu}{\sigma}
\end{aligned}
$$

When x is 14,

$$Z = \frac{(14-15)}{1} = -1$$

When x is 16,

$$Z = \frac{(16-15)}{1} = 1$$

Use the unit normal distribution method with tables given in the Engineering Probability and Statistics section of the *NCEES Handbook.* The illustration shows the unit normal distribution curve with the parameter Z along its x-axis. The area between −1 and 1 represents the rainfall from 14 in to 16 in.

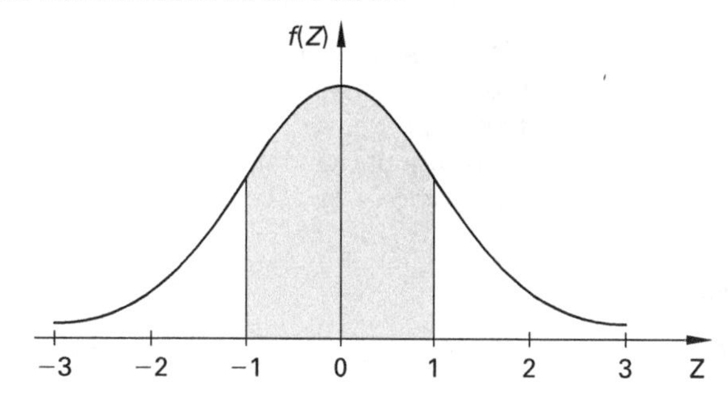

From the table, the shaded area is

$$F(1,-1) = W(1) = 0.6827 \quad (0.68)$$

The answer is (C).

11. Ethics are a set of guidelines, rules, philosophical concepts, customs, norms, and aspirations for a licensee to follow. The Code of Engineering Ethics articulates the ways in which moral and ethical principles apply to unique situations encountered in professional practice. It indicates to others that the professionals are seriously concerned about responsible and professional conduct. In some cases, it is impossible to comply with every aspect of the Code. Therefore, ethics are also called a set of aspirations that a licensed engineer should aim for.

However, ethics are not subject to the law. As the late Chief Justice of the U.S. Supreme Court Earl Warren put it, "Society would come to grief without ethics, which is unenforceable in the courts and cannot be made part of law ... Not only does law in a civilized society presuppose ethical commitment, it presupposes the existence of a broad area of human conduct controlled only by ethical norms and not subject to law at all."

The answer is (A), (B), and (C).

12. A licensee's first and foremost responsibility in performance of professional service is to the public welfare.

The answer is (D).

13. The first three options concern whether the design engineer can be sued. The last two options concern the liability of the engineer. An injured third party can file their claim against the consultant even though the consultant had no privity of contract with the injured party. The privity of contract is not required. The courts have allowed the right to sue for injuries or damages to third parties' beneficiaries, and even to innocent bystanders. Options A and C are incorrect and option B is correct. One of the questions in such claims is whether the engineer met the standard of care or not. If not, then the engineer is likely negligent and can be found liable. An opinion by another engineer or design firm does not by itself make a design professional liable. Option D is incorrect and option E is correct.

The answer is (B) and (E).

14. Options A, B, C, and D are correct. A licensee found violating the professional code of ethics may be asked to pay a fine. The amount of the fine imposed may vary with the nature of the violation and the jurisdiction the licensee practices in. Therefore, option E is incorrect.

The answer is (E).

15. The retained amount is 10% of the construction work claimed.

$$\begin{aligned}\text{retained} &= (\$2{,}000{,}000)(95\%)(10\%)\\ &= \$190{,}000\end{aligned}$$

Although, the project is substantially complete and only 5% work is remaining (which amounts to \$100,000), the owner is entitled to retain the amount agreed upon in the contract. The term *substantial completion* should not be confused with *full completion* since each pertains to a different level of completion.

The answer is (C).

16. The formula to determine future value given present value is provided in the Engineering Economics section of the *NCEES Handbook*.

$$\left(\frac{F}{P},\ i\%,\ n\right) = (1+i)^n$$

In this case, the present worth, P, is 24. The interest rate as a decimal is 0.07, and there are 390 compounding periods. The future worth in 2016 of the original \$24 investment is

$$\begin{aligned}\left(\frac{F}{P},\ i\%,\ n\right) &= (1+i)^n\\ F &= P(1+0.07)^{390}\\ &= \$24\left(2.882\times 10^{11}\right)\\ &= \$6.916\times 10^{12}\\ &= \$6.916\ \text{trillion}\quad (\$7\ \text{trillion})\end{aligned}$$

The answer is (D).

17. The firm would earn a uniform amount, A, of \$100,000 per year from now to perpetuity. The amount, A, is assumed to be inflation adjusted. It would not increase since there is no growth. The rate of return, i, is 10% or 0.1. The Engineering Economics section of the *NCEES Handbook* provides the capitalized costs formula. Using the formula, the present value, P, of the firm is

$$P = \frac{A}{i} = \frac{\$100{,}000}{0.1} = \$1{,}000{,}000$$

The answer is (B).

18. The statements in options A, B, and C are correct. The sunk cost is defined as a cost that an entity has incurred and that it can no longer recover. In considering the feasibility of a new project, a sunk cost should be considered a cost item for the new project.

The answer is (D).

19. The method of considering more than the initial cost of a facility is called the life-cycle costing method. The service life of a facility is measured in the number of years the facility is expected to be in service. The life-cycle cost includes the initial cost and the periodic repair costs to maintain the facility over its service lifetime. It can be described in the form

$$\text{life-cycle cost} = \text{initial cost} + \text{present value of periodic repair costs and operating costs}$$

The present value of future payments is calculated using the formulas given in the *NCEES Handbook*. The life-cycle cost can also be expressed as an equivalent uniform annual cost.

The answer is (C).

20. The cost grows linearly.

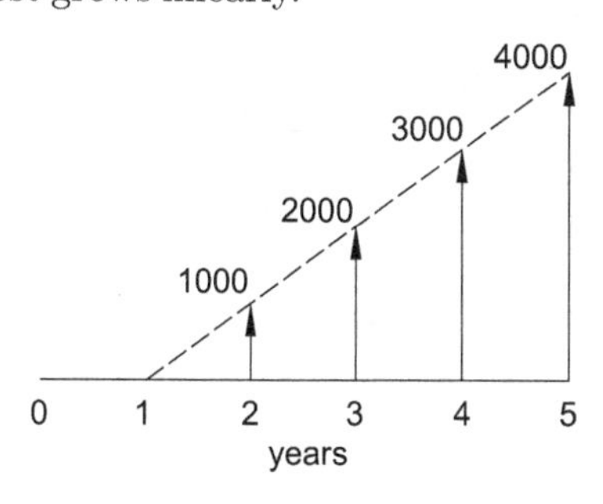

The growth increment, g, is \$1000. From Interest Rate Tables in the *NCEES Handbook*,

$$\left(\frac{A}{g}, 8\%, 5\right) = 1.8465$$

The annual equivalent repair cost is

$$\begin{aligned} A &= \$1000(1.8465) \\ &= \$1846.50 \quad (\$1850) \end{aligned}$$

The answer is (A).

21. A payoff, or reward, is expressed in monetary units. It represents the net change in the total wealth of the decision maker as a result of his or her decision and the actual state of the world. The decision could be to either do something or do nothing at all. For example, in this problem, Gr8 has two options: drill or don't drill. The payoff is specific to an event or outcome. The payoff occurs if the event occurs. The probability of the event happening is irrelevant. Payoffs are always expressed in net terms rather than gross terms. A payoff can be either positive or negative.

The payoff or reward depends on whether the decision is made to drill or not drill, and then whether or not oil is present. There are two choices, and each choice has two outcomes, so there are four scenarios to consider. This question concerns the decision to drill, which has two outcomes. The payoff depends on what the outcome is.

If Gr8 strikes oil, the oil will be worth \$13,000,000, and the cost of drilling will be \$3,000,000.

$$\text{net payoff, } R_1 = \$13{,}000{,}000 - \$3{,}000{,}000 = \$10{,}000{,}000$$

If Gr8 does not strike oil, it will incur the \$3,000,000 in drilling cost and gain nothing.

$$\text{net payoff, } R_2 = -\$3{,}000{,}000$$

Comparing the two outcomes, reward R_1 is larger.

The answer is (C).

22. A force is a vector possessing magnitude and direction. The forces are added according to the parallelogram law. A parallelogram with sides A and B represents the resultant force vector. Option D shows the vectors and the correct resultant.

The answer is (D).

23. A Newton is defined as a force that gives an acceleration of 1 m/s^2 to a mass of 1 kg. The weight of a body with a mass of 1 kg is

$$\begin{aligned} W = mg &= (1\text{ kg})\left(9.81\ \frac{\text{m}}{\text{s}^2}\right) \\ &= 9.81\text{ N} \quad (10\text{ N}) \end{aligned}$$

The answer is (C).

24. It is given that the barge's direction should be along the x-axis. Therefore, the resultant, R, is directed along the x-axis. Use the triangle rule and draw the vector in tip-to-toe fashion.

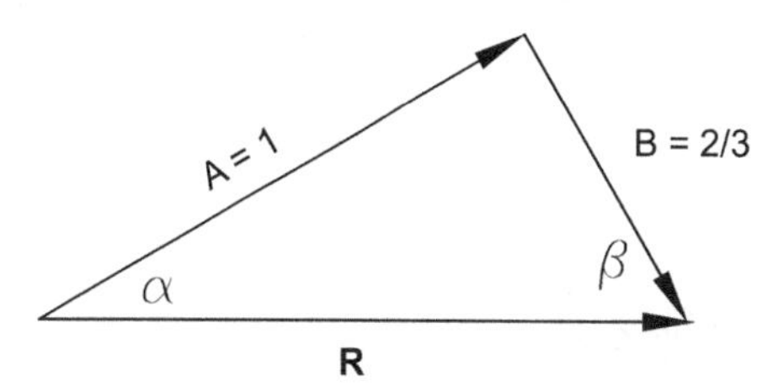

See the Mathematics section of the *NCEES Handbook* for the law of sines. The law of sines is applicable and it states that

$$\frac{a}{\sin A} = \frac{b}{\sin B}$$

a and b are the sides of a triangle and A and B are the angles in front of lines a and b, respectively. In this case, let the sides be

$$a = 1N$$

$$b = \frac{2}{3}N$$

The angle α is

$$\alpha = \angle B = 30°$$

The angle β is

$$\beta = \angle A = \text{unknown}$$

Therefore, the equation is

$$\frac{1}{\sin\beta} = \frac{2/3}{\sin\alpha}$$

$$\sin\beta = \frac{3}{2}\sin\alpha = \frac{3}{2}(\sin 30°) = \frac{3}{2}(0.5) = 0.75$$

Solve for the angle.

$$\measuredangle\beta = \sin^{-1}0.75 = 48.6° \quad (49°)$$

The answer is (D).

25. Assume that bar OB is in tension and the force in brace OA is compressive.

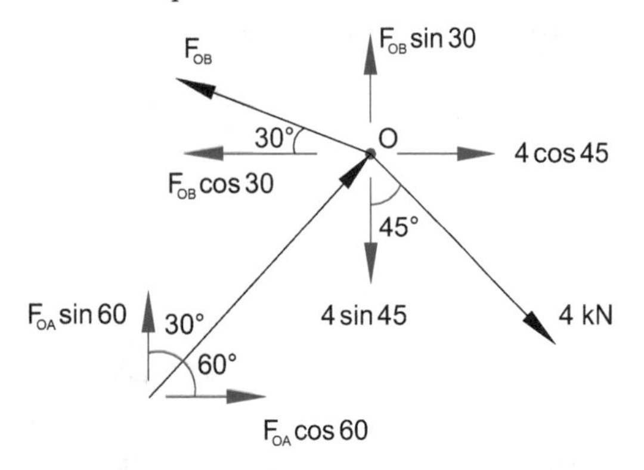

From equilibrium at point O, the sum of force components in both the x and y directions is zero.

$$\sum F_x = F_{OA}\cos 60° - F_{OB}\cos 30° + 4\cos 45° = 0$$

$$\sum F_y = F_{OA}\sin 60° + F_{OB}\sin 30° - 4\sin 45° = 0$$

Solving the equations simultaneously yields

$$F_{OA} = 1.04 \text{ kN}$$

$$F_{OB} = 3.86 \text{ kN}$$

Since the signs are positive for both forces, both assumptions are correct.

Shortcut

Note that the applied force makes an angle of 15° with the element OB, and the elements OA and OB are perpendicular to each other. Resolve the applied force in the direction of the truss elements.

$$F_{OB} = 4\cos 15° = 3.86 \text{ kN}$$

$$F_{OA} = 4\sin 15° = 1.04 \text{ kN}$$

The answer is

	member	
	OA	OB
tension or compression	compression	tension
force magnitude (kN)	1.04	3.86

26. To simplify the problem from 3-D to 2-D, consider that the moment is equal to the force multiplied by the perpendicular distance. The cable is oriented neither in the x-y plane nor in the x-z plane.

Consider the inclined plane in which the cable and the cantilever lie, and resolve the cable force into two directions: along the longitudinal axis of the cantilever (P_{axial}) and perpendicular to the longitudinal axis.

Since P_{axial} is along the member axis, its moment about A is zero.

$$P_\perp = P\sin\theta = \frac{P\sqrt{y^2+z^2}}{\sqrt{x^2+y^2+z^2}} = \frac{(9 \text{ kN})\sqrt{(4 \text{ m})^2+(5 \text{ m})^2}}{\sqrt{(10 \text{ m})^2+(4 \text{ m})^2+(5 \text{ m})^2}} = 4.85 \text{ kN}$$

A moment is defined in the *NCEES Handbook* as the cross product of the radius vector and the force. In a planar case, the moment calculations can be simplified by using the expression

$$\text{moment} = (\text{force})(\text{perpendicular distance})$$

The moment about A is

$$M_A = P_\perp(\overline{AB}) = (4.85 \text{ kN})(10 \text{ m}) = 48.5 \text{ kN-m} \quad (49 \text{ kN-m})$$

Since the rotation about A is counterclockwise, the moment is positive.

The answer is (C).

27. The centroid of an area can be found by integration.

$$\bar{x} = \frac{\int x \, dA}{A} = \frac{\sum x_i A_i}{\sum A_i}$$

Divide the L-section into two rectangles (labeled as 1 and 2 in the illustration).

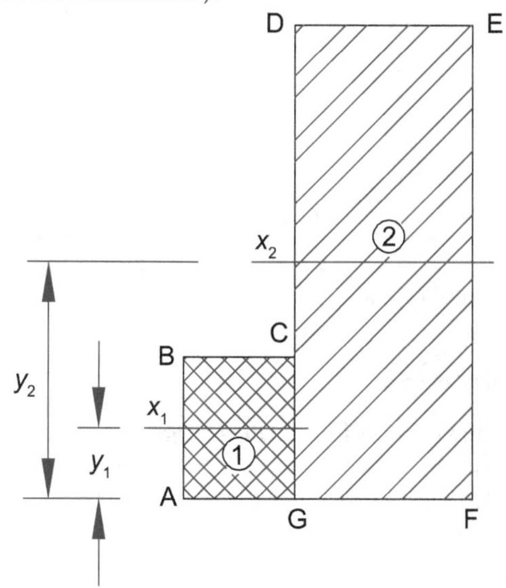

The areas of the rectangles are

$$A_1 = (4 \text{ in})(8 \text{ in}) = 32 \text{ in}^2$$
$$A_2 = (8 \text{ in})(24 \text{ in}) = 192 \text{ in}^2$$

Each rectangle has its own centroidal axis located at its middepth (4 in for rectangle 1 and 12 in for rectangle 2 from the base).

The distance to the centroid of the composite section is

$$\bar{y} = \frac{\sum A_i y_i}{\sum A_i} = \frac{(4 \text{ in})(32 \text{ in}^2) + (12 \text{ in})(192 \text{ in}^2)}{32 \text{ in}^2 + 192 \text{ in}^2} = 10.86 \text{ in} \quad (11 \text{ in})$$

The answer is (B).

28. The I-shaped beam can be considered in terms of three rectangular shapes.

$$\text{I-shaped section} = \text{rectangle A} - \text{rectangle B} - \text{rectangle C}$$

$$\text{I-shaped section} = \text{rectangle A} - (\text{rectangle B} + \text{C})$$

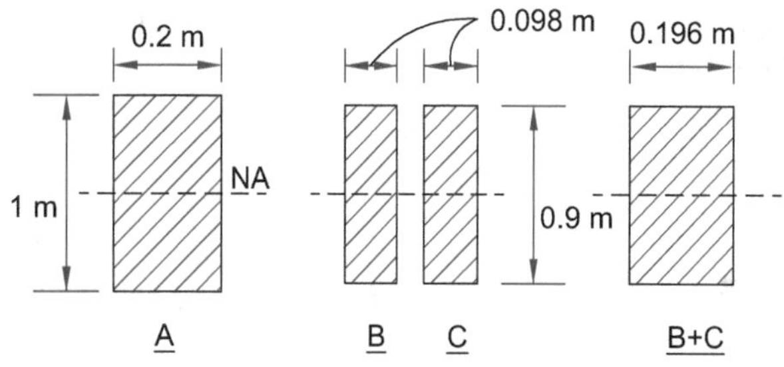

The moment of inertia of a rectangular section about its axis is determined using the formula

$$I = \frac{bd^3}{12}$$

Substituting the sizes of the two rectangles into the equation, the moment of inertia is

$$I = \frac{(0.2 \text{ m})(1 \text{ m})^3}{12} - \frac{(0.196 \text{ m})(0.9 \text{ m})^3}{12} = 0.00476 \text{ m}^4 \quad \left(48 \times 10^{-4} \text{ m}^4\right)$$

The answer is (C).

29. Draw a free-body diagram of the block.

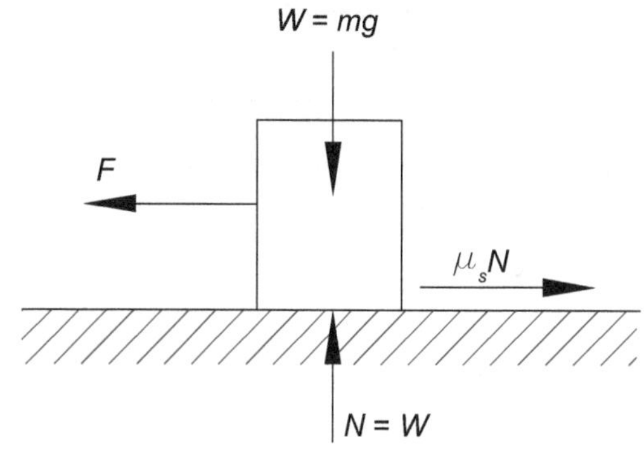

The largest friction force is called the limiting friction and is given by

$$F = \mu_s N$$

The normal force at the interface is

$$W = mg = (100 \text{ kg})\left(9.81 \ \frac{\text{m}}{\text{s}^2}\right) = 981 \text{ N}$$

The coefficient of static friction is 0.3. The limiting friction is

$$F = \mu_s N = 0.3(981 \text{ N}) = 294.3 \text{ N} \quad (290 \text{ N})$$

The answer is (D).

30. Since the ball is thrown against gravity, its deceleration is constant. Use the kinematic constant acceleration equation given in the Dynamics section of the *NCEES Handbook*.

$$v^2 = v_0^2 + 2\ a_0(s - s_0)$$

The deceleration, a_0, is equal and opposite to gravity.

$$a_0 = -32.2\ \text{ft/sec}^2$$

The initial velocity and height are

$$\begin{aligned} v_0 &= 60\ \text{ft/sec} \\ s_0 &= 5\ \text{ft} \end{aligned}$$

The final speed, v, is zero when the ball reaches its maximum height, s, as shown in the illustration.

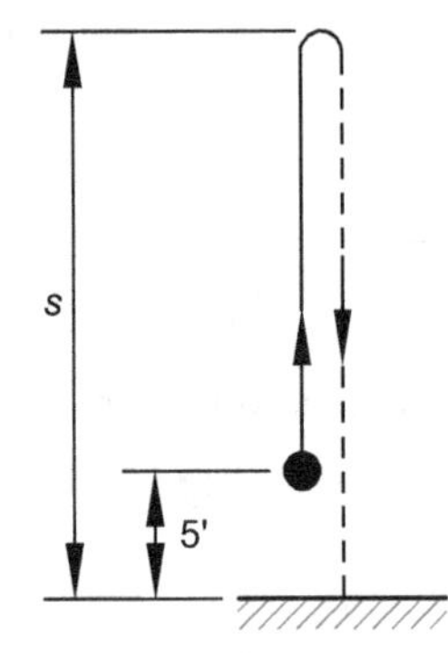

Substitute the values into the kinematic equation and solve for the final height.

$$\begin{aligned} v^2 &= v_0^2 + 2a_0(s - s_0) \\ s &= \frac{(v^2 - v_0^2)}{2a} + s_0 \\ &= \frac{0^2 - \left(60\ \frac{\text{ft}}{\text{sec}}\right)^2}{(2)\left(-32.2\ \frac{\text{ft}}{\text{sec}^2}\right)} + 5\ \text{ft} \\ &= 60.9\ \text{ft} \quad (61\ \text{ft}) \end{aligned}$$

The answer is (B).

31. The change in kinetic energy is the work done in accelerating the flywheel from ω_1 to ω_2. The work-energy formula is

$$\begin{aligned} (\text{torque})(\text{angle turned}_{\text{rad}}) &= \text{increase}_{\text{KE}} \\ T \times \theta &= \frac{1}{2}\left(I\omega_2^2 - I\omega_1^2\right) \end{aligned}$$

It is given that

$$T = 0.1\ \text{N·m} = 0.1\ \frac{\text{kg·m}^2}{\text{s}^2}$$

The speed of the flywheel is

$$\begin{aligned} \theta &= (10\ \text{turns})\left(2\pi\ \frac{\text{rad}}{\text{turns}}\right) = 20\pi\ \text{rad} \\ \omega_1 &= 0 \\ T\theta &= \frac{I\omega_2^2}{2} \\ \omega_2 &= \sqrt{\frac{2T\theta}{I}} = \sqrt{\frac{2(0.1\ \text{kg·m}^2/\text{s}^2)(20\pi\text{rad})}{3.14\ \text{kg·m}^2}} \\ &= 2.0\ \text{rad/s} \end{aligned}$$

The answer is (B).

32. The natural period of the system is

$$\tau_n = 2\pi\sqrt{\frac{I}{k_t}} = 1\ \text{sec}$$

The mass moment of inertia for a disk rotating about the center is

$$\begin{aligned} I &= \frac{1}{2}MR^2 = \frac{1}{2}\frac{W}{g}R^2 = \frac{1}{2}\left(\frac{50\ \text{lbf}}{32.2\ \frac{\text{ft}}{\text{sec}^2}}\right)\left(\frac{9\ \text{in}}{12\ \frac{\text{in}}{\text{ft}}}\right)^2 \\ &= 0.44\ \text{ft-lbf-sec}^2 \end{aligned}$$

Rearrange the natural period equation to solve for the torsional stiffness of the rod.

$$\begin{aligned} \tau_n &= 1\ \text{sec} = 2\pi\sqrt{\frac{0.44\ \text{ft-lbf-sec}^2}{k_t}} \\ k_t &= 17.24\ \text{ft-lbf/rad} \quad (17\ \text{ft-lbf/rad}) \end{aligned}$$

The answer is (D).

33. Use Newton's second law of motion.

$$\sum F_y = ma$$

Alternatively,

$$T - N = ma$$

T is the tension in the cable and N is the weight of the elevator. The tension in the cable is

$$T = N + ma = 1000 \text{ lbf} + \left(\frac{1000 \text{ lbf}}{32.2 \ \frac{\text{ft}}{\text{sec}^2}} \right) \left(10 \ \frac{\text{ft}}{\text{sec}^2} \right)$$
$$= 1311 \text{ lbf} \quad (1320 \text{ lbf})$$

The answer is (B).

34. The applied force, F, is proportional to the spring deformation, x. The force-deformation relationship is

$$F \propto x$$
$$F = kx$$
$$k = \frac{F}{x} = \frac{1000 \text{ lbf}}{4 \text{ in}} = 250 \text{ lbf/in}$$

The work done is the force multiplied by the distance traveled by the spring from 4 in to 8 in.

$$W = \int_4^8 F\,dx = \int_4^8 kx\,dx = \left(250 \ \frac{\text{lbf}}{\text{in}}\right)\left(\frac{x^2}{2}\right)\Bigg|_4^8$$
$$= 6000 \text{ lbf-in} = 6 \text{ in-kips}$$

The answer is (C).

35. By symmetry, the reactions at both supports equal 10 kips.

The shear force between points A and B is 10 kips.

The shear force between points B and C is

$$10 \text{ kips} - 10 \text{ kips} = 0 \text{ kips}$$

The beam shear force diagram is

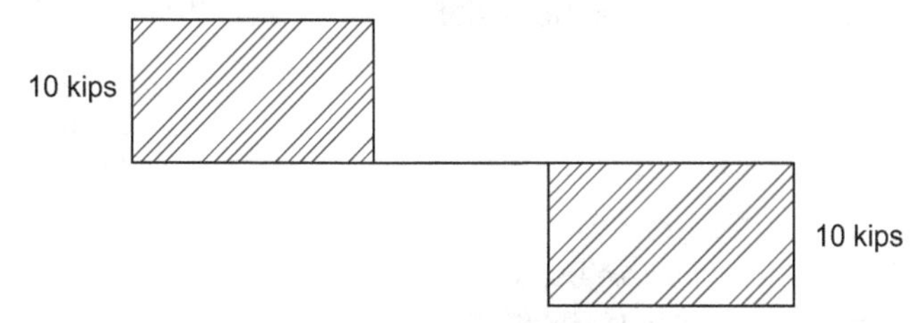

The answer is (A).

36. Determine the reactions at supports A and D. Support D is a roller and is incapable of resisting any horizontal thrust. Therefore, the entire applied load of 44 kips is resisted by support A. To determine the vertical reactions, take the moment about A.

$$\sum M_A = R_D(10 \text{ ft} + 12 \text{ ft}) - 44 \text{ kips}(16 \text{ ft}) = 0$$
$$R_D = 32 \text{ kips}$$
$$R_A = -32 \text{ kips}$$

The reactions for the frame can be seen by

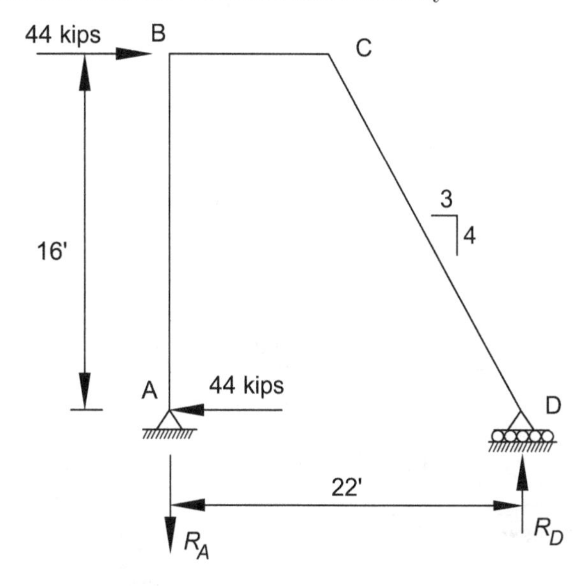

The moment at point B is

$$M_B = (44 \text{ kips})(16 \text{ ft}) = 704 \text{ ft-kips}$$

The moment at point C is

$$M_C = (32 \text{ kips})(12 \text{ ft})$$
$$= 384 \text{ ft-kips}$$

The bending moment diagram for the frame is

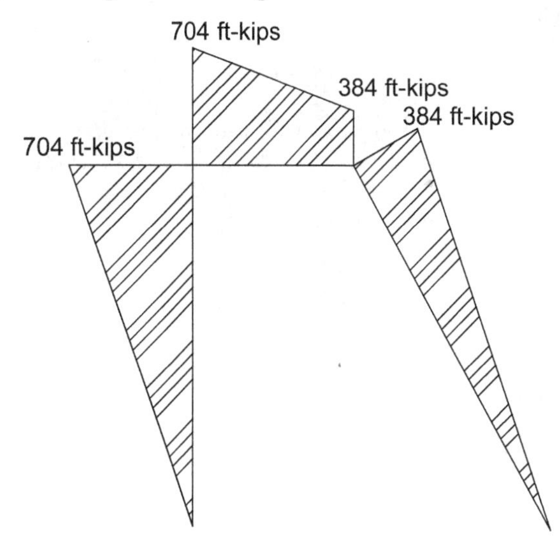

The answer is (C).

37. The material properties of steel are given in the Typical Material Properties table in the Mechanics of Materials section of the *NCEES Handbook*. From the Conversion Factors table,

$$1\ \text{Pa} = 1\ \frac{\text{N}}{\text{m}^2}$$
$$1\ \text{kPa} = 1\ \frac{\text{kN}}{\text{m}^2}$$

Young's modulus of steel is

$$E = 200\ \text{GPa} = 2 \times 10^8\ \text{kPa} = 2 \times 10^8\ \text{kN/m}^2$$
$$A = \frac{\pi D^2}{4} = \frac{\pi (0.03\ \text{m})^2}{4} = 7.07 \times 10^{-4}\ \text{m}^2$$

The elongation of the bar is

$$\Delta L = \frac{PL}{AE} = \frac{(300\ \text{kN})(1\ \text{m})}{(7.07 \times 10^{-4}\ \text{m}^2)\left(2 \times 10^8\ \frac{\text{kN}}{\text{m}^2}\right)} = 2.12\ \text{mm} \quad (2\ \text{mm})$$

The answer is (B).

38. This problem describes a uniaxial case of Hooke's law. The bar under tension is

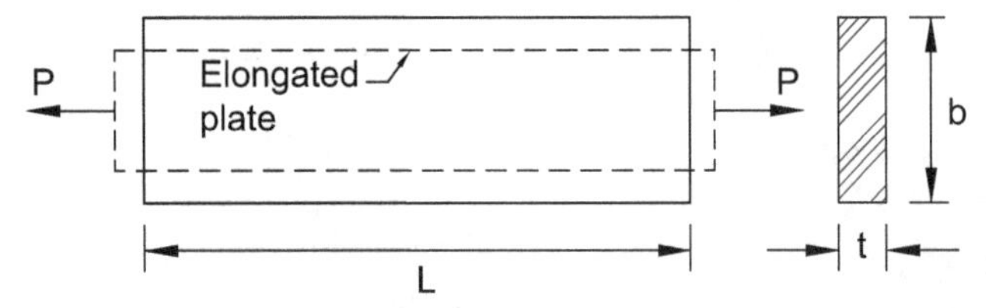

As the bar elongates, it decreases in width. The decrease in width is a function of lateral strain and the plate width.

$$\Delta w = \epsilon \times b$$

The material properties of aluminum are given in the Typical Material Properties table in the Mechanics of Materials section of the *NCEES Handbook*. From the Conversion Factors table,

$$1\ \text{Pa} = 1\ \frac{\text{N}}{\text{m}^2}$$
$$1\ \text{kPa} = 1\ \frac{\text{kN}}{\text{m}^2}$$

Young's modulus of elasticity for aluminum is

$$E = 69\ \text{GPa} = 69 \times 10^9\ \text{Pa} = 69 \times 10^9\ \frac{\text{N}}{\text{m}^2}$$

The area of the bar is

$$A = b \times t = (20\ \text{cm})(20\ \text{mm})\left(\frac{1\ \text{cm}}{10\ \text{mm}}\right) = 40\ \text{cm}^2 = 4 \times 10^{-3}\ \text{m}^2$$

The Poisson ratio for aluminum is

$$\nu = 0.33$$

The stress and strain in the bar along its longitudinal axis is

$$\sigma = \frac{P}{A} = \frac{300\ \text{kN}}{4 \times 10^{-3}\ \text{m}^2} = 75{,}000\ \frac{\text{kN}}{\text{m}^2} = 75 \times 10^6\ \frac{\text{N}}{\text{m}^2}$$
$$\epsilon_x = \frac{\sigma}{E} = \frac{75 \times 10^6\ \frac{\text{N}}{\text{m}^2}}{69 \times 10^9\ \frac{\text{N}}{\text{m}^2}} = 1.09 \times 10^{-3}$$

The lateral strain is

$$\epsilon_y = \nu\epsilon_x = 0.33\left(1.09 \times 10^{-3}\right) = 0.36 \times 10^{-3}$$

The decrease in width is

$$\Delta w = b\epsilon_y = 0.2\ \text{m}\left(0.36 \times 10^{-3}\right) = 72\ \mu\text{m} \quad (70\ \mu\text{m})$$

The answer is (B).

39. The modular ratio is defined in the *NCEES Handbook* as

$$n = \frac{E_1}{E_2}$$

For a steel-concrete beam, the steel modulus of elasticity is generally used as the numerator.

$$n = \frac{E_{\text{steel}}}{E_{\text{concrete}}} = \frac{29{,}000\ \text{ksi}}{3600\ \text{ksi}} = 8.06 \quad (8)$$

The answer is (A).

40. The UDL, w, is 2.50 kN/m (2500 N/m).

The maximum deflection is at midspan.

$$y_{max} = \frac{5wL^4}{384EI}$$

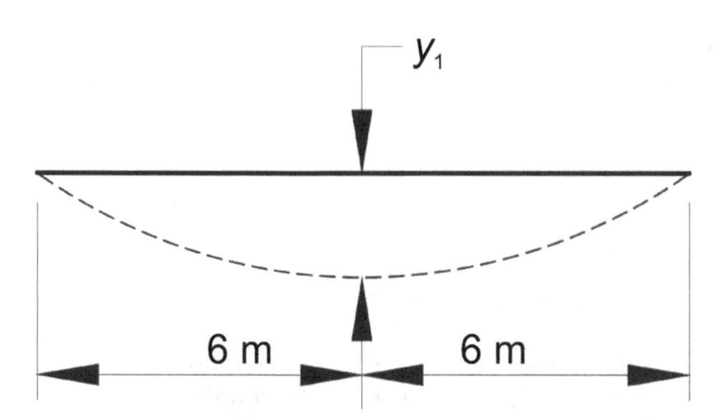

The moment of inertia of a hollow shaft is given in the Statics section of the *NCEES Handbook*, in the of Area Moment of Inertia table.

$$I = \frac{\pi\left(a^4 - b^4\right)}{4}$$

The outer radius is

$$r_{\text{out}} = \frac{400 \text{ mm}}{2} = 200 \text{ mm} = 0.2 \text{ m}$$

The inner radius is

$$\begin{aligned} r_{\text{in}} = r_{\text{out}} - t_{\text{pipe}} &= 200 \text{ mm} - 20 \text{ mm} \\ &= 180 \text{ mm} = 0.18 \text{ m} \end{aligned}$$

The mass deflection can be found using Young's modulus and the mass moment of inertia.

$$\begin{aligned} E &= 200 \text{ GPa} = 200 \times 10^9 \text{ Pa} = 200 \times 10^9 \ \frac{\text{N}}{\text{m}^2} \\ I &= \frac{\pi\left((0.2 \text{ m})^4 - (0.18 \text{ m})^4\right)}{4} = 0.000432 \text{ m}^4 \\ y_{max} &= \frac{5\left(2500 \ \frac{\text{N}}{\text{m}}\right)(12 \text{ m})^4}{384\left(200 \times 10^9 \ \frac{\text{N}}{\text{m}^2}\right)\left(0.000432 \text{ m}^4\right)} \\ &= 0.0078 \text{ m} \quad (8 \text{ mm}) \end{aligned}$$

The answer is (B).

41. The radius of Mohr's circle is

$$R = \sqrt{\left(\frac{\sigma_x - \sigma_y}{2}\right)^2 + \tau_{xy}^2}$$

In this case,

$$\begin{aligned} \sigma_x &= \sigma_y = 5 \text{ ksi} \\ \tau_{xy} &= 0 \end{aligned}$$

Therefore,

$$R = \sqrt{\left(\frac{5 \text{ ksi} - 5 \text{ ksi}}{2}\right)^2 + 0^2} = 0 \,\text{ksi}$$

The answer is (A).

42. For a pinned column loaded concentrically, use the Euler equation.

$$P_{cr} = \frac{\pi^2 EI}{L^2}$$

I is the moment inertia of the section about its weak axis (i.e., the y-axis). The moment of inertia of a rectangular section bending about its y-axis, given in Statics section of the *NCEES Handbook*, is

$$\begin{aligned} I_y &= \frac{b^3 h}{12} = \frac{(6 \text{ cm})^3(6 \text{ cm})}{12} \\ &= 108 \text{ cm}^4 = 1.08 \times 10^{-6} \text{ m}^4 \end{aligned}$$

Young's modulus for steel is

$$\begin{aligned} E &= 200 \text{ GPa} = 200 \times 10^9 \text{ Pa} \\ &= 200 \times 10^9 \ \frac{\text{N}}{\text{m}^2} \end{aligned}$$

The buckling capacity is

$$\begin{aligned} P_{cr} &= \frac{\pi^2\left(200 \times 10^9 \ \frac{\text{N}}{\text{m}^2}\right)\left(1.08 \times 10^{-6} \text{ m}^4\right)}{(2 \text{ m})^2} \\ &= 533 \times 10^3 \text{ N} = 533 \text{ kN} \quad (500 \text{ kN}) \end{aligned}$$

The answer is (C).

43. A phase diagram of the asphalt mix is shown.

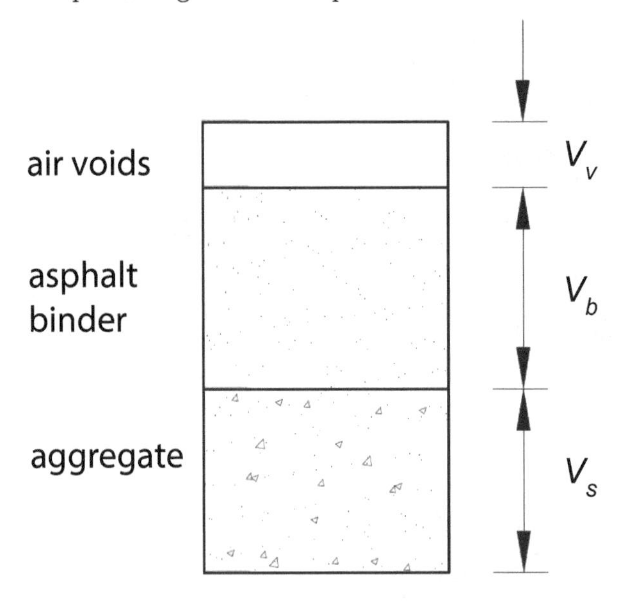

Consider a mix sample of 1 ft^3. The sample volume is

$$V_m = 1 \text{ ft}^3$$

The sample weight is 152 lbf.

The binder weight in the sample is

$$W_{\text{binder}} = (5\%)(152 \text{ lbf}) = 7.6 \text{ lbf}$$

The binder volume is

$$V_b = \frac{W_{\text{binder}}}{\rho_{\text{binder}}} = \frac{7.6 \text{ lbf}}{64 \ \frac{\text{lbf}}{\text{ft}^3}} = 0.12 \text{ ft}^3$$

The percentage of binder in the mix (by volume) is

$$\begin{aligned} V_{b,\%} &= \frac{V_b}{V_m}(100\%) = \frac{0.12 \text{ ft}^3}{1 \text{ ft}^3}(100\%) \\ &= 12\% \end{aligned}$$

The answer is (C).

44. Consider a 1 ft^3 mix volume. The mix has air voids, which have volume, but no weight.

$$V_{\text{voids}} = V_{\text{total}} - V_{\text{aggregate}} - V_{\text{binder}}$$

The weight and volume of the aggregate are

$$\begin{aligned} W_{\text{aggregate}} &= (95\%)(152 \text{ lbf}) = 144.4 \text{ lbf} \\ V_{\text{aggregate}} &= \frac{144.4 \text{ lbf}}{172 \ \frac{\text{lbf}}{\text{ft}^3}} = 0.84 \text{ ft}^3 \end{aligned}$$

The binder volume is 0.10 ft^3. The volume of the air voids is

$$\begin{aligned} V_{\text{voids}} &= V_{\text{total}} - V_{\text{aggregate}} - V_{\text{binder}} \\ &= 1.0 \text{ ft}^3 - 0.84 \text{ ft}^3 - 0.10 \text{ ft}^3 \\ &= 0.06 \text{ ft}^3 \end{aligned}$$

Since the volume of the sample is 1 ft^3, the air void ratio is

$$\text{air/void ratio} = \frac{V_{\text{voids}}}{V_{\text{total}}}(100\%) = \left(\frac{0.06 \text{ ft}^3}{1 \text{ ft}^3}\right)(100\%) = 6\%$$

The answer is (B).

45. The water-cement ratio is by weight. A concrete mix is generally described by the number of "sacks" of Portland cement needed to produce a cubic yard of concrete. A sack of cement weighs 94 lbf. Similarly, the water requirement is stated as the number of gallons of water per sack of concrete. Recall that one gallon of water weighs 8.345 lbf.

$$\begin{aligned} \frac{W_{\text{water}}}{W_{\text{cement}}} &= 0.4 \\ W_{\text{water}} &= (0.4)(W_{\text{cement}}) = (0.4)(94 \text{ lbf}) = 37.6 \text{ lbf} \end{aligned}$$

The volume of water is

$$\begin{aligned} V_{\text{water}} &= \frac{37.6 \text{ lbf}}{8.345 \ \frac{\text{lbf}}{\text{gal}}} \\ &= 4.51 \text{ gal} \quad (4.5 \text{ gal}) \end{aligned}$$

4.5 gal of water is needed per sack of cement to maintain the water-cement ratio.

The answer is (B).

46. The concrete compressive strength is defined as the ultimate compressive stress a concrete cylinder can bear. It is the ultimate strength over the area of the cylinder.

$$f'_c = \frac{P_{\text{ult}}}{A}$$

The area of the cylinder is

$$A = \frac{\pi d^2}{4} = \frac{\pi(6 \text{ in})^2}{4} = 28.26 \text{ in}^2$$

The concrete compressive strength is

$$f_c' = \frac{255{,}000 \text{ lbf}}{28.26 \text{ in}^2} = 9023 \text{ lbf/in}^2 \quad (9000 \text{ psi})$$

The answer is (D).

47. The stress at which a material will experience permanent deformation is called the yield stress. The answer is shown in the illustration.

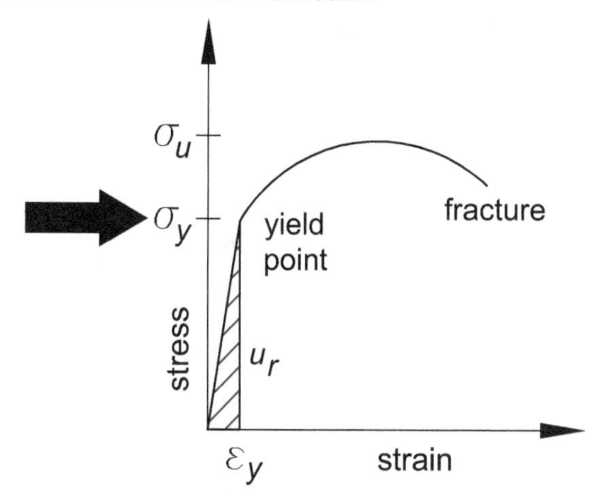

48. The term $h_{f,\text{fitting}}$ stands for the head loss due to pipe fittings, pipe expansions, and contractions. It does not represent the frictional head loss in the pipe which is represented by h_f. Manning's equation applies to the open channel flow, not to the pipe fittings. The fitting loss is expressed as

$$h_{f,\text{fitting}} = C\frac{\text{v}^2}{2g}$$

The answer is (D).

49. The maximum hydrostatic pressure at depth, h, from the top is

$$P_h = \gamma_{\text{water}}(h)$$

γ_{water}, the density of water is the specific weight of the water, 62.4 lbf/ft^3. h is the height of the water column.

The illustration shows the corresponding lateral force per unit length.

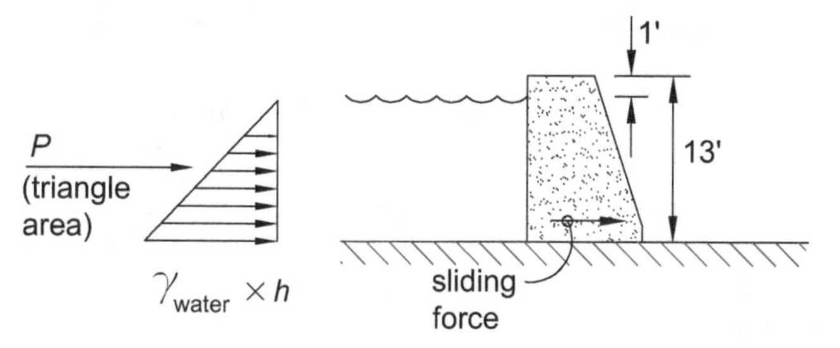

The corresponding hydrostatic force per unit width of the wall is

$$P_h = \frac{1}{2}(\gamma_{\text{water}})(h^2)$$

The force equals the area under the hydrostatic pressure diagram. The maximum lateral force is at the base, as shown in the illustration. The water height above the base is

$$h = 13 \text{ ft} - 1 \text{ ft} = 12 \text{ ft}$$

The pressure at the base is

$$P_{\text{base}} = \left(\frac{1}{2}\right)\left(62.4 \ \frac{\text{lbf}}{\text{ft}^3}\right)(12 \text{ ft})^2 = 4493 \text{ lbf/ft}$$

For the 100 ft length of the dam, the total lateral pressure at the base is

$$P_{\text{total}} = (100 \text{ ft})\left(4493 \ \frac{\text{lbf}}{\text{ft}}\right) = 449{,}300 \text{ lbf} \quad (450 \text{ kips})$$

The answer is (C).

50. The elevation head is the difference between the elevations of the water surface levels of the tanks.

$$\text{elevation at A} - \text{elevation at B} = 10 + 5 + 5 = 20 \text{ ft.}$$

The answer is (B).

51. The pressure at the base is

$$P_{\text{base}} = \left(\frac{1}{2}\right)\left(62.4 \ \frac{\text{lbf}}{\text{ft}^3}\right)(12 \text{ ft})^2 = 4493 \text{ lbf/ft} = 4.493 \text{ kips/ft}$$

The resultant hydrostatic force acts at $h/3$ above the base. The maximum overturning moment is at the base.

$$\text{overturning moment} = \frac{12 \text{ ft}}{3}(4.493 \text{ kips/ft}) = 17.97 \text{ ft-kips/ft}$$

For the 100 ft length of the dam,

$$\begin{array}{c}\text{total overturning} \\ \text{moment}\end{array} = \left(17.97 \ \frac{\text{ft-kips}}{\text{ft}}\right)(100 \text{ ft}) = 1797 \text{ ft-kips} \quad (1800 \text{ ft-kips})$$

The answer is (B).

52. The Reynolds number is dimensionless and is expressed as

$$Re = \frac{vD}{\nu}$$

The pipe diameter, D, is 2.54 cm or 0.0254 m. Use the kinematic viscosity and flow velocity to calculate the Reynolds number.

$$\begin{aligned} Re &= \frac{vD}{\nu} = \frac{\left(10\ \frac{\text{m}}{\text{s}}\right)(0.0254\ \text{m})}{0.0003\ \frac{\text{m}^2}{\text{s}}} \\ &= 846 \end{aligned}$$

The answer is (B).

53. From the Units and Conversion Factors section of the *NCEES Handbook*, use the relation 1 cfs = 0.646317 MGD. The oil flow rate in cubic feet per second is

$$Q = (3\ \text{MGD})\left(\frac{1\ \text{cfs}}{0.646317\ \text{MGD}}\right) = 4.69\ \text{cfs}$$

The area of the pipe is

$$A = \pi r^2 = \pi(0.5\ \text{ft})^2 = 0.785\ \text{ft}^2$$

The average velocity in the pipe is

$$\begin{aligned} \text{v} &= \frac{Q}{A} = \frac{4.69\ \frac{\text{ft}^3}{\text{sec}}}{0.785\ \text{ft}^2} \\ &= 5.97\ \text{ft/sec} \quad (6.0\ \text{ft/sec}) \end{aligned}$$

The answer is (B).

54. Draw the triangle. Side b is 80 ft, side c is 100 ft, and angle A is 60°.

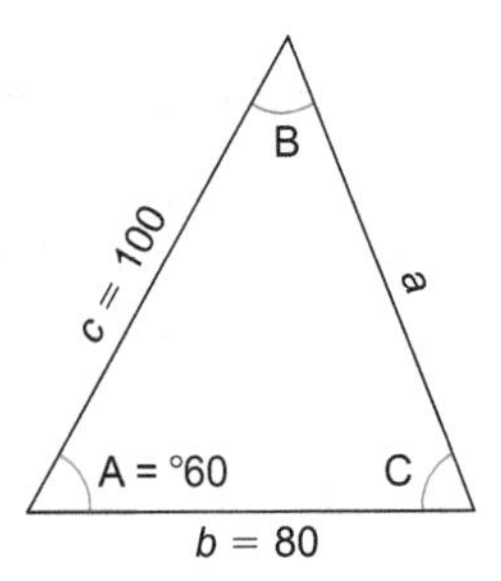

Use the law of cosines to determine the length of side a.

$$\begin{aligned} a^2 &= b^2 + c^2 - 2bc\ \cos A \\ a &= \sqrt{b^2 + c^2 - 2bc\ \cos A} \\ &= \sqrt{(80\ \text{ft})^2 + (100\ \text{ft})^2 - 2(80\ \text{ft})(100\ \text{ft})\cos 60^\circ} \\ &= 91.65\ \text{ft} \quad (92\ \text{ft}) \end{aligned}$$

The answer is (C).

55. Since the three sides of the triangle are given, use the law of sines to determine the associated angles.

$$\frac{a}{\sin A} = \frac{b}{\sin B} = \frac{c}{\sin C}$$

Using the side lengths, the equations become

$$\frac{80\ \text{ft}}{\sin B} = \frac{100\ \text{ft}}{\sin C} = \frac{91.65\ \text{ft}}{\sin 60^\circ} = \frac{91.65\ \text{ft}}{0.866} = 105.83$$

Therefore,

$$\begin{aligned} \frac{100}{\sin C} &= 105.83 \\ \sin C &= 0.945 \\ \measuredangle C &= 70.91^\circ \end{aligned}$$

Since the sum of the three angles is 180°, angle B is

$$\begin{aligned} \measuredangle B &= 180^\circ - 60^\circ - 70.91^\circ \\ &= 49.09^\circ \end{aligned}$$

Therefore, the largest angle is most nearly 71°.

The answer is (B).

56. For a polygon of n sides, the internal angle is expressed as

$$\theta = (\pi\ \text{radians})\left(1 - \frac{2}{n}\right) = 180^\circ\left(\frac{n-2}{n}\right)$$

Therefore,

$$n\theta = 180^\circ(n-2)$$

Since the traverse has six sides, n is six.

The sum of the interior angles is

$$\sum \measuredangle_{\text{interior}} = (6-2)180^\circ = 720^\circ$$

The sum of the four known angles is

$$\sum \measuredangle_{\text{known}} = 4\ (90^\circ) = 360^\circ$$

The sum of the two unknown angles is

$$\sum \measuredangle_{\text{unknown}} = 720° - 360° = 360°$$

The illustration shows one of the possible configurations for the traverse.

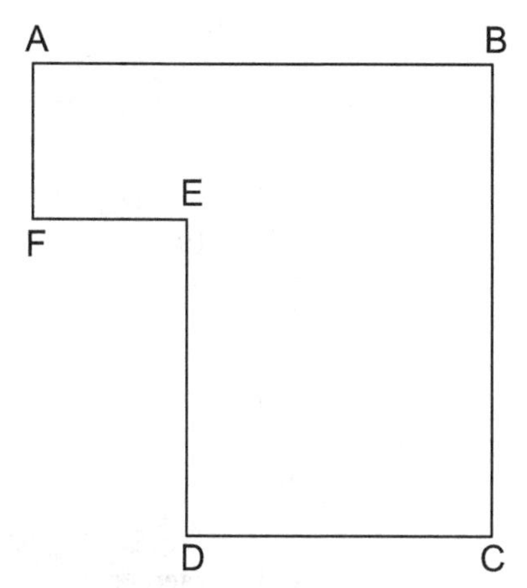

The answer is (D).

57. Generally, if the properties of two end sections are given from which the end areas can be readily computed, the volume equals the average cross-sectional area multiplied by the height. This is called the average end area method. The volume is expressed as

$$V = \frac{L(A_1 + A_2)}{2}$$

The pile base is a circle with a 100 ft diameter. Therefore, its base area, A_1, is

$$A_1 = \frac{\pi d_1^2}{4} = \frac{\pi(100\text{ ft})^2}{4} = 7854\text{ ft}^2$$

The top area, A_2, is

$$A_2 = \frac{\pi d_2^2}{4} = \frac{\pi(25\text{ ft})^2}{4} = 491\text{ ft}^2$$

The volume of the piled material is

$$\begin{aligned} V &= \frac{L(A_1 + A_2)}{2} = \frac{25\text{ ft}\left(7854\text{ ft}^2 + 491\text{ ft}^2\right)}{2} \\ &= 104{,}312\text{ ft}^3 = 104{,}312\text{ ft}^3\left(\frac{1\text{ yd}^3}{27\text{ ft}^3}\right) \\ &= 3863\text{ yd}^3 \quad (3860\text{ yd}^3) \end{aligned}$$

The answer is (B).

58. While azimuths are typically expressed from the north, some navigation systems use south as the reference plane. The rotation of the azimuth is always clockwise. In this case, the bearing needs to be subtracted from 360° to determine the azimuth from the south.

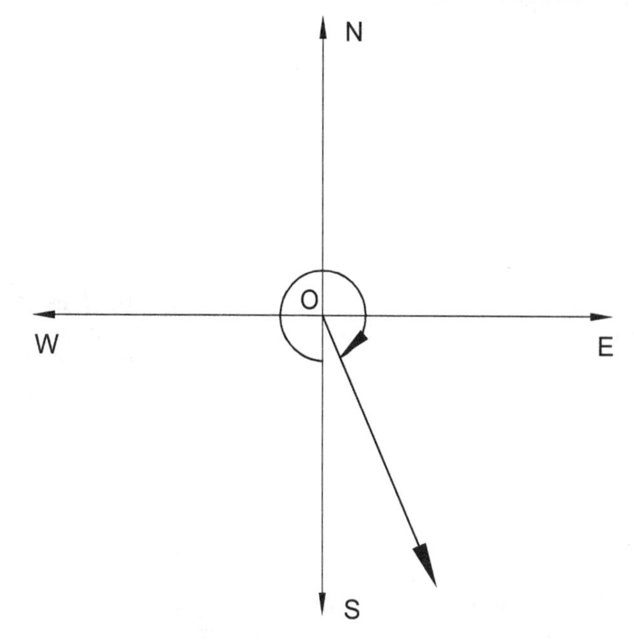

The answer is (D).

59. Options B, C, and D are correct. Option A is incorrect.

The answer is (A).

60. In general, runoff is the rainfall minus the losses.

Given that the curve number is 60 and the precipitation amount is 5 in, the surface storage is

$$\begin{aligned} S &= \frac{1000}{\text{CN}} - 10 \\ &= \frac{1000}{60} - 10 \\ &= 6.67\text{ in} \end{aligned}$$

The watershed runoff is

$$\begin{aligned} Q &= \frac{(P - 0.2S)^2}{P + 0.8S} \\ &= \frac{\left(5 - (0.2)(6.67\text{ in})\right)^2}{5 + (0.8)(6.67\text{ in})} \\ &= \frac{13.44\text{ in}^2}{10.33\text{ in}} \\ &= 1.3\text{ in} \end{aligned}$$

Therefore, out of 5 in of total precipitation, only 1.3 in is runoff. The remaining 3.7 in of the rainfall is lost (i.e., it is absorbed in the soil, evaporated, or trapped in locally depressed areas).

The runoff volume is

$$\begin{aligned} V &= (\text{runoff depth})(\text{drainage area}) \\ &= (1.3 \text{ in})\left(\frac{1 \text{ ft}}{12 \text{ in}}\right)(300 \text{ ft})(400 \text{ ft}) \\ &= 13{,}000 \text{ ft}^3 \end{aligned}$$

The answer is (B).

61. The statement in option A is correct. It pertains to one unit of effective rainfall occurring uniformly in space and time over a unit period of time. Hydrographs reflect all features of the watershed basin; however, the basin features are fixed and do not change from one storm to another. The theory of superposition is applicable in using a unit hydrograph. The storm characteristics have a significant effect on the shape of hydrographs.

The answer is (D).

62. The maximum height, h, that can be used in siphoning is limited by the absolute pressure needed to avoid separation of dissolved gases. In this case, separation occurs at an absolute pressure of 8 ft of water head, which is less than the atmospheric pressure of 34 ft of water. The difference can be used to raise the pipe elevation above point A. Apply Bernoulli's energy equation to points A and C.

Energy at point A equals energy at point C plus losses between A and C.

$$\frac{p_{\text{A}}}{\gamma} + \frac{v_{\text{A}}^2}{2g} + z_{\text{A}} = \frac{p_{\text{C}}}{\gamma} + \frac{v_{\text{C}}^2}{2g} + z_{\text{C}} + h_f + h_{f,\text{fitting}}$$

Express the height as the difference between the heights of point C and A.

$$h = z_{\text{C}} - z_{\text{A}}$$

To maximize the siphoning height, h, the water pressure at point C should be reduced as much as possible without causing separation of the gases from the water.

The water velocity at point A is practically zero ($v_{\text{A}} = 0$), and at point C it is 3 ft/sec.

The pipe friction loss and entry loss are given. The energy equation reduces to

$$\begin{aligned} 34 \text{ ft} + 0 + z_{\text{A}} &= 8 \text{ ft} + \frac{\left(3 \ \frac{\text{ft}}{\text{sec}}\right)^2}{(2)\left(32.2 \ \frac{\text{ft}}{\text{sec}^2}\right)} \\ &\quad + z_{\text{C}} + 15 \text{ ft} + 1 \text{ ft} + 0.1 \text{ ft} \\ z_{\text{A}} &= z_{\text{C}} - 9.76 \text{ ft} \\ h &= z_{\text{C}} - z_{\text{A}} \\ &= 9.76 \text{ ft} \quad (10 \text{ ft}) \end{aligned}$$

The answer is (A).

63. The pumping power equation is

$$\begin{aligned} \dot{W} &= \frac{Q\gamma h}{\eta} \\ Q &= 0.5 \text{ ft}^3/\text{sec} \end{aligned}$$

The pipe diameter is 6 in (or 0.5 ft). The pipe area is

$$A = \frac{\pi D^2}{4} = \frac{(\pi)(0.5 \text{ ft})^2}{4} = 0.2 \text{ ft}^2$$

The water velocity is

$$v = \frac{Q}{A} = \frac{0.5 \ \frac{\text{ft}^3}{\text{sec}}}{0.2 \text{ ft}^2} = 2.5 \text{ ft/sec}$$

The pump efficiency is 0.8.

The head of the water being lifted, h, is the friction loss plus the elevation difference in reservoir elevations (100 ft).

The friction head loss is determined by the Darcy-Weisbach equation.

$$h_f = f\frac{L}{D}\frac{v^2}{2g}$$

The friction factor is

$$\begin{aligned} f &= 0.028 \\ L &= 1 \text{ mi} = 5280 \text{ ft} \end{aligned}$$

$$h_f = f\frac{L}{D}\frac{v^2}{2g} = (0.028)\left(\frac{5280 \text{ ft}}{0.5 \text{ ft}}\right)\left(\frac{\left(2.5 \ \frac{\text{ft}}{\text{sec}}\right)^2}{2\left(32.2 \ \frac{\text{ft}}{\text{sec}^2}\right)}\right) = 28.7 \text{ ft}$$

The total head is

$$h = 28.7 \text{ ft} + 100 \text{ ft} = 128.7 \text{ ft}$$

$$\begin{aligned}\dot{W} &= \frac{\left(0.5\ \frac{\text{ft}^3}{\text{sec}}\right)\left(62.4\ \frac{\text{lbf}}{\text{ft}^3}\right)(128.7\ \text{ft})}{0.8} \\ &= 5019\ \frac{\text{ft-lbf}}{\text{sec}} \\ &= \frac{(5019\ \text{ft})(1\ \text{hp})}{550\ \frac{\text{ft-lbf}}{\text{sec}}} \\ &= 9.1\ \text{hp}\quad (9\ \text{hp})\end{aligned}$$

The answer is (C).

64. A smooth pipe is defined as a pipe with a roughness factor of

$$e\ (\text{or}\ \epsilon) = 0$$

See the Moody diagram. The friction head loss is given by the Darcy-Weisbach equation.

$$h_f = f\frac{L}{D}\frac{v^2}{2g}$$

The friction factor, f, depends on the Reynolds number and the relative roughness of the pipe.

D = diameter of the pipe

L = length over which the pressure drop occurs

e = pipe roughness factor

$g = 9.81\ \text{m}^2/\text{s}$

For a smooth pipe, $e = 0$.

The equation for the Reynolds number is

$$\text{Re} = \frac{D\text{v}}{\upsilon}$$

It is given that given that the diameter is 10 cm (0.1 m).

The velocity, v, is 2 m/s.

The kinematic viscosity of water is

$$\upsilon = 0.000000326\ \frac{\text{m}^2}{\text{s}}$$

The Reynolds number is

$$\text{Re} = \frac{D\text{v}}{\upsilon} = \frac{(0.1\ \text{m})\left(2\ \frac{\text{m}}{\text{s}}\right)}{0.000000326\ \frac{\text{m}^2}{\text{s}}} = 6.13 \times 10^5$$

Use the Moody Diagram. For the Re number obtained from the smooth pipe curve.

$$f = 0.0125$$

Therefore,

$$\begin{aligned}h_f &= f\frac{L}{D}\frac{v^2}{2g} = (0.0125)\left(\frac{100\ \text{m}}{0.1\ \text{m}}\right)\left(\frac{\left(2\ \frac{\text{m}}{\text{s}}\right)^2}{2\left(9.81\ \frac{\text{m}}{\text{s}^2}\right)}\right) \\ &= 2.55\ \text{m}\quad (2.5\ \text{m})\end{aligned}$$

The answer is (C).

65. For a hydraulic jump to occur, the depth upstream must be less than the critical depth. The associated specific energy is at the minimum level. There is a loss of energy at the jump which can be calculated as the difference between specific energy levels before and after the hydraulic jump. Options A, C, and D are incorrect.

The answer is (B).

66. The curve number (CN) for a basin varies between 1 and 100. A CN value of 100 indicates a soil that has no basin retention value, so the entire rainfall results in the runoff. It is validated by the formula

$$S = \frac{1000}{\text{CN}} - 10$$

For a CN of 100, the basin retention value is

$$S = \frac{1000}{100} - 10 = 0$$

The answer is (A).

67. *Transmissivity* is defined as the product of hydraulic conductivity, K, and the thickness of the aquifer. Hydraulic conductivity depends on two factors: permeability of soil and properties of the fluid. Permeability alone is used in the statement to define K; it is insufficient to define hydraulic conductivity. Therefore, option A is incorrect.

Option B states that the effective hydraulic conductivity, $\overline{K}$, is the weighted average of the hydraulic conductivity of all layers in the aquifer with water flowing parallel to stratification; the statement is not valid for the flow perpendicular to the stratification. Therefore, option B is incorrect.

Option C states that the effective hydraulic conductivity, $\overline{K}$, for the flow is perpendicular to stratification: it does not pertain to the flow parallel to the stratification. Therefore, option C is incorrect.

A confined aquifer occurs where groundwater is under greater-than-atmospheric pressure due to an overlaying confined layer of a relatively impermeable medium. Therefore, option D is correct.

The answer is (D).

68. See the Flow in Noncircular Conduits section of the *NCEES Handbook*. The hydraulic radius is defined as the ratio

$$R_H = \frac{\text{cross-sectional area}}{\text{wetted perimeter}}$$

In this case, the wetted perimeter is an arc length. The cross-sectional area of the flow is the area of a partial circle. The diameter of the pipe is 36 in, and the depth of flow is 27 in. Use the Hydraulic Elements Graph for Circular Sewers given in the *NCEES Handbook*.

The ratio of water depth to pipe diameter is

$$\frac{d}{D} = \frac{3}{4} = 0.75$$

From the hydraulic element graph, for a ratio of water depth to pipe diameter of 0.75, the ratio is

$$\begin{aligned} \frac{R}{R_f} &= 1.21 \\ R &= (1.21)(R_f) \end{aligned}$$

The factor, R_f, is the hydraulic radius of the pipe running full. From the Fluid Mechanics section of the *NCEES Handbook*,

$$\begin{aligned} R_H &= \frac{\text{cross-sectional area}}{\text{wetted perimeter}} = \frac{\pi r^2}{2\pi r} = \frac{r}{2} \\ &= \frac{D}{4} \\ &= \frac{36 \text{ in}}{4} \\ &= 9 \text{ in} \end{aligned}$$

Therefore, the hydraulic radius for this pipe is

$$\begin{aligned} R_H &= (1.21)(9 \text{ in}) \\ &= 10.89 \text{ in} \quad (11 \text{ in}) \end{aligned}$$

The answer is (B).

69. This type of growth is called log growth, exponential growth, or geometric growth. (See the Population Modeling section of the *NCEES Handbook*.)

Bacteria multiply or double during the growth phase without any decay or microbial death. The population is

$$\begin{aligned} P &= P_0(2^n) \\ P_0 &= 1000 \text{ cells/L} \\ n &= 20 \end{aligned}$$

Therefore,

$$\begin{aligned} P &= \left(1000 \ \frac{\text{cells}}{\text{L}}\right)(2^{20}) \\ &= 1.05 \times 10^9 \text{ cells/L} \quad (1{,}000{,}000{,}000 \text{ cells/L}) \end{aligned}$$

The answer is (D).

70. The horizontal velocity, or approach velocity, is

$$V_h = \frac{Q}{A_x}$$

The design flow rate, Q, is 20 ft^3/sec.

The cross-sectional area is

$$A_x = (6 \text{ ft})(8 \text{ ft}) = 48 \text{ ft}^2$$

The approach velocity is

$$\begin{aligned} V_h &= \frac{20 \ \frac{\text{ft}^3}{\text{sec}}}{48 \text{ ft}^2} \\ &= 0.42 \text{ ft/sec} \end{aligned}$$

The answer is (A).

71. See the Stability, Determinacy, and Classification of Structures section of the *NCEES Handbook* for conditions for determinacy. An arch is considered a frame. It is given that the arch has three hinges—two at supports points A and B, and the third at the crown point, C. Thus, the moment is zero at point A, point B, and point C. The arch has the following properties.

$$\begin{aligned} \text{number of members, } m &= 2 \\ \text{number of independent reaction components, } r &= \left(\frac{2 \text{ reactions}}{\text{support}}\right)(2 \text{ supports}) \\ &= 4 \text{ reactions} \\ \text{number of joints, } j &= 3 \end{aligned}$$

The number of condition equations, c, based on known internal moments or forces, such as internal moment of zero, is 1.

As described in the Stability, Determinacy, and Classification of Structures section of the *NCEES Handbook*, the stability and determinacy of an arch depends on whether $3m + r$ is greater than, equal to, or less than $3j + c$.

$$\text{left-hand side}: 3m + r = 3(2) + 4 = 10$$
$$\text{right-hand side}: 3j + c = 3(3) + 1 = 10$$

This means that

$$3m + r = 3j + c$$

Therefore, the arch ABC is stable and statically determinate.

The answer is (B).

72. The slopes and deflections of a simply supported beam under a uniformly distributed load (UDL) are given in the Simply Supported Beam Slopes and Deflections table in the *NCEES Handbook*. The maximum downward deflection is at the midspan, and maximum curvature occurs at supports B and C

$$\theta_B = \frac{wL^3}{24EI}$$

There is no load applied over segment AB. Therefore, segment AB remains a straight line with point A deflecting up distance δ_A.

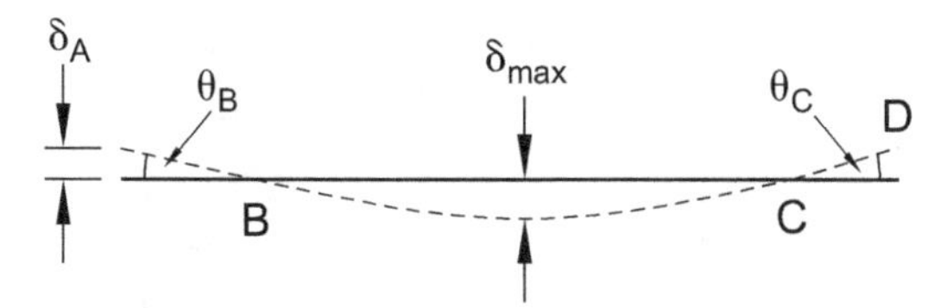

The deflection at A is

$$\delta_A = (\text{slope at B})(\text{distance AB})$$

Steel has the following properties

$$E = 29{,}000 \text{ ksi}$$
$$I = 500 \text{ in}^4$$

The span length is

$$L = 20 \text{ ft} = 20 \text{ ft}\left(12 \ \frac{\text{in}}{\text{ft}}\right) = 240 \text{ in}$$

The UDL is

$$w = 2 \ \frac{\text{kips}}{\text{ft}} = 2 \ \frac{\text{kips}}{\text{ft}}\left(\frac{1 \text{ ft}}{12 \text{ in}}\right) = 0.167 \text{ kips/in}$$

The maximum angle of deflection is

$$\theta_B = \frac{wL^3}{24EI} = \frac{\left(0.167 \ \frac{\text{kips}}{\text{in}}\right)(240 \text{ in})^3}{24\left(29{,}000 \ \frac{\text{kips}}{\text{in}^2}\right)(500 \text{ in}^4)}$$
$$= 0.0066 \text{ rad}$$
$$AB_{\text{distance}} = 5 \text{ ft} = 60 \text{ in}$$

The maximum deflection is

$$\delta_A = (\theta_B)(AB_{\text{distance}}) = (0.0066 \text{ rad})(60 \text{ in})$$
$$= 0.396 \text{ in} \quad (0.40 \text{ in})$$

The answer is (B).

73. The properties of a W12 × 79 section are given in the W Shapes Dimensions and Properties table of the *NCEES Handbook*. Use the smaller of the two radii of gyration given for the section. For pinned-pinned end conditions,

$$K = 1$$
$$L = 240 \text{ in}$$
$$A = 23.2 \text{ in}^2$$
$$r_y = 3.05 \text{ in}$$
$$\frac{KL}{r} = \frac{(1)(240 \text{ in})}{3.05 \text{ in}} = 79$$

From AISC Table 4-14, given in the Civil Engineering section of the *NCEES Handbook*, use the allowable critical stress for the compression member for the slenderness ratio.

$$\phi F_{cr} = 28.5 \text{ ksi}$$
$$\phi P_n = (\phi F_{cr})(A) = \left(28.5 \ \frac{\text{kips}}{\text{in}^2}\right)(23.2 \text{ in}^2)$$
$$= 661.2 \text{ kips} \quad (660 \text{ kips})$$

The answer is (C).

74. A planar truss is defined as a truss that lies in a two-dimensional plane. A planar truss is comprised of slender bars that are connected by frictionless pin joints with no moment connections. In order to have static determinacy, all truss supports should not be restrained against translation. The pinned truss joints are capable of transferring an axial load only. Thus, all elements of a planar truss are subjected to axial loads only.

The answer is (D).

75. Take the moment about joint B due to the cantilever load.

$$M_B = (24 \text{ kips})(5 \text{ ft}) = 120 \text{ ft-kips} \quad \text{(counterclockwise)}$$

This moment is distributed between members BC and BD. The carryover factor is 0.5.

The equation for flexural stiffness is given as

$$k = \frac{4EI}{L}$$

In this case, members BC and BD have equal *EI*, but length of BC is twice the length of BD.

Since the stiffness, *k*, is inversely proportional to the member length, the stiffness of BD is twice the stiffness of BC. In relative terms, if k_{BC} is 1, then k_{BD} is 2.

The distribution factor for member BC is

$$DF_{BC} = \frac{k_i}{\sum k_i} = \frac{1}{1+2} = 0.333$$

The distributed moment at the near end for member BC is

$$M_{BC} = (0.333)(120 \text{ ft-kips}) = 40 \text{ ft-kips}$$

The carryover moment at joint C is

$$\begin{aligned} M_C &= (0.5)(40 \text{ ft-kips}) \\ &= 20 \text{ ft-kips} \end{aligned}$$

The answer is (B).

76. The applied moment, M_A, looks like

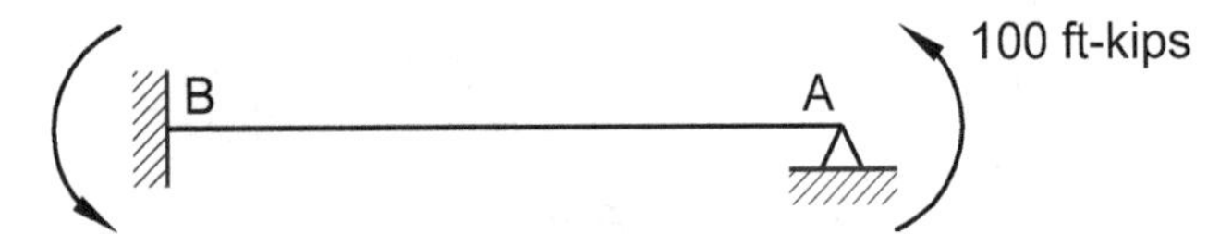

Support B is the fixed end of the propped cantilever. The induced moment at support B, M_B, is called the carryover moment. The associated carryover factor is 0.5.

$$M_B = (0.5)(M_A) = (0.5)(100 \text{ ft-kips}) = 50 \text{ ft-kips}$$

The answer is (B).

77. To determine the influence line for A_y, take the moment about point B. Let clockwise moments be positive, and let x be the distance from point C to the point load.

$$\begin{aligned} \sum M_B &= 0 = A_y(20 \text{ ft}) - (1 \text{ kip})(30 \text{ ft} - x) \\ A_y &= 1.5 - \frac{x}{20} \text{ kips} \end{aligned}$$

The equation represents a straight line, with 1.5 kips as the intercept when x is 0.

Use the above equation to determine the reaction A_y when the unit load is at point G.

$$A_y = 1.5 \text{ kips} - \frac{40}{20} = 1.5 - 2 = -0.5 \text{ kips}$$

This is shown in option C.

Option D represents the influence line for B_y.

Support B has a roller, which can translate in the horizontal direction and thus is incapable of developing a horizontal reaction. Option A shows the influence line for reactions A_x and B_x, which both equal zero.

Option B is the influence diagram for A_x and B_x if the frame has a hinge instead of a roller support at B and a rotational hinge at midspan E.

The answer is (C).

78. Use AISC Tables 1-1 and 3-2, as given in the Civil Engineering section in the *NCEES Handbook*. For a W21 × 44 shape,

$$\begin{aligned} S_x &= 81.6 \text{ in}^3 \\ L_p &= 4.45 \text{ ft} \\ L_r &= 13.0 \text{ ft} \\ \phi_b M_{px} &= 358 \text{ ft-kips} \\ \phi_b M_{rx} &= 214 \text{ ft-kips} \\ \phi_b &= 0.9 \end{aligned}$$

Since $L_p < L_b \le L_r$, use the limit state of lateral-torsional buckling equation given in the *NCEES Handbook*.

$$M_n = C_b\left[M_p - (M_p - 0.7F_yS_x)\left(\frac{L_b - L_p}{L_r - L_p}\right)\right] \le M_p$$

In the M_n equation,

$$\begin{aligned} 0.7F_yS_x &= (0.7)(50 \text{ ksi})(81.6 \text{ in}^3) \\ &= 2856 \text{ in-kips} = 238 \text{ ft-kips} \end{aligned}$$

Using ft-kips units in the M_n equation,

$$M_n = \begin{pmatrix} 358 \text{ ft-kips} - (358 \text{ ft-kips} - 238 \text{ ft-kips}) \\ \times \left(\dfrac{10 \text{ ft} - 4.45 \text{ ft}}{13.0 \text{ ft} - 4.45 \text{ ft}} \right) \end{pmatrix} = 280 \text{ ft-kips}$$

The moment capacity is

$$\phi_b M_n = (0.9)(280 \text{ ft-kips}) = 252 \text{ ft-kips} \quad (250 \text{ ft-kips})$$

The answer is (A).

79. A composite action requires full resistance to sliding at the slab-beam interface. Casting a concrete slab over a steel beam creates a cold joint that is incapable of resisting sliding at the slab-beam interface. Therefore, there is no composite action between the slab and beam. For the beam to be braced, its compression flange should be restrained against lateral-torsional rotation under vertical loads. A cold joint as shown does not provide the required restraint. To ascertain that the slab and beam act together compositely, shear connectors are needed to transfer the horizontal shear at the interface.

The answer is (A).

80. The frame is unstable without the brace. The force, P, that the frame can resist depends on capacity of the diagonal member. Determine the member force, F_{AC}. Assume member AC is in tension. Use equilibrium at point D.

$$\theta = \tan^{-1}\left(\frac{12}{30}\right) = \arctan(0.4) = 21.8°$$

$$F_{AC} \cos\theta = P$$

$$F_{AC}(\cos 21.8°) = 0.93 \; F_{AC} = P$$

The ultimate load, P, depends on the tensile capacity of the diagonal member. The formulas for tension members are given in the *NCEES Handbook*.

For a W10 × 45 section, the cross-sectional area is

$$A_g = 13.3 \text{ in}^2$$

For tension,

$$\phi_y = 0.90$$

Available tensile strength is

$$\phi T_n = \phi_y F_y A_g = (0.90)\left(50 \; \frac{\text{kips}}{\text{in}^2}\right)(13.3 \text{ in}^2) = 598.5 \text{ kips}$$

At yield,

$$F_{AC} = \phi T_n = 598.5 \text{ kips}$$

Therefore,

$$P = 0.93 \; F_{AC} = (0.93)(598.5 \text{ kips}) = 557 \text{ kips} \quad (560 \text{ kips})$$

The answer is (C).

81. The L-beam is subjected to an eccentric loading and is therefore in torsion. Determine the center of gravity of the beam. The magnitude of torsion depends upon the force and the eccentricity with respect to the beam's centroid.

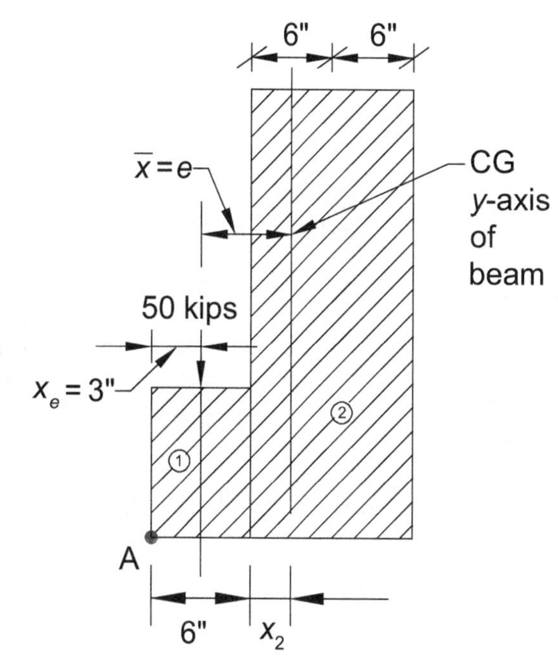

Torsional moment, or torque, is the load multiplied by the eccentricity.

The eccentricity is defined as the shortest distance between the load and the centroid (CG) axis of the L-shaped beam. Use the formula to compute the CG of the entire beam section.

$$x_c = \frac{\int x \, dA}{A}$$

Divide the beam into two rectangles. The equation is

$$\bar{x} = \frac{A_1 \bar{x}_1 + A_2 \bar{x}_2}{A_1 + A_2}$$

Calculate the beam section centroid distance, $\bar{x}$, from the bottom-left corner, A.

	B_i	D_i	$A_i = (B_i)(D_i)$	$\bar{x}_i$	$(A_i)(\bar{x}_i)$
area 1	6 in	10 in	60 in^2	$\frac{6\text{ in}}{2} = 3$ in	180 in^3
area 2	12 in	34 in	408 in^2	$6\text{ in} + \frac{12\text{ in}}{2} = 12$ in	4896 in^3
		$\Sigma =$	468 in^2		5076 in^3
				$\bar{x} =$	10.85 in

The eccentricity is the load offset distance from the beam CG. In this case,

$$e = 10.85\text{ in} - 3\text{ in} = 7.85\text{ in}$$

Torsional moment is

$$\begin{aligned} T &= (P)(e) = (50\text{ kips})(7.85\text{ in}) \\ &= 392.5\text{ in-kips} \quad (390\text{ in-kips}) \end{aligned}$$

The answer is (C).

82. See the Statics section of the *NCEES Handbook.* Using the equilibrium condition, the tension force in the steel rod is equal to the compressive force on the concrete area.

$$P_{\text{steel}} = A_s\sigma_s = (4\text{ in}^2)\left(55\ \frac{\text{kips}}{\text{in}^2}\right) = 220\text{ kips}$$

The concrete area is

$$A_c = \frac{\pi}{4}\left(D_o^2 - D_i^2\right) = \frac{\pi}{4}\left((24\text{ in})^2 - (4\text{ in})^2\right) = 440\text{ in}^2$$

The concrete stress is

$$\sigma_{\text{concrete}} = \frac{220\text{ kips}}{440\text{ in}^2} = 0.5\text{ ksi}$$

The answer is (B).

83. The plasticity index (PI) is the liquid limit (LL) minus the plastic limit (PL).

$$\text{PI} = \text{LL} - \text{PL} = 52 - 35 = 17$$

Use the ASTM D2487-11 Standard Practice for Classification of Soils for Engineering Purposes (Unified Soil Classification System) table and/or graph given in the *NCEES Handbook.* For classification of fine-grained soils, use the USCS plasticity graph.

1. For a PI of 17 and an LL of 52, the choice narrows to two soil types: OH and MH. The symbol "M" is for silty soils, and the symbol "O" is for organic soils.

2. The problem statement describes the soil as a silty type. There is no indication that it is organic silt. So, by process of elimination, the category for the soil noted in the problem is MH. The group symbol "MH" stands for inorganic silts with high plasticity.

The answer is (D).

84. It is correct that the coefficient of permeability has units of velocity. However, when water flows through soil, its velocity is small. Therefore, the velocity head can be ignored in determining permeability, and option B is incorrect. Options A, C, and D are correct.

The answer is (B).

85. The illustration shows the phase diagram.

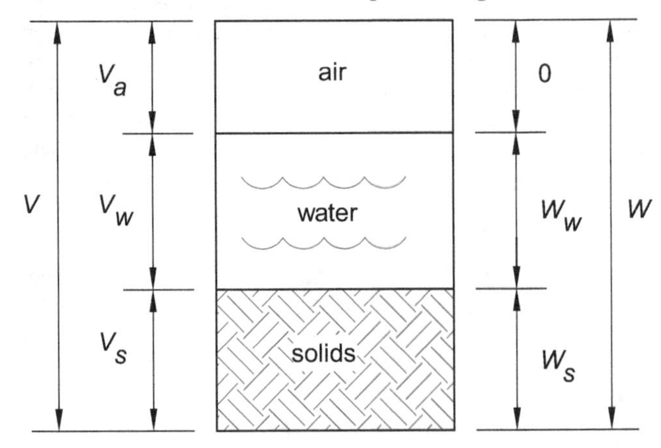

A soil consists of three parts: solids, water, and air voids between them, all combined and arranged in a complex manner. The phase diagram simplifies the complex arrangement by showing the three parts contiguously. The total volume of a soil mass equals the volume of the solids plus the volume of the water and air.

$$V = V_s + V_w + V_a$$

The weight of a soil mass equals the weight of the solids plus the weight of water.

$$W = W_s + W_w$$

The formulas are based on the weight-volume or mass-volume relationships. Options A, C, and D are correct.

The answer is (B).

86. The water content is the ratio of the weight of water, W_w, in a given mass of soil to the weight of soil solids, W_s, in the same mass. The moisture content and degree of saturation are not the same. The water content percentage is

$$\omega(\%) = \frac{W_w}{W_s}(100)$$

The moist sample weight is

$$W_w = 222 \text{ g}$$

The dry sample weight is

$$W_s = 185 \text{ g}$$

The weight of the water in the sample is

$$W_w = 222 \text{ g} - 185 \text{ g} = 37 \text{ g}$$

Therefore,

$$\omega = \frac{W_w}{W_s} = \frac{37 \text{ g}}{185 \text{ g}} = 20\%$$

The answer is (A).

87. The total vertical stress is the vertical normal stress on a horizontal plane. It is the sum of the effective vertical stress and the pore pressure.

The total stress is the effective stress plus the pore water pressure.

$$\sigma = \sigma' + u$$

The total effective stress is

$$\begin{aligned}
\sigma' &= \sigma - u = \sigma - h_u\gamma_w \\
\sigma &= \text{weight of water above point} \\
&\quad + \text{weight of saturated soil above point} \\
\sigma &= \gamma_w h_w + \gamma_{\text{sat}} h_s \\
&= \left(62.4 \ \frac{\text{lbf}}{\text{ft}^3}\right)(20 \text{ ft}) + \left(120 \ \frac{\text{lbf}}{\text{ft}^3}\right)(2 \text{ ft}) \\
&= 1488 \text{ lbf/ft}^2
\end{aligned}$$

The pore pressure is

$$\begin{aligned}
u &= h_w\gamma_w = (2 \text{ ft})\left(62.4 \ \frac{\text{lbf}}{\text{ft}^3}\right) \\
&= 124.8 \text{ lbf/ft}^2 \quad (125 \text{ lbf/ft}^2)
\end{aligned}$$

The soil's effective vertical stress at the footing's bottom level is

$$\begin{aligned}
\sigma' &= \sigma - u = \sigma - h_u\gamma_w = 1488 \ \frac{\text{lbf}}{\text{ft}^2} - 125 \ \frac{\text{lbf}}{\text{ft}^2} \\
&= 1363 \text{ lbf/ft}^2 \quad (1360 \text{ psf})
\end{aligned}$$

The answer is (C).

88. The soil density is 120 pcf. The retaining wall is shown.

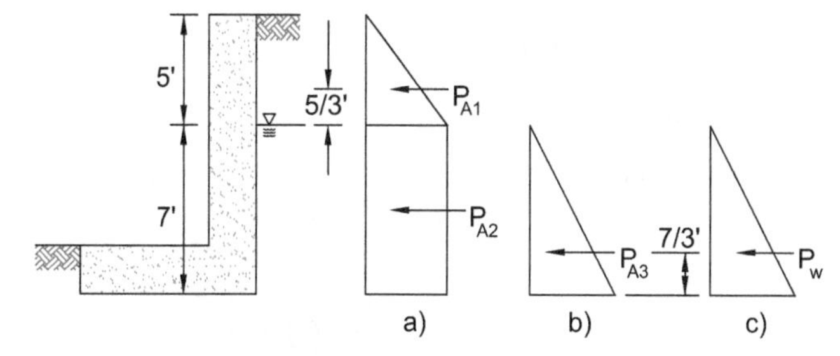

The Rankine coefficient for the resultant active lateral force (lbf) behind the wall is

$$K_A = \tan^2\left(45° - \frac{\phi}{2}\right) = \tan^2\left(45° - \frac{30}{2}\right) = \tan^2(30°) = \frac{1}{3}$$

Given that the heights are 5 ft, 7 ft, and 12 ft, the pore pressure is 62.4 pcf, the submerged unit weight is

$$\gamma' = 120 \ \frac{\text{lbf}}{\text{ft}^3} - 62.4 \ \frac{\text{lbf}}{\text{ft}^3} = 57.6 \text{ lbf/ft}^3$$

The lateral pressure on the wall consists of two parts: the effective horizontal soil pressure and the pore water pressure. The effective part can be further divided into three pressure blocks, as shown in the illustration above.

The total pressure is

$$P_{A(\text{total})} = P_w + \sum P_{Ai}$$

The resultant lateral force on each pressure block is

$$\begin{aligned}
P_{A1} &= \frac{1}{2}\gamma K_A H_1^2 = \left(\frac{1}{2}\right)\left(120 \ \frac{\text{lbf}}{\text{ft}^3}\right)\left(\frac{1}{3}\right)(5 \text{ ft})^2 \\
&= 500 \text{ lbf/ft}
\end{aligned}$$

$$\begin{aligned}
P_{A2} &= \gamma K_A H_1 H_2 \\
&= \left(120 \ \frac{\text{lbf}}{\text{ft}^3}\right)\left(\frac{1}{3}\right)(5 \text{ ft})(7 \text{ ft}) \\
&= 1400 \text{ lbf/ft}
\end{aligned}$$

$$
\begin{aligned}
P_{A3} &= \frac{1}{2}\gamma' K_A H_2^2 \\
&= (0.5)\left(57.6\ \frac{\text{lbf}}{\text{ft}^3}\right)\left(\frac{1}{3}\right)(7\ \text{ft})^2 \\
&= 470\ \text{lbf/ft}
\end{aligned}
$$

$$
\begin{aligned}
P_w &= \frac{1}{2}\gamma_w H_2^2 \\
&= (0.5)\left(62.4\ \frac{\text{lbf}}{\text{ft}^3}\right)(7\ \text{ft})^2 \\
&= 1529\ \text{lbf/ft}
\end{aligned}
$$

The total pressure is

$$
\begin{aligned}
P_{A(\text{total})} &= P_w + \sum P_{Ai} \\
&= 1529\ \frac{\text{lbf}}{\text{ft}} + 500\ \frac{\text{lbf}}{\text{ft}} \\
&\quad + 1440\ \frac{\text{lbf}}{\text{ft}} + 470\ \frac{\text{lbf}}{\text{ft}} \\
&= 3899\ \text{lbf/ft} \quad (3900\ \text{lbf/ft})
\end{aligned}
$$

The answer is (B).

89. The key is generally called a shear key. It increases resistance to sliding, but does not increase the resistance to overturning of the dam. Since the shear key creates an obstruction to the flow in the permeable medium, the total flow decreases and head drop increases. A flow net is shown with flow lines and equipotential lines. Options A, C, and D are correct.

The answer is (B).

90. Since the sample was tested under uniaxial loading, it implies that the sample was tested under unconfined compression. The cross-sectional area of the sample is

$$
A = \frac{\pi D^2}{4} = \frac{\pi(3\ \text{in})^2}{4} = 7.1\ \text{in}^2
$$

$$
q_U = \frac{P}{A} = \frac{499\ \text{lbf}}{7.1\ \text{in}^2} = 70.28\ \text{lbf/in}^2
$$

For clays, unconfined compressive strength is

$$
q_U = 2c
$$

Therefore,

$$
\begin{aligned}
c &= \frac{q_U}{2} = \frac{70.28\ \frac{\text{lbf}}{\text{in}^2}}{2} \\
&= 35.14\ \text{lbf/in}^2 \quad (35\ \text{psi})
\end{aligned}
$$

The answer is (B).

91. The terms *caisson* and *drilled pier* are used interchangeably.

The area of the cross section is

$$
A = \frac{\pi D^2}{4} = \frac{\pi(4\ \text{ft})^2}{4} = 12.57\ \text{ft}^2
$$

The allowable soil pressure is 20 ksf. The caisson allowable bearing capacity is

$$
P_{\text{allow}} = (\sigma_{\text{allow}})(A) = \left(20\ \frac{\text{kips}}{\text{ft}^2}\right)(12.57\ \text{ft}^2) = 251.3\ \text{kips}
$$

The caisson ultimate bearing capacity is

$$
\begin{aligned}
P_{\text{ult}} &= \text{FS} \times P_{\text{allow}} = (3)(251.3\ \text{kips}) \\
&= 753.9\ \text{kips} \quad (750\ \text{kips})
\end{aligned}
$$

The answer is (D).

92. The range of recompression applies to overconsolidated soils. The range of virgin compression applies to normally consolidated soils. The recompression index, C_R, is also known as the swelling index. According to the Phase Relationships section of the *NCEES Handbook*, its slope is $1/6$ of C_C. It is true that as the pressure is increased on a soil sample, the void ratio, e, decreases. However, the decrease is nonlinear, and its curve is called the *e-log p* curve.

The answer is (D).

93. The embankment is shown here.

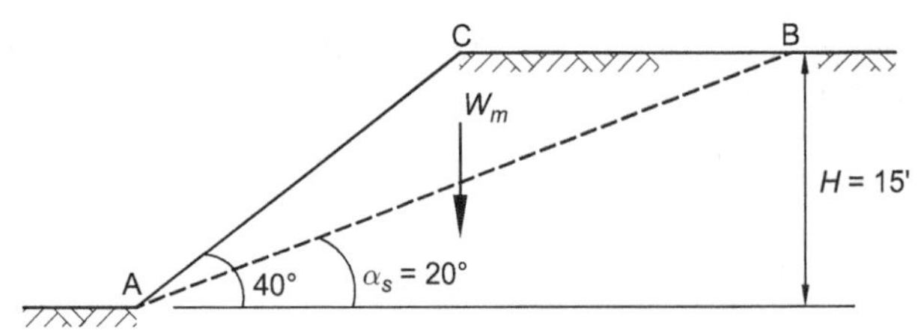

The factor of safety, FS, is

$$
\begin{aligned}
\text{FS} &= \frac{\text{available shearing resistance along slip surface}}{\text{mobilized shear force along slip surface}} \\
&= \frac{T_{\text{FF}}}{T_{\text{MOB}}}
\end{aligned}
$$

The slip angle and slope angle are

$$
\alpha_{\text{slip}} = 20^\circ
$$

$$
\alpha_{\text{slope}} = 40^\circ
$$

The slip plane length is

$$
L_S = \frac{H}{\sin \alpha_S} = \frac{15\ \text{ft}}{\sin 20^\circ} = \frac{15\ \text{ft}}{0.34} = 43.86\ \text{ft}
$$

The triangle ABC in the diagram shows the volume of the soil wedge above the slip plane AB. The triangle ABC is an isosceles triangle. Two equal angles in the triangle are

$$\measuredangle \text{CAB} = \measuredangle \text{CBA} = 20°$$

$$\text{base AB} = 43.86 \text{ ft}$$

$$\begin{aligned} A_{\Delta\text{ABC}} &= \frac{1}{2}(\text{AB})(\perp\text{C to AB}) \\ &= \frac{1}{2}(43.86 \text{ ft})\left(\frac{43.86 \text{ ft}}{2}\right)\tan(20°) \\ &= 175.04 \text{ ft}^2 \end{aligned}$$

The soil wedge mass is

$$\begin{aligned} W_M &= A_{\Delta\text{ABC}} \text{ (soil density)} = 175.04 \text{ ft}^2\left(120 \ \frac{\text{lbf}}{\text{ft}^3}\right) \\ &= 21{,}005 \text{ lbf/ft length} \end{aligned}$$

The available shearing resistance along the slip surface is

$$\begin{aligned} T_{\text{FF}} &= cL_S + W_M \cos\alpha_S \tan\phi \\ &= \left(300 \ \frac{\text{lbf}}{\text{ft}^2}\right)(43.86 \text{ ft}) + \left(21{,}005 \ \frac{\text{lbf}}{\text{ft}}\right)(\cos 20°)(\tan 30°) \\ &= 24{,}554 \text{ lbf/ft} \end{aligned}$$

The mobilized shear force along the slip surface is

$$T_{\text{MOB}} = W_M \sin\alpha_S = 21{,}005 \ \frac{\text{lbf}}{\text{ft}}(\sin 20°) = 7184 \text{ lbf/ft}$$

The factor of safety is

$$\text{FS} = \frac{24{,}554}{7184} = 3.4$$

The answer is (D).

94. The roadway can be shown as

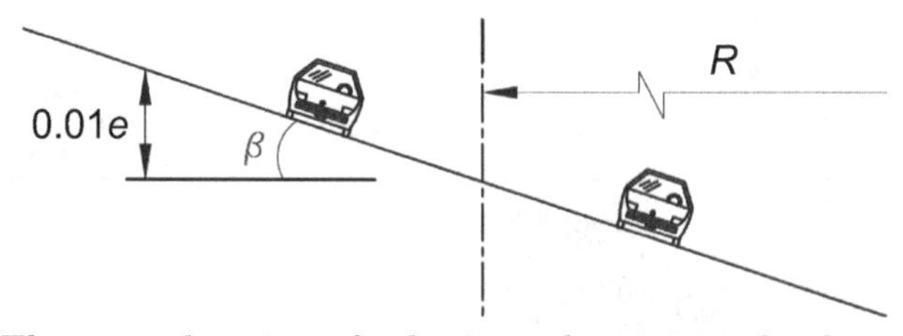

The superelevation of a horizontal curve can be determined by the formula given in the Horizontal Curves table of the *NCEES Handbook*.

$$0.01e + f = \frac{V^2}{15R}$$

In this formula, the variables have been adjusted so that the car speed, V, is in miles per hour, and the horizontal curve radius is in feet. The term e is the percentage superelevation.

The following were given in the problem statement.

- vehicle velocity, $V = 55$ mph
- curve radius, $R = 1000$ ft
- friction factor, $f = 0.10$

Therefore,

$$\begin{aligned} e &= \frac{\frac{V^2}{15R} - f}{0.01} = \frac{\frac{\left(55 \ \frac{\text{mi}}{\text{hr}}\right)^2}{(15)(1000 \text{ ft})} - 0.10}{0.01} \\ &= 10\% \end{aligned}$$

This means that the superelevation slope is 10%.

The answer is (D).

95. From the Stopping Sight Distance section in the *NCEES Handbook*, the formula for the stopping sight distance, SSD, is

$$\text{SSD} = 1.47Vt + \frac{V^2}{30\left(\left(\frac{a}{32.2}\right) \pm G\right)}$$

$$\begin{aligned} V &= 55 \text{ mph} \\ t &= 2.5 \text{ sec} \\ G &= -0.04 \end{aligned}$$

The deceleration rate, a, is

$$a = 10 \text{ ft/sec}^2$$

The stopping sight distance is

$$\begin{aligned} \text{SSD} &= 1.47\left(55 \ \frac{\text{mi}}{\text{hr}}\right)(2.5 \text{ sec}) \\ &\quad + \frac{\left(55 \ \frac{\text{mi}}{\text{hr}}\right)^2}{30\left(\left(\frac{10 \ \frac{\text{ft}}{\text{sec}^2}}{32.2 \ \frac{\text{ft}}{\text{sec}^2}}\right) - 0.04\right)} \\ &= 202 \text{ ft} + 372 \text{ ft} \\ &= 574 \text{ ft} \quad (575 \text{ ft}) \end{aligned}$$

The answer is (B).

96. In mountainous terrains that have long and large continuous uphill slopes, heavier vehicles do not have adequate horsepower to maintain the speed that lighter vehicles can. Therefore, the design speed for heavier vehicles is reduced in such areas. The statement that the posted speed is increased as the grade change becomes longer and larger is incorrect. Options A, B, and D are correct.

The answer is (C).

97. In the AASHTO pavement design method, a single axle load of 18,000 lbf is defined as a single unit called a *load equivalency factor* (LEF). LEFs are tabulated in the Highway Pavement Design section of the *NCEES Handbook.*

The truck factor, TF, is defined as the number of LEFs caused by a single passage of a vehicle.

$$\text{TF} = \sum \begin{pmatrix} \text{(number of axles)} \\ \text{(load equivalency factor per axle)} \end{pmatrix}$$

The LEFs for various axle loadings are tabulated in the *NCEES Handbook.*

For 18,000 lbf single axles, LEF = 1.0.

For 30,000 lbf tandem axles, LEF = 0.658.

For 20,000 lbf single axles, LEF = 1.51.

The truck factor is

$$\begin{aligned} \text{TF} &= 1.0 + (2)(0.658) + 1.51 \\ &= 3.83 \quad (3.8) \end{aligned}$$

The answer is (C).

98. The levels of service, designated A through F, were determined subjectively. Level of service A denotes free flow. Level of service F denotes forced or breakdown flow, which is the most congested flow.

The answer is (C).

99. See the Transportation section of the *NCEES Handbook.* The length of the red clearance interval is the time needed for a vehicle to clear an intersection after it enters the intersection. To clear the intersection, the entire length of the vehicle must be out of the intersection, as shown in the illustration.

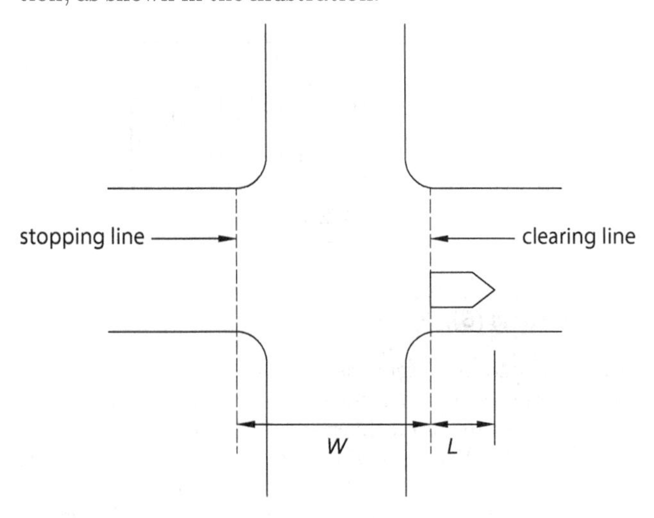

The red clearance time interval, r, can be computed as

$$r = \frac{W + L}{v}$$

The vehicular speed is

$$v = 35 \text{ mph} = \left(35 \ \frac{\text{mi}}{\text{hr}}\right)\left(1.47 \ \frac{\frac{\text{ft}}{\text{sec}}}{\frac{\text{mi}}{\text{hr}}}\right) = 51.5 \text{ ft/sec}$$

The red clearance time interval is

$$\begin{aligned} r &= \frac{60 \text{ ft} + 19 \text{ ft}}{51.5 \ \frac{\text{ft}}{\text{sec}}} \\ &= 1.53 \text{ sec} \quad (1.5 \text{ sec}) \end{aligned}$$

The answer is (B).

100. The traffic flow, q, is defined as the equivalent hourly rate at which vehicles pass through a point on a roadway. It is expressed as

$$q = \frac{(n)(60)}{t}$$

n is the number of vehicles passing a point in t minutes.

The traffic flow is

$$q = \frac{(15 \text{ vehicles})\left(60 \ \frac{\text{min}}{\text{hour}}\right)}{10 \text{ min}} = 90 \text{ vehicles per hour}$$

The answer is (B).

101. The traffic density, k, is the number of vehicles traveling over a unit length (generally one mile) of roadway at a selected time. It is expressed as

$$\begin{aligned} k &= \frac{(\text{number of vehicles in the section length})\left(5280 \ \frac{\text{ft}}{\text{mi}}\right)}{\text{length of section}_{\text{ft}}} \\ &= \frac{(6 \text{ vehicles})\left(5280 \ \frac{\text{ft}}{\text{mi}}\right)}{200 \text{ ft}} \\ &= 158 \text{ vehicles per mile} \quad (160 \text{ vehicles per mile}) \end{aligned}$$

The answer is (B).

102. Average daily traffic (ADT) and average annual daily traffic (AADT) are correctly defined. They are used in evaluating current demand and existing traffic flow. Therefore, options A and B are correct.

Peak hour volume (PHV) is the maximum number of vehicles that pass a point during a period of 60 consecutive minutes. PHV is used for capacity analysis and design of geometric characteristics of a highway, such as signalization or channelization. Therefore, option D is correct.

Vehicle miles of travel (VMT) is a measure of travel along a section of a road. It is defined as

$$\begin{aligned}\text{VMT} &= (\text{traffic volume}) \\ &\quad \times \begin{pmatrix}\text{length of roadway in miles} \\ \text{to which the volume is applicable}\end{pmatrix}\end{aligned}$$

VMT is used as a base for allocating resources for highway maintenance and improvement. The definition of VMT in option C is incorrect.

The answer is (C).

103. Various defaults or delays of construction are possible during the construction phase. For example, a bidder may realize that he or she has underbid the job and may not start construction, or a contractor may not have means to complete the job on time. The concept of a guarantee allows an owner to protect itself against defaults or delays caused by a bidder or contractor. The guarantee is called a *bond*. The owner is called an obligee. The contractor is called the principal, and the surety company is the guarantor or obligor. The engineer working on the project is not a party to the bond.

The answer is (D).

104. *Retainage* is defined as a portion of the agreed upon contract price deliberately withheld until the work is substantially complete, to assure that the contractor or subcontractor will satisfy its obligations and complete a construction project.

The retained amount is 10% of the construction work claimed.

$$A_{\text{retained}} = \$2{,}000{,}000(95\%)(10\%) = \$190{,}000$$

The cost to complete the 5% remaining work is

$$C_{5\%} = (100\% - 95\%)\$2{,}000{,}000 = \$100{,}000$$

At this stage, the expected cost to complete the project is \$100,000, and the owner is holding \$190,000, which is more than the amount needed to complete the work without additional retainage. However, the owner is entitled to retain the amount agreed upon in the contract. For the remainder of the work, the additional retainage required is

$$A_{\text{extra}} = 10\% \text{ of } \$100{,}000 = \$10{,}000$$

The answer is (B).

105. According to 29 CFR 1926, the top edge of top rail or equivalent guardrail system member will be 42 in (within 3 in) above the working or walking level. Thus, the required height ranges from 39 in to 45 in.

The answer is (D).

106. The measures to reduce formwork costs include options A through C. Commercially available forms are less expensive to use than a specially developed form. Repeated use of a form also reduces the cost. Similarly, use of available lumber sizes is economical. For example, the actual size of a 2 × 4 is

$$1\frac{1}{2} \text{ in} \times 3\frac{1}{2} \text{ in}$$

Specifying a 1.5 in wide shear key so that the commercially available size can be used is less expensive than specifying a 2 in wide shear key. Designing member sizes for which formwork is commercially unavailable would require detailing and developing the new formwork, which would in turn add unnecessary cost to the project.

The answer is (D).

107. A critical path is defined as the longest path that connects activities from start to finish. Float is always zero for activities on the critical path. In a CPM diagram, an activity is represented by an arrow, and an event is represented by a node. For an event to happen, all activities culminating at the node must be completed. The terms *activity* and *event* are not synonymous.

The answer is (B).

108. The illustration shows the concrete wall dimensions.

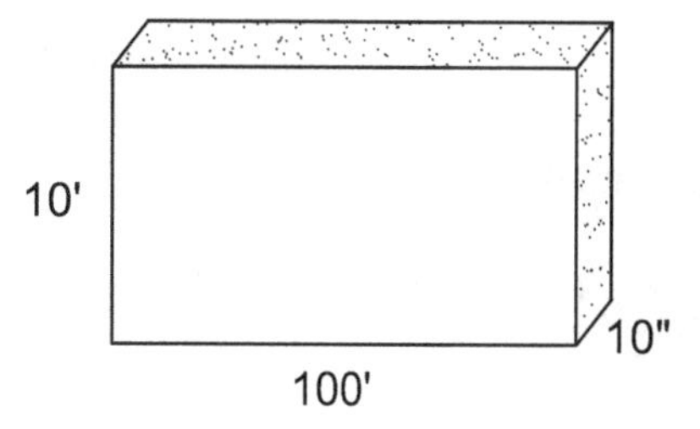

The volume of the wall is

$$v_{\text{wall}} = (100 \text{ ft})(10 \text{ ft})\left(\frac{10 \text{ in}}{12 \frac{\text{in}}{\text{ft}}}\right)\left(\frac{1 \text{ yd}^3}{27 \text{ ft}^3}\right) = 30.86 \text{ yd}^3$$

Considering 10% waste, the volume is

$$v_{10\%} = \left(30.86\ \text{yd}^3\right)(1.10) = 33.95\ \text{yd}^3$$

The cost of the concrete is

$$\begin{aligned}\text{cost} &= \left(33.95\ \text{yd}^3\right)\left(\frac{\$100}{\text{yd}^3}\right) \\ &= \$3395 \quad (\$3400)\end{aligned}$$

The answer is (B).

109. From the ASTM Standard Reinforcement Bars table in the *NCEES Handbook*, the weight of a #9 rebar is

$$W_{\#9} = 3.4\ \text{lbf/ft}$$

The weight of all the vertical rebar is

$$\begin{aligned}W_{\text{all}} &= \left(\frac{8\ \text{bars}}{\text{col}}\right)(100\ \text{columns}) \\ &\quad \times (12.75\ \text{ft})\left(3.4\ \frac{\text{lbf}}{\text{ft}}\right)\left(\frac{1\ \text{ton}}{2000\ \text{lbf}}\right) \\ &= 17.34\ \text{tons} \quad (17.3\ \text{tons})\end{aligned}$$

The answer is (B).

110. Since the delay time is 10 min every hour, the time available for hauling is 50 min per hour.

The time taken to complete one hauling trip is 20 min.

The number of trips in 50 min is

$$\text{trips}_{\text{hr}} = \frac{50\ \frac{\text{min}}{\text{hr}}}{20\ \frac{\text{min}}{\text{trip}}} = 2.5\ \text{trips/hr}$$

Considering the delay time, a truck makes 2.5 trips per hour.

The number of trips per day is

$$\text{trips}_{\text{day}} = \left(2.5\ \frac{\text{trips}}{\text{hr}}\right)\left(8\ \frac{\text{hr}}{\text{day}}\right) = 20\ \text{trips/day}$$

The answer is (B).

Solutions

Exam 2

111. Geothermal energy is generated when water comes in contact with heated underground rocks. The heat turns the water into steam, which is used to generate power. In this problem, the thermal gradient, m, is constant because the thermal variation along the depth is linear. Calculate the slope, m.

$$\begin{aligned} m &= \frac{y_2 - y_1}{x_2 - x_1} = \frac{\text{temperature difference}}{\text{depth difference}} \\ &= \frac{90°\text{C} - 20°\text{C}}{2000\ \text{m} - 0\ \text{m}} \\ &= 0.035°\text{C/m} \end{aligned}$$

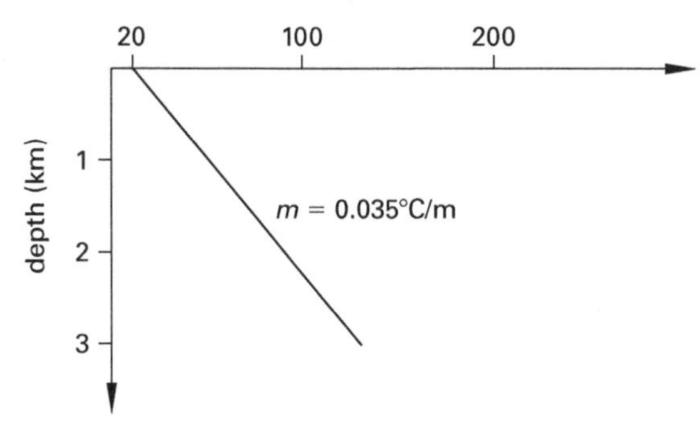

Use the point-slope form of the straight-line equation.

$$\begin{aligned} y - y_1 &= m(x - x_1) \\ T_{3800} - T_0 &= m(3800\ \text{m} - 0\ \text{m}) \\ T_{3800} - 20°\text{C} &= (0.035°\text{C/m})(3800\ \text{m} - 0\ \text{m}) \end{aligned}$$

The temperature at a depth of 3800 m is

$$\begin{aligned} T_{3800} &= (0.035°\text{C/m})(3800\ \text{m}) + 20°\text{C} \\ &= 153°\text{C} \end{aligned}$$

The answer is 153°C.

112. The equations are solved using Cramer's rule, which uses determinants. First, write the coefficients of the variables as they appear in the system of three equations as determinant A.

$$A = \begin{vmatrix} 1 & 2 & -1 \\ 2 & 3 & 2 \\ 1 & -2 & -2 \end{vmatrix}$$

The determinant of the matrix is 17. According to Cramer's rule, if A is not zero, replace the first column in A with the system answers (the column on the right side of the equation signs) as determinant B.

$$\begin{aligned} B &= \begin{vmatrix} -7 & 2 & -1 \\ -3 & 3 & 2 \\ 3 & -2 & -2 \end{vmatrix} \\ &= -7\begin{vmatrix} 3 & 2 \\ -2 & -2 \end{vmatrix} - 2\begin{vmatrix} -3 & 2 \\ 3 & -2 \end{vmatrix} - 1\begin{vmatrix} -3 & 3 \\ 3 & -2 \end{vmatrix} \\ &= (-7)\big((3)(-2) - (2)(-2)\big) - (2)\big((-3)(-2) - 3(2)\big) \\ &\quad + (-1)\big((-3)(-2) - (3)(3)\big) \\ &= 17 \end{aligned}$$

$$x = \frac{B}{A} = \frac{17}{17} = 1$$

The answer is 1.

113. The line represented by the equation is shown.

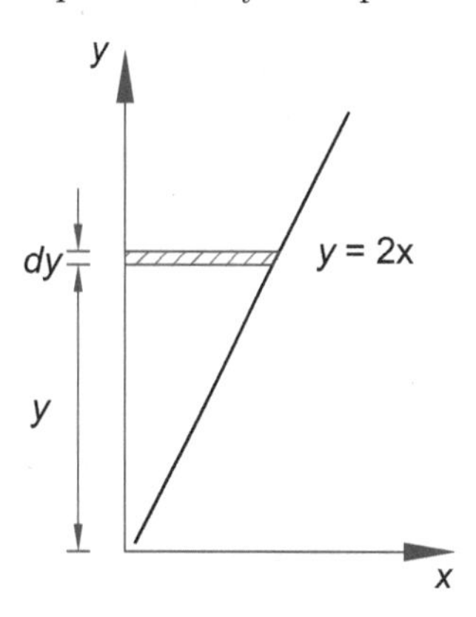

The rotation of the line about its y-axis would generate a conical shape. The area of a slice of width x rotated about the y-axis is

$$A = \pi x^2$$

The volume of the slice is

$$\begin{aligned} dV &= \pi x^2 dy = \pi\left(\frac{y}{2}\right)^2 dy \\ &= \frac{\pi y^2}{4} dy \end{aligned}$$

The volume between $y = 0$ and $y = 10$ is

$$V = \int dV = \int_0^{10} \frac{\pi y^2}{4} dy$$
$$= \left(\frac{\pi}{4}\right)\left(\frac{y^3}{3}\right)\Bigg|_0^{10}$$
$$= \left(\frac{\pi}{12}\right)(10^3 - 0^3)$$
$$= 250\pi/3$$

Shortcut: The solution can be arrived at by using the cone volume formula given in the Mensuration of Areas and Volumes section of the *NCEES Handbook*.

The answer is (C).

114. Find the lot area.

$$\text{present lot area, } A = (50 \text{ m})(30 \text{ m}) = 1500 \text{ m}^2$$

$$\text{proposed lot area} = 2A = 3000 \text{ m}^2$$

The size is doubled by adding strips of equal widths to each dimension.

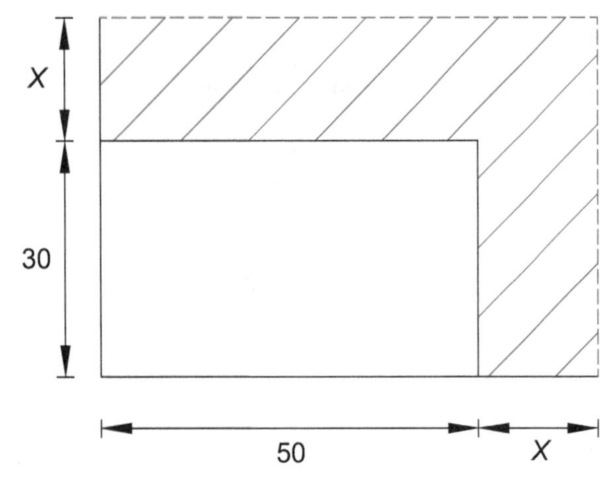

The new length, L, equals $50 + x$.

The new width, W, equals $30 + x$.

Find the value of x from the area equation with new dimensions.

$$(50 + x)(30 + x) = 3000$$
$$x^2 + 80x - 1500 = 0$$
$$x = 15.68 \text{ and } -95.68$$

Since the strips have positive widths, the value of x is

$$x = 15.68 \text{ m} \quad (16 \text{ m})$$

The answer is 16 m.

115. To determine whether the series is a geometric progression, divide each number starting with the second number by the number preceding it. If the quotients are equal, the series is geometric.

$$\text{second number/first number} = (-1/3)/1 = -1/3$$

$$\text{third number/second number} = (1/9)/(-1/3) = -1/3$$

The quotients are equal. Therefore, it is a geometric progression series.

The first term, a, is 1.

$$\text{common ratio, } r = \frac{i^{\text{th}} \text{ term}}{(i-1)^{\text{th}} \text{ term}} = \left(-\frac{1}{3}\right) \div (1) = -\frac{1}{3}$$

For $r < 1$, the sum is

$$S_n = \frac{a}{1-r} = \frac{1}{1-\left(-\frac{1}{3}\right)} = \frac{3}{4}$$

The answer is 3/4.

116. The velocity, v, is defined as the distance travelled per unit time.

$$\text{v} = \frac{ds}{dt} = \frac{d}{dt}\left(\frac{t^2+1}{t+1}\right)$$

Use the quotient rule.

$$\frac{d\left(\frac{u}{\text{v}}\right)}{dx} = \frac{\text{v}\left(\frac{du}{dx}\right) - u\left(\frac{d\text{v}}{dx}\right)}{\text{v}^2}$$

Let $u = t^2 + 1$ and $\text{v} = t + 1$.

$$\frac{du}{dt} = \frac{d(t^2+1)}{dt} = 2t$$

$$\frac{d\text{v}}{dt} = \frac{d(t+1)}{dt} = 1$$

Substitute these values into the velocity equation.

$$\text{v} = \frac{(t+1)(2t) - (t^2+1)(1)}{(t+1)^2} = \frac{(t^2+2t-1)}{(t+1)^2}$$

For $t = 2.5$ sec, find the velocity.

$$\text{v} = \frac{((2.5 \text{ sec})^2 + 2(2.5 \text{ sec}) - 1)}{(2.5 \text{ sec} + 1)^2} = \frac{10.25}{12.25} = 0.837 \quad (0.84)$$

The answer is (B).

117. The three vectors are

$$\vec{A} = 0\,\mathbf{i} + 4\,\mathbf{j} + 0\mathbf{k}$$

$$\vec{B} = 4\,\mathbf{i} + 8\,\mathbf{j} + 0\mathbf{k}$$

$$\vec{C} = 1\,\mathbf{i} + 2\,\mathbf{j} + 3\mathbf{k}$$

The vectors and parallelepiped bounded by the vectors are shown.

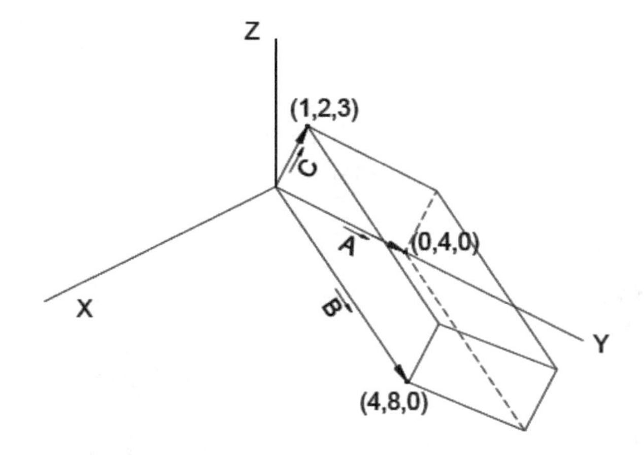

The volume, V, of the parallelepiped is height × base area.

$$V = |\vec{A} \cdot (\vec{B} \times \vec{C})|$$

Use cross product and then dot product. The two steps are combined.

$$\vec{B} \times \vec{C} = \begin{vmatrix} i & j & k \\ b_x & b_y & b_z \\ c_x & c_y & c_z \end{vmatrix}$$

$$\begin{aligned} V = \vec{A} \cdot (\vec{B} \times \vec{C}) &= \begin{vmatrix} a_x & a_y & a_z \\ b_x & b_y & b_z \\ c_x & c_y & c_z \end{vmatrix} = \begin{vmatrix} 0 & 4 & 0 \\ 4 & 8 & 0 \\ 1 & 2 & 3 \end{vmatrix} \\ &= |-(4)(4(3) - 0(1))| \\ &= 48 \end{aligned}$$

The answer is 48.

118. The order in which workers are combined is not considered in this group formation, so this is a combination problem: n objects taken r at a time. The combination is expressed as

$$C(n,r) = \frac{n!}{(n-r)!\,r!}$$

In this case, one man is selected from eight men, so the combination is $C(8,1)$. Four women are selected from 12 women, so the combination is $C(12,4)$. Find the number of combinations.

$$\begin{aligned} N = C(8,1) \times C(12,4) &= \left(\frac{8!}{(8-1)!\,1!}\right)\left(\frac{12!}{(12-4)!\,4!}\right) \\ &= \left(\frac{8!}{7!\,1!}\right)\left(\frac{12!}{8!\,4!}\right) \\ &= 3960 \end{aligned}$$

The answer is 3960.

119. Use the weighted arithmetic mean formula given in the Engineering Probability and Statistics section of the *NCEES Handbook*. The weighted arithmetic mean, $\overline{X}_w$, is the average wage.

$$\begin{aligned} \overline{X}_w &= \frac{\sum w_i X_i}{\sum w_i} \\ &= \frac{(10)\left(\frac{\$10}{\text{hr}}\right) + (10)\left(\frac{\$30}{\text{hr}}\right) + (15)\left(\frac{\$25}{\text{hr}}\right) + (20)\left(\frac{\$12}{\text{hr}}\right) + (20)\left(\frac{\$15}{\text{hr}}\right) + (25)\left(\frac{\$20}{\text{hr}}\right)}{10 + 10 + 15 + 20 + 20 + 25} \\ &= \$18.15/\text{hr} \quad (\$18/\text{hr}) \end{aligned}$$

The answer is (B).

120. The coefficient of variation is generally expressed as a percentage rather than in terms of the units of the particular data. The sample coefficient of variation, CV, is

$$\begin{aligned} \text{CV} &= \frac{s}{\overline{X}} = \frac{6.6}{30} \\ &= 0.22 \quad (22\%) \end{aligned}$$

The answer is (C).

121. Ethics are not fixed, but vary with time, location, and culture.

The answer is (A), (B), (C), and (D).

122. In some states, an oral contract for design services is unenforceable, but in this case the oral contract between owners and engineers is deemed valid. In general, it is good business practice to insist on a written contract. The engineer's report should be objective and truthful, even if it goes against the client's interest.

The answer is (D).

123. Meeting the minimum requirements of a building code may not yield sound design documents. The documents must conform to the accepted engineering standards in order to safeguard the life, health, property, and welfare of the public. According to Model Rules, Section 240.15, Rules of Professional Conduct, licensees shall sign and seal only those plans, surveys, and other documents that conform to accepted engineering and surveying standards and that safeguard the health, safety, and welfare of the public.

Further, according to Model Law, Section 150.10, Grounds for Disciplinary Action—Licensees and Interns, "The board shall have the power to suspend, revoke, place on probation, fine, recover costs, and/or reprimand, or to refuse to issue, restore, or renew a license or intern certification to any licensee or intern that is found guilty of signing, affixing, or permitting the licensee's seal or signature to be affixed to any specifications, reports, drawings, plans, plats, design information, construction documents or calculations, surveys, or revisions thereof which have not been prepared by the licensee or under the licensee's responsible charge."

The answer is (B) and (E).

124. Distribution of calendars showing photos of completed projects to clients is generally considered an acceptable form of advertising. An exception is using a project that the client requires the design firm to keep confidential or not use in the firm's advertisement material. The problem description does not allude to any restrictions placed on the firm to publish the photos of the projects. Use of self-laudatory language is permissible in advertising. Social media advertising is now considered the most effective form of advertising. However, luring in prospective clients by offering tickets to the Super Bowl is a violation. NCEES code prohibits giving any commission, gift, or other valuable consideration in order to secure work.

The answer is (C).

125. An offer to provide repair services does not need to be in writing, so the verbal offer was valid when made. The general rule is that the person making an offer can withdraw the offer before it is accepted. In this case, an offer was made, but there was no acceptance. Therefore, the parties did not enter into a binding contract. The contractor was under no obligation to keep the offer open. The homeowner did not accept the offer, and the contractor withdrew it.

The answer is (B).

126. The problem involves a single lump sum investment of \$100 now. The investment would grow at a rate of 10% compounded annually. Therefore, the problem requires converting the present payment value, P, to its future worth, F, after n interest periods at an interest rate of i over the entire period.

$$\begin{aligned} P &= \$100 \\ n &= 25 \text{ years} \\ i &= 10\% \text{ per year} \end{aligned}$$

Use the relationship below and the factor table in the Engineering Economics section of the *NCEES Handbook*:

$$\begin{aligned} F &= P(F/P, 10\%, 25) = (\$100)(10.8347) \\ &= \$1084 \quad (\$1100) \end{aligned}$$

The answer is (C).

127. In this transaction, 80% of the car price is being financed.

$$\text{loan amount} = (0.8)(\$30{,}000) = \$24{,}000$$

The monthly installment, A, is \$599. Find the A/P factor.

$$\left(\frac{A}{P}, i, 60\right) = \frac{\$599}{\$24{,}000} = 0.025$$

The interest, i, is the monthly interest. Determine the interest rate where the A/P factor equals 0.025 for $n = 60$.

$$\left(\frac{A}{P}, 1\%, 60\right) = 0.0222$$

$$\left(\frac{A}{P}, 1.5\%, 60\right) = 0.0254$$

Find the interest rate by interpolation.

$$\begin{aligned} i &= 1\% + \left(\frac{0.025 - 0.0222}{0.0254 - 0.0222}\right) \times (1.5\% - 1\%) \\ &= 0.0144 \\ &= 1.44\% \end{aligned}$$

$$\text{nominal annual rate} = 12 \times 1.44\% = 17.28\%$$

$$\begin{aligned} \text{effective annual rate} &= (1+i)^{12} - 1 = \left(1 + \frac{0.1728}{12}\right)^{12} - 1 \\ &= 18.72\% \quad (19\%) \end{aligned}$$

The answer is (D).

128. Assume that the concrete mix contains A doses of admixture A and B doses of admixture B. The mix cost is

$$W = \left(\frac{\$4}{\text{dose}}\right)(A \text{ doses of admixture A}) + \left(\frac{\$12}{\text{dose}}\right)(B \text{ doses of admixture B})$$

The mix requires both admixtures. To keep the mix cost to a minimum, three conditions must be satisfied.

$$3 \leq A \leq 9$$
$$B \geq 8$$
$$A + B \leq 16$$

Graph the conditions as shown in the illustration. The shaded area bound by the three points marked 1, 2, and 3 represents the limits that satisfy the inequalities. The mix cost, W, lies within or at the boundaries of the shaded area. The value of W closest to the origin (0,0) would yield the minimum mix cost. Compute the cost for the three corners of the shaded area. By inspection, the corner (3,8) is located closest to the origin (0,0) and provides the least-expensive mix design.

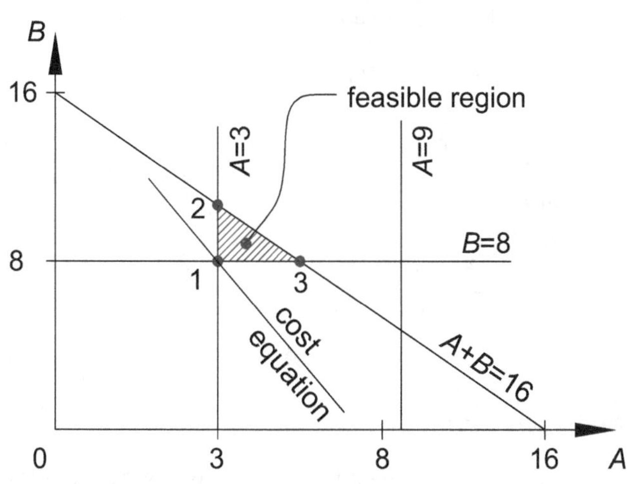

The minimum cost of the mix is

$$W_{\text{min}} = (3)(\$4) + (8)(\$12) = \$108$$

The answer is (B).

129. This is an optimization problem in which the effect of each toll fee increment must be evaluated. From the Differential Calculus section of Mathematics in the *NCEES Handbook*,

$$y = f(x) \text{ is maximum for } x = a, \text{if } f'(a) = 0 \text{ and } f''(a) < 0$$

Let x be the number of \$0.25 fee increases needed to fully optimize the revenue. The new fee is

$$\text{new fee} = \$3.00 + (\$0.25)x$$

The number of motorists is

$$\text{number of motorists} = 20{,}000 - 1000x$$

Find the toll fee increase from the equation for new income, I.

$$\begin{aligned} I &= (\$3.00 + (\$0.25)x)(20{,}000 - 1000x) \\ &= 60{,}000 - 3000x + 5000x - 250x^2 \\ &= 60{,}000 + 2000x - 250x^2 \\ f'(I) &= \frac{dI}{dx} \\ 0 &= 0 + 2000 - 500x \\ x &= 4 \text{ quarters} \quad (\$1.00) \end{aligned}$$

Check the second condition.

$$f''(I) = -500 < 0$$

Both maxima conditions are met. The income is maximized when the toll is increased by a dollar. A graph of the optimal toll increase versus income/revenue is shown.

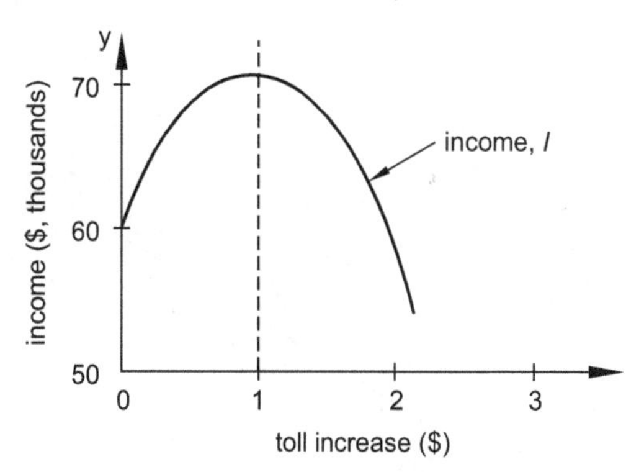

The answer is (A).

130. The problem involves a single lump sum investment of \$100,000 now. The investment would lose its value at a rate of 10% compounded annually. Therefore, the problem requires converting the present value, P, to its future worth, F, after n interest periods at an interest rate of i per period. The interest rate is negative here and the value would decline since the equipment is aging and depreciating.

$$\begin{aligned} P &= \$100{,}000 \\ n &= 10 \text{ yrs} \\ i &= -10\% \text{ per yr} \\ &= -0.1 \end{aligned}$$

There is no factor table provided in the *NCEES Handbook* for depreciation or declining rates. Therefore, use the relationship given and find the future worth.

$$\begin{aligned} F &= P(F/P, -10\%, 10) = P(1+i)^n \\ &= 100{,}000(1-0.1)^{10} \\ &= \$34{,}868 \quad (\$35{,}000) \end{aligned}$$

The answer is (B).

131. If Gr8 decides not to drill, it will sell its right to drill to the wildcatter for a fixed sum of \$1,000,000, and receive another \$1,000,000 contingent on presence of oil.

If oil is struck, then Gr8 would collect a total of \$2,000,000 from the wildcatter.

If no oil is struck, then Gr8 would collect \$1,000,000 for sale of its drilling rights. Its contingency clause would be worthless.

The expected maximum payoff is \$2,000,000.

The answer is \$2,000,000.

132. A force is a vector and is defined by its magnitude and direction. The vectors are added according to the parallelogram law. A parallelogram with sides A and B is drawn.

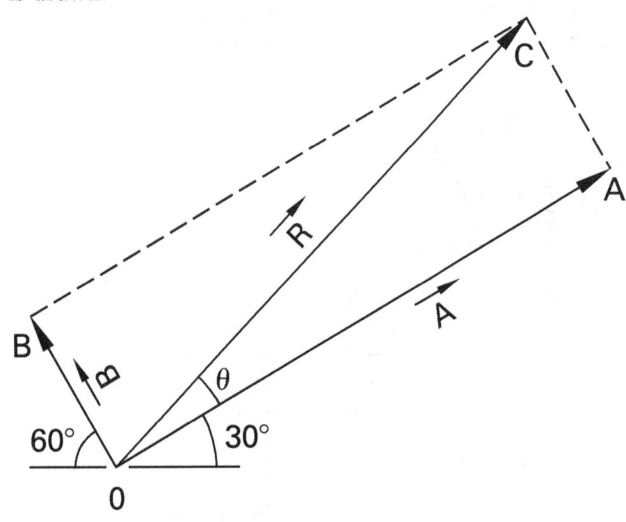

From the data, the included angle between $\vec{A}$ and $\vec{B}$ is 90°, and triangle OAC is a right triangle. The magnitude of the resultant R is computed.

$$\begin{aligned} R^2 &= A^2 + B^2 \\ \vec{R} &= \sqrt{(100\ \text{N})^2 + (30\ \text{N})^2} = 104.4\ \text{N} \quad (104\ \text{N}) \end{aligned}$$

Find the angle of the resultant force from the x-axis.

$$\begin{aligned} \tan\theta &= \frac{30\ \text{N}}{100\ \text{N}} = 0.3 \\ \theta &= \tan^{-1} 0.3 \\ &= 16.7^\circ \text{ from } \vec{A} \\ \text{clockwise angle from } x\text{-axis} &= 30^\circ + 16.7^\circ \\ &= 46.7^\circ \quad (47^\circ) \end{aligned}$$

The answer is 104 N and 47° from the x-axis.

133. See the Statics section of the *NCEES Handbook.* Two conditions must be met for equilibrium.

$$\sum F_n = 0$$
$$\sum M_n = 0$$

The block weighs 1200 lbf. The weight acts vertically down at the centroid of the block, which is located midway between points A and B with the block in a level position, as shown. The weight is equally split between the two corners A and B. Corner A bears against the surface below, and corner B is carried by the crowbar.

Therefore, the load on the tip of the crowbar at B is

$$F_B = (0.5)(1200\ \text{lbf}) = 600\ \text{lbf}$$

The crowbar has two arms, BC and CD. Assuming counterclockwise moment is positive, consider the moment at point C.

$$\begin{aligned} \sum M_C &= 0 \\ F_B(\text{BC}) - F_D(\text{CD}) &= (600\ \text{lbf})(2\ \text{ft}) - F_D(6\ \text{ft}) = 0 \\ F_D &= 200\ \text{lbf} \end{aligned}$$

The answer is (B).

134. A force with its arrow pointing toward a joint denotes a compressive force, and a force with its arrow pointing away from a joint denotes a tensile force. Assume that forces F_2 and F_3 are positive. From equilibrium, the sums of force components in the x- and y-directions are zero.

$$\sum F_x = F_1\cos\theta - F_4\cos\theta + F_3 = 0$$
$$\sum F_y = F_1\sin\theta - F_4\sin\theta - F_2 = 0$$

The angle θ is determined from the member slope of 3:4.

$$\sin\theta = \frac{3}{5}$$
$$\cos\theta = \frac{4}{5}$$

Substitute the values into the force equation for the x-axis.

$$(10\ \text{kN})\left(\frac{4}{5}\right) - (20\ \text{kN})\left(\frac{4}{5}\right) + F_3 = 0$$
$$F_3 = 8\ \text{kN}$$

The calculations show that force F_3 is positive and therefore tensile.

Similarly, substitute the values into the force equation for the y-axis.

$$(10\ \text{kN})\left(\frac{3}{5}\right) - (20\ \text{kN})\left(\frac{3}{5}\right) - F_2 = 0$$
$$F_2 = -6\ \text{kN}$$

The negative sign shows that F_2 is a compressive force.

The answer is

	force F_2 (kN)	force F_3 (kN)
tension or compression?	compression	tension
magnitude	6	8

135. The moment is given by the equation

$$M = (\text{force})(\text{lever arm})$$

The force, F, is 200 N. The lever arm, d, is the distance from the base to the center of the billboard where the total wind force acts.

$$d = 5\ \text{m} + \frac{1\ \text{m}}{2} = 5.5\ \text{m}$$

Find the moment at the base of the pole.

$$M = Fd = (200\ \text{N})(5.5\ \text{m}) = 1100\ \text{N}\cdot\text{m} \quad (1.1\ \text{kN}\cdot\text{m})$$

The answer is (C).

136. Resolve the applied load F at point C into its x- and y-components, as shown.

$$F_x = F\cos 30°$$
$$F_y = F\sin 30°$$

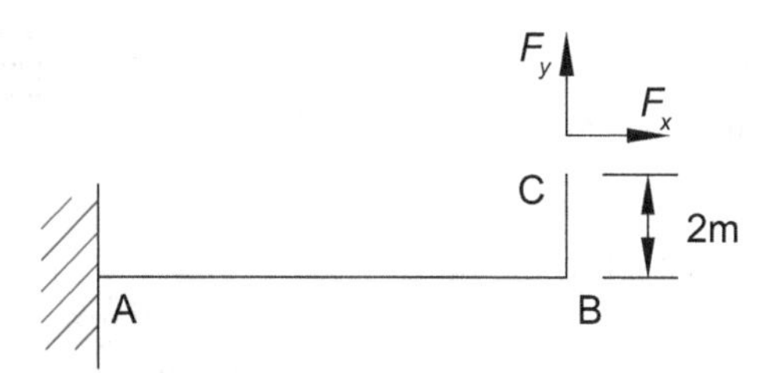

Move the force system from point C to point B as shown in the illustration. Use the principle that states that any force acting at point C can be moved to point B provided that a couple is added with a moment equal to the moment of force F about B.

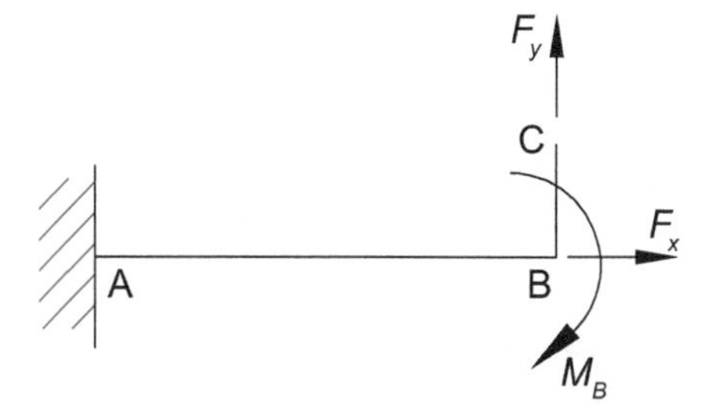

Let counterclockwise moments be positive. The associated moment is

$$M_B = F_x(-2\ \text{m}) + F_y(0) = -2F_x\ \text{kN}\cdot\text{m}$$

The moment at B is negative, which means the moment is acting in a clockwise direction. This is shown above.

Move the force system from point B to support A, as shown in the following illustration.

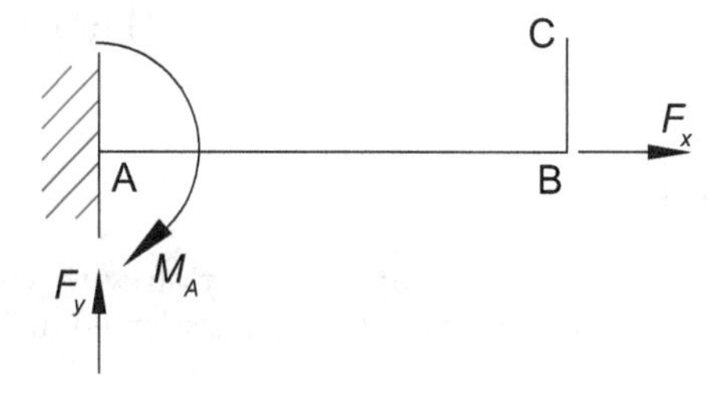

The resulting moment at support A is

$$\begin{aligned} M_A &= F_y(9\ \text{m}) + M_B = (9\ \text{m})F_y - (2\ \text{m})F_x \\ &= (9\ \text{m})F\sin 30° - (2\ \text{m})F\cos 30° \\ &= F(9\sin 30° - 2\cos 30°) \\ &= (2.77\ \text{m})F \\ &= (2.77\ \text{m})(10\ \text{kN}) \\ &= 27.7\ \text{kN}\cdot\text{m} \quad (28\ \text{kN}\cdot\text{m}) \end{aligned}$$

The moment at point A is positive. This means that it is in counterclockwise direction, as shown in the illustration above.

The answer is

	moment
clockwise or counterclockwise	counterclockwise
magnitude (kN·m)	28

137. Material properties such as strength and modulus of elasticity of each material are not considered in computing the centroid of a composite section. The centroid can be found by summation.

$$\bar{y} = \frac{\sum y_i A_i}{\sum A_i}$$

The areas of the components are

$$\begin{aligned} A_{\text{beam}} &= 14.1 \text{ in}^2 \\ A_{\text{slab}} &= (6 \text{ in})(60 \text{ in}) \\ &= 360 \text{ in}^2 \end{aligned}$$

The centroid of each component is at its mid-depth.

$$\begin{aligned} y_{\text{beam}} &= \frac{13.8 \text{ in}}{2} = 6.9 \text{ in} \\ y_{\text{slab}} &= 13.8 \text{ in} + \left(\frac{6 \text{ in}}{2}\right) \\ &= 16.8 \text{ in} \end{aligned}$$

The distance to the centroid of the composite section is

$$\begin{aligned} \bar{y} &= \frac{\sum y_i A_i}{\sum A_i} = \frac{(6.9 \text{ in})(14.1 \text{ in}^2) + (16.8 \text{ in})(360 \text{ in}^2)}{(14.1 \text{ in}^2) + (360 \text{ in}^2)} \\ &= 16.40 \text{ in} \quad (16 \text{ in}) \end{aligned}$$

The answer is 16.

138. The MOI of a rectangular section with width, b, and depth, d, about its axis is determined using the formula

$$I_1 = \frac{bd^3}{12}$$

For the welded bars, the depth increases to $2d$, as shown.

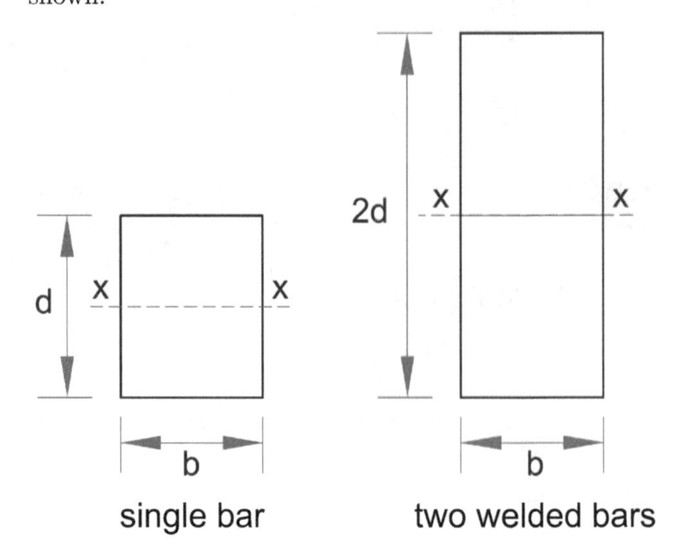

Therefore, the MOI of the welded bars is

$$I_2 = \frac{b(2d)^3}{12} = \frac{8bd^3}{12} = 8I_1$$

The ratio, I_2/I_1, of the MOI of the welded bars to the MOI of a single bar is 8.

The answer is (C).

139. For the weight, W, consider equilibrium along the x- and y-axes at impending motion between the block and the plane.

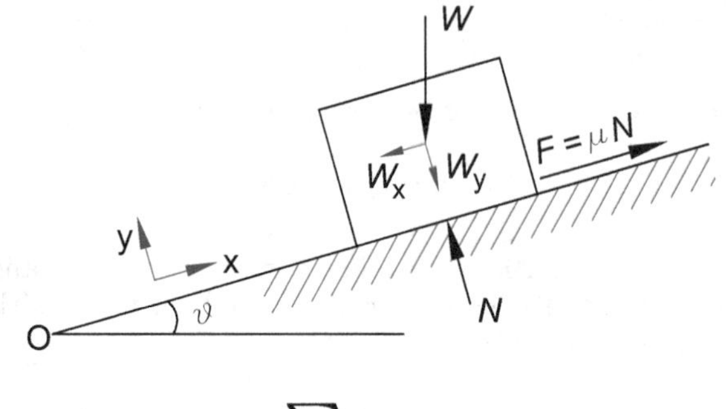

$$\begin{aligned} \sum F_y &= 0 \\ W_y - N &= 0 \\ W\cos\theta - N &= 0 \\ N &= W\cos\theta \end{aligned}$$

$$\begin{aligned} \sum F_x &= 0 \\ F - W_x = F - W\sin\theta &= 0 \end{aligned}$$

At impending motion, the friction force, F, is given by

$$\begin{aligned} F &= \mu_s N \\ \mu_s N - W\sin\theta &= 0 \end{aligned}$$

Substituting the value of N and rearranging the equation, find the sloping angle.

$$\begin{aligned}\mu_s W\cos\theta - W\sin\theta &= 0\\ \mu_s &= \tan\theta\\ \tan\theta &= 0.3\\ \theta &= 16.7^\circ \quad (17^\circ)\end{aligned}$$

The answer is (A).

140. The stone's initial velocity, u, is zero. To determine the distance travelled, s, use the equation

$$s = ut + \tfrac{1}{2}at^2$$

In this case,

$$\begin{aligned}u &= 0\ \text{m/s}\\ a = g &= 9.81\ \text{m/s}^2\\ t &= 2.9\ \text{s}\end{aligned}$$

Find the cliff height above the ground.

$$\begin{aligned}s &= \left(0\ \frac{\text{m}}{\text{s}}\right)(2.9\ \text{s}) + \left(\frac{1}{2}\right)\left(9.81\ \frac{\text{m}}{\text{s}^2}\right)(2.9\ \text{s})^2\\ &= 41.25\ \text{m} \quad (40\ \text{m})\end{aligned}$$

The answer is (C).

141. This is a plastic case wherein the vehicles stick together after the collision. Use the impact equation for conservation of momentum given in the Dynamics section of the *NCEES Handbook.*

$$m_1 v_1 + m_2 v_2 = m_1 v_1' + m_2 v_2' = v_{\text{comb}}(m_1 + m_2)$$

In this case, the velocities are equal, but in opposite directions.

Therefore, $v_2 = -v_1$. Assume the eastbound velocity, v_1, is positive.

$$\begin{aligned}v_1(m_1 - m_2) &= v_{\text{comb}}(m_1 + m_2)\\ v_{\text{comb}} &= v_1\left(\frac{m_1 - m_2}{m_1 + m_2}\right) = v_1\left(\frac{g(w_1 - w_2)}{g(w_1 + w_2)}\right)\\ &= v_1\left(\frac{6000\ \text{lbm} - 15{,}000\ \text{lbm}}{6000\ \text{lbm} + 15{,}000\ \text{lbm}}\right)\\ &= \frac{\left(30\ \frac{\text{mi}}{\text{hr}}\right)(-9\ \text{lbm})}{21\ \text{lbm}}\\ &= -12.9\ \text{mph} \quad (13\ \text{mph, westbound})\end{aligned}$$

The answer is (B) and (D).

142. Since the mass is uniformly distributed, the flywheel is a disk-like wheel. The mass moment of inertia of a hollow disk about its centroidal axis is

$$\begin{aligned}I &= \frac{1}{2}MR^2\\ M &= \frac{W}{g}\\ &= \frac{1000\ \text{lbf}}{32.2\ \frac{\text{ft}}{\text{sec}^2}}\\ &= 31.06\ \text{slugs}\\ R &= 18\ \text{in} = 1.5\ \text{ft}\\ I &= \left(\frac{1}{2}\right)\left(31.06\ \frac{\text{lbf-sec}^2}{\text{ft}}\right)(1.5\ \text{ft})^2\\ &= 34.94\ \text{slugs-ft}^2 \quad (35\ \text{slugs-ft}^2)\end{aligned}$$

The answer is (B).

143. Use ft-lbf units. The spring undamped frequency, ω_n, can be determined in terms of the static deflection, δ_{st}, of the system.

$$\omega_n = \sqrt{\frac{k}{m}}$$

The stiffness, k, is the force required to stretch a spring per unit length.

Use the static spring-displacement formula for constant force.

$$mg = k\delta_{\text{st}}$$

Since $mg = W$, the formula reduces to

$$k = \frac{W}{\delta_{\text{st}}} = \left(\frac{4\ \text{lbf}}{6\ \text{in}}\right)\left(\frac{12\ \text{in}}{1\ \text{ft}}\right) = 8\ \text{lbf/ft}$$

$$\begin{aligned}g &= 32.2\ \text{ft/sec}^2\\ m &= \frac{W}{g} = \frac{4\ \text{lbf}}{32.2\ \frac{\text{ft}}{\text{sec}^2}} = \frac{1}{8}\ \text{slugs} = 0.125\ \text{slugs}\\ \omega_n &= \sqrt{\frac{8\ \frac{\text{lbf}}{\text{ft}}}{0.125\ \frac{\text{lbf-sec}^2}{\text{ft}}}} = 8\ \text{rad/sec}\end{aligned}$$

The answer is (C).

144. The acceleration, α, of a particle rotating at a distance i from a fixed point has two components, tangential and radial.

$$\text{tangential acceleration, } \alpha_t = r\alpha$$
$$\text{radial acceleration, } \alpha_r = r\omega^2$$
$$\alpha = \sqrt{\alpha_t{}^2 + \alpha_r{}^2}$$

In this case, the shaft is rotating at a constant velocity, so the angular acceleration is zero. Therefore, its tangential acceleration, $\alpha_t = 0$.

The radial acceleration is

$$\alpha_r = r\omega^2 = (0.5 \text{ m})\left(20 \ \frac{\text{rad}}{\text{s}}\right)^2 = 200 \text{ m/s}^2$$

$$\begin{aligned}\alpha &= \sqrt{\alpha_r{}^2 + \alpha_t{}^2} \\ &= \alpha_r \\ &= 200 \text{ m/s}^2\end{aligned}$$

The answer is (D).

145. By symmetry, the reactions at both supports equal 10 kips. The shear force diagram is shown.

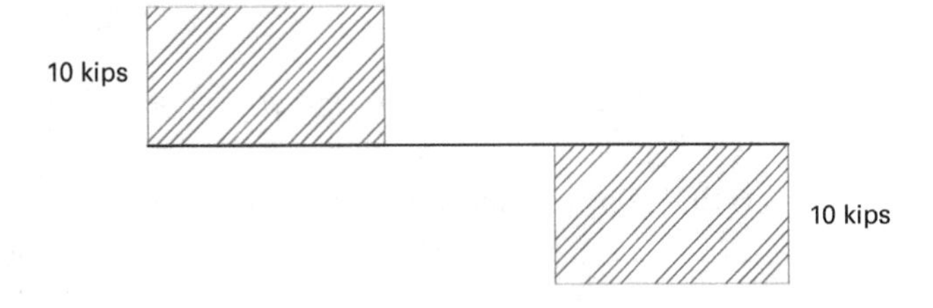

By definition, the maximum bending moment occurs at the section where shear force is zero. The shear force is zero in segment BC. Therefore, the bending moment is at maximum in that segment.

The exact value of the bending moment in segment BC is not needed for this problem but can be found. The change in bending moment between the two points equals the area of the shear force diagram between the two points.

$$\begin{aligned}M_B &= M_A + (10 \text{ kips})(8 \text{ ft}) \\ &= 0 \text{ ft-kips} + 80 \text{ ft-kips} = 80 \text{ ft-kips}\end{aligned}$$

The moment diagram is shown.

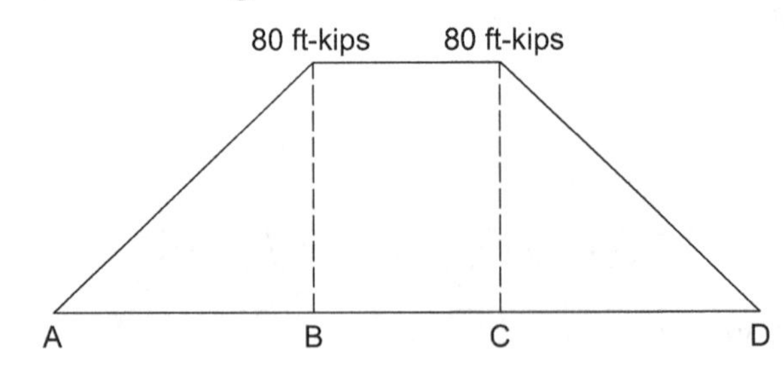

The answer is

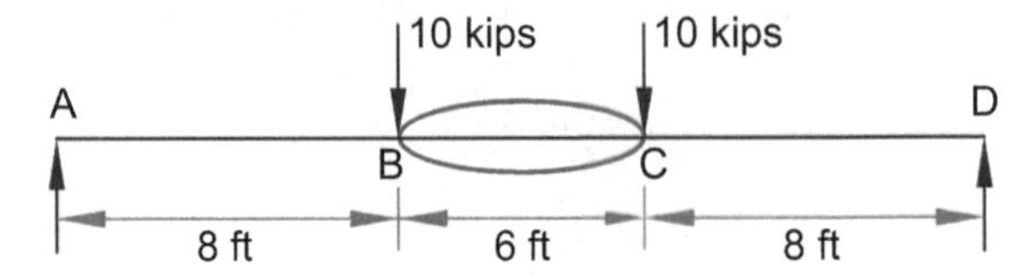

146. The bolt is in double shear. The bolt shear is distributed over two cross-sectional areas of the bolt. Find the shear stress.

$$\tau = \frac{\text{shear force}}{\text{total cross-sectional area}} = \frac{V}{2A}$$

$$A = \frac{\pi d^2}{4} = \frac{\pi(1 \text{ in})^2}{4} = 0.785 \text{ in}^2$$

$$\begin{aligned}\tau &= \frac{16 \text{ kips}}{(2)(0.785 \text{ in}^2)} \\ &= 10.19 \text{ ksi} \quad (10 \text{ ksi})\end{aligned}$$

The answer is (B).

147. The reaction block load-deformation configuration is shown in the problem statement.

To compute the deformation under shear loading, use the shear modulus for steel. The material properties of steel are given in the Typical Material Properties table in the Mechanics of Materials section of the *NCEES Handbook*. From the table, the shear modulus, G, of steel is 11.5 Mpsi.

The shear stress-strain relationship is given by

$$\text{shear strain, } \gamma = \frac{\text{shear stress, } \tau}{\text{shear modulus, } G}$$

$$G = 11.5 \text{ Mpsi} = 11{,}500 \text{ ksi}$$

$$\begin{aligned}\tau &= \frac{\text{shear force}}{\text{area of cross-section}} \\ &= \frac{24 \text{ kips}}{(1 \text{ in})(12 \text{ in})} \\ &= 2 \text{ ksi}\end{aligned}$$

Substitute the values into the shear stress-strain relationship equation and find the shear strain.

$$\gamma = \frac{2 \text{ ksi}}{11{,}500 \text{ ksi}} = 1.74 \times 10^{-4} \text{ in/in}$$

Since the shear stress is uniform along the height of the block, the shear strain will also be uniform. By definition, the shear deformation is the sum of strains over the length of the member. The deformation is

$$\Delta = \gamma h = \left(1.74 \times 10^{-4} \frac{\text{in}}{\text{in}}\right)(12 \text{ in}) = 2.1 \times 10^{-3} \text{ in}$$

The answer is (D).

148. The modulus of rigidity, G, is

$$G = \frac{E}{2(1+\nu)} = \frac{200 \text{ GPa}}{(2)(1+0.3)} = 76.9 \text{ GPa} \quad (77 \text{ GPa})$$

The answer is (B).

149. The error induced is

$$\begin{aligned} \Delta L &= e_{\text{th}} L\,(\Delta T) \\ \Delta T &= 98°\text{F} - 68°\text{F} \\ &= 30°\text{F} \\ e_{\text{th}} &= 6.5 \times 10^{-6} \frac{\text{in}}{\text{in}} \text{ per } °\text{F} \\ \Delta L &= \left(6.5 \times 10^{-6} \frac{\text{in}}{\text{in-}°\text{F}}\right)(3001.20 \text{ ft})(30°\text{F}) \\ &= 0.59 \text{ ft} \end{aligned}$$

Due to the high temperature, the tape expanded 0.59 ft. Therefore, the measured length is less than the true length. The true distance is

$$L_{\text{true}} = 3001.20 \text{ ft} + 0.59 \text{ ft} = 3001.79 \text{ ft}$$

The answer is (C).

150. The biaxial stress is shown.

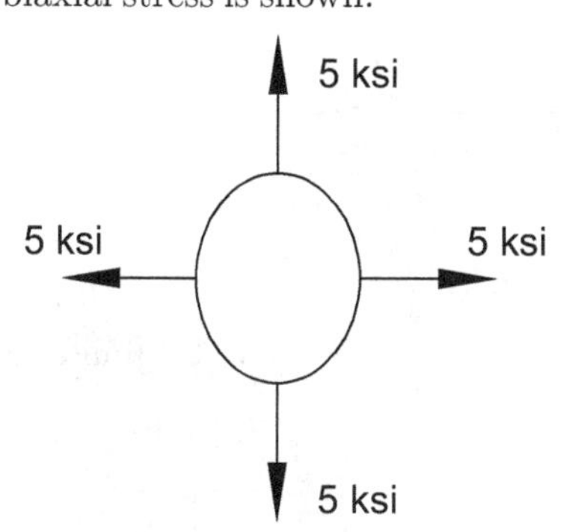

The maximum inplane shear stress is $\tau_{\text{in}} = R$.

The maximum shear stress given by Mohr's circle is

$$\tau_{\max} = \sqrt{\left(\frac{\sigma_x - \sigma_y}{2}\right)^2 + \tau_{xy}^2}$$

In this case,

$$\begin{aligned} \sigma_x &= \sigma_y = 5 \text{ ksi} \\ \tau_{xy} &= 0 \text{ ksi} \end{aligned}$$

The maximum shear stress is

$$\tau_{\max} = \sqrt{\left(\frac{5 \text{ ksi} - 5 \text{ ksi}}{2}\right)^2 + (0 \text{ ksi})^2} = 0 \text{ ksi}$$

The answer is (A).

151. The modular ratio, n, is defined as the ratio of moduli of elasticity of the member's materials. The modulus ratio is expressed as more than unity. As such, the weaker modulus of the elasticity is in the denominator of the ratio.

$$n = \frac{\text{modulus of elasticity of steel}}{\text{modulus of elasticity of concrete}}$$

The answer is (B).

152. Shaft ABC has a step-up section. Each segment is subjected to a different torsional force. The rotation at C is the sum of the rotations in segments AB and BC.

$$\phi_C = \left(\frac{TL}{GJ}\right)_{\text{AB}} + \left(\frac{TL}{GJ}\right)_{\text{BC}}$$

$$\begin{aligned} T_{\text{BC}} &= 0.2 \text{ kN}\cdot\text{m} \\ T_{\text{AB}} &= 0.2 \text{ kN}\cdot\text{m} + 0.1 \text{ kN}\cdot\text{m} \\ &= 0.3 \text{ kN}\cdot\text{m} \\ J_{\text{AB}} &= \frac{\pi d^4}{32} \\ &= \frac{\pi(0.06 \text{ m})^4}{32} \\ &= 1.27 \times 10^{-6} \text{ m}^4 \\ J_{\text{BC}} &= \frac{\pi(0.04 \text{ m})^4}{32} \\ &= 0.25 \times 10^{-6} \text{ m}^4 \end{aligned}$$

The shear modulus, G, for steel is given in the Typical Material Properties table of the Mechanics of Materials section in the *NCEES Handbook.*

$$G = 80 \text{ GPa} = 80 \times 10^6 \text{ kPa} = 80 \times 10^6 \text{ kN/m}^2$$

Find the shaft's total rotation at C.

$$\phi_C = \frac{1}{G}\left(\left(\frac{TL}{J}\right)_{AB} + \left(\frac{TL}{J}\right)_{BC}\right)$$

$$= \left(\frac{1}{80 \times 10^6 \ \frac{\text{kN}}{\text{m}^2}}\right) \times \left(\frac{(0.3 \text{ kN}\cdot\text{m})(1 \text{ m})}{1.27 \times 10^{-6} \text{ m}^4} + \frac{(0.2 \text{ kN}\cdot\text{m})(2 \text{ m})}{0.25 \times 10^{-6} \text{ m}^4}\right)$$

$$= 0.0229 \text{ rad}$$

$$= \frac{(180°)(0.0229 \text{ rad})}{\pi}$$

$$= 1.31°$$

The answer is (D).

153. Portland cement, water, and fine and coarse aggregates are four necessary ingredients.

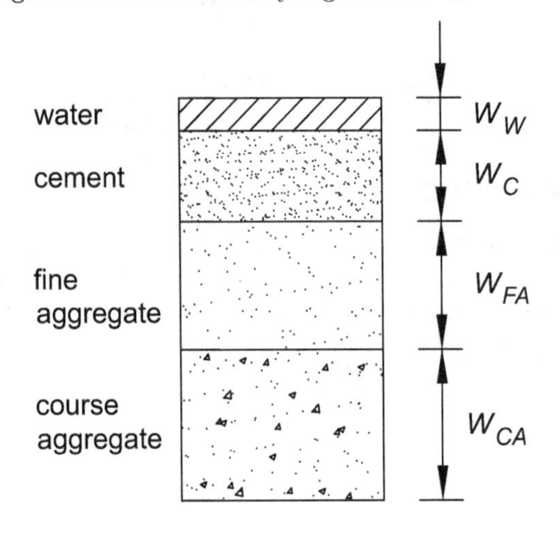

Concrete is composed principally of aggregates, Portland cement, and water. Admixtures are often added to improve its workability and curing, and long-term performance. Concrete is known to have low tensile strength and is prone to cracking. The freeze and thaw cycles aggravate the crack propagation because the water present in concrete voids and capillary pores freezes. As water freezes to ice, it expands. The expansion produces stress that is far in excess of the tensile strength of concrete.

Air-entraining admixtures are primarily used to stabilize tiny air bubbles in concrete, produced by mixing, in order to protect the concrete against damage from repeated freezing and thawing. Concrete contains some amount of entrapped air in form of voids. Entrained air should not be confused with entrapped air. Concrete setting time depends on the ambient temperature the concrete is being placed in. The time of concrete setting increases as the ambient temperature drops. A retarding admixture is used in hot weather to retard or to slow down the chemical setting process and thus increase the time of setting. On the other hand, an accelerating admixture is used in cold weather to accelerate the chemical setting process and thus reduce the time of setting. Neither a retarding admixture nor an accelerating admixture is intended to provide the air void system needed to entrain air in concrete. Therefore, options B, C, and D are incorrect.

The answer is (A).

154. The mix designation 1:x:y denotes one part of cement, x parts of fine aggregate, and y parts of coarse aggregate. The ratio is by weight. The coarse aggregate is generally a saturated surface-dried aggregate.

The answer is (C) and (E).

155. The total volume and the weight of the concrete mix is

$$\text{volume} = 1 \text{ yd}^3 = 27 \text{ ft}^3$$

$$\text{weight} = \left(\frac{27 \text{ ft}^3}{1 \text{ yd}^3}\right)\left(145 \ \frac{\text{lbm}}{\text{ft}^3}\right) = 3915 \text{ lbm/yd}^3$$

Use mix ratio of 1:2:3. The following proportion (by weight) is needed.

$$\begin{aligned} \text{cement} &= 1 \text{ lbm} \\ \text{fine aggregate} &= 2 \text{ lbm} \\ \text{coarse aggregate} &= 3 \text{ lbm} \\ \text{w/c ratio} &= 0.4 \\ \text{water} &= (0.4)(1 \text{ lbm of cement}) = 0.4 \text{ lbm} \end{aligned}$$

The weight of the mix produced by 1 lbm of cement is

$$1 \text{ lbm} + 2 \text{ lbm} + 3 \text{ lbm} + 0.4 \text{ lbm} = 6.4 \text{ lbm}$$

1 lbm of cement will produce 6.4 pounds of concrete. 1 yd^3 of concrete weighs 3915 pounds. The weight, W_c, of cement needed to produce 1 yd^3 (3915 lbm) of concrete is

$$W_c = \frac{3915 \ \frac{\text{lbm}}{\text{yd}^3}}{6.4 \ \frac{\text{lbm of concrete}}{\text{lbm of cement}}} = 611.7 \text{ lbm of cement/yd}^3$$

The weight of one sack of cement is 94 lbm.

n = the number of cement sacks needed to produce 1 yd^3 of concrete.

$$n = \frac{611.7 \ \frac{\text{lbm of cement}}{\text{yd}^3}}{94 \text{ lbm per sack}} = 7 \text{ sacks/yd}^3$$

The answer is (C).

156. The tensile yield strength of grade 60 steel, f_y, is 60,000 psi, as shown in the Interaction Diagram for reinforced concrete columns in the *NCEES Handbook.*

The ratio of steel strength to concrete tensile strength is

$$\frac{f_y}{f'_c} = \frac{60{,}000 \ \frac{\text{lbf}}{\text{in}^2}}{5000 \ \frac{\text{lbf}}{\text{in}^2}} = 12$$

A schematic comparison of the strengths is shown.

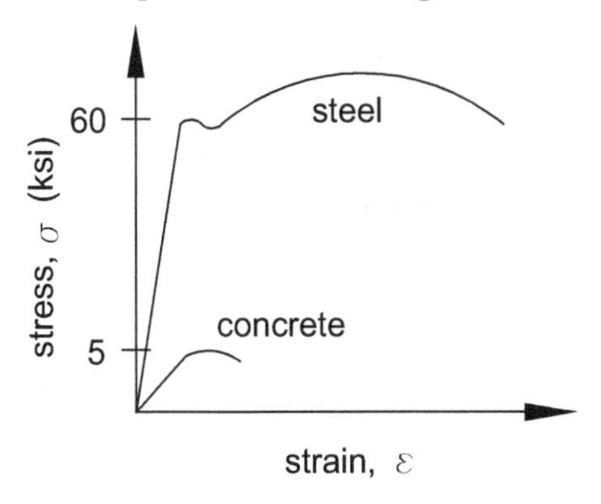

The answer is (C).

157. The ability of a material to undergo deformation without failure under high tensile stress is called ductility.

The answer is (C).

158. Steady incompressible flow in conduits and pipes is expressed by Bernoulli's energy equation.

$$\frac{p_1}{\gamma} + \frac{v_1^2}{2g} + z_1 = \frac{p_2}{\gamma} + \frac{v_2^2}{2g} + z_2 + h_f + h_{f,\text{fitting}}$$

For static fluid in a tank, the fluid velocity is zero. Head losses that occur as the fluid flows through pipe fittings are also neglected. Thus, solving the problem only involves elevations and pressures, so the Bernoulli equation can be simplified as

$$\frac{p_1}{\gamma} + z_1 = \frac{p_2}{\gamma} + z_2$$

The gauge pressure decreases as the head reduces (i.e., height about the base increases).

$$\text{For } z_1 = 5 \text{ m}, \ p_1 = 80 \text{ kPa} = 80{,}000 \text{ Pa}$$
$$\text{For } z_2 = 8 \text{ m}, \ p_2 = 57.4 \text{ kPa} = 57{,}400 \text{ Pa}$$

Substitute the values into the Bernoulli energy equation and find the unit weight of the liquid.

$$\left(\frac{80{,}000 \text{ Pa}}{\gamma}\right) + 5 \text{ m} = \left(\frac{57{,}400 \text{ Pa}}{\gamma}\right) + 8 \text{ m}$$
$$\gamma = 7530 \ \text{kg/m}^3$$

The answer is (D).

159. In SI units, the specific gravity, SG, is

$$\text{SG} = \frac{\gamma}{\gamma_w} = \frac{\rho}{\rho_w}$$
$$\gamma = \rho g$$

Find the specific weight of the liquid.

$$\begin{aligned}\gamma_l &= (\text{SG})\rho_w g \\ &= (2.7)(1000 \text{ kg/m}^3)\left(9.81 \ \frac{\text{m}}{\text{s}^2}\right) \\ &= 26{,}487 \text{ kg/m}^2\cdot\text{s}^2 \ \ (27{,}000 \text{ kg/m}^2\cdot\text{s}^2)\end{aligned}$$

The answer is (D).

160. The hydrostatic pressure acts perpendicular to the wall. It exerts outward pressure on the wall, and therefore is tensile in nature on the weld used on the plate. Also, the circular wall is under hoop stress, which is also tensile in nature.

The answer is (B).

161. The magnitude of the total force that the weld is required to resist equals the hydrostatic pressure exerted on the entire area of the plate. The pressure intensity varies with the depth of water. The pressure distribution diagram is shown.

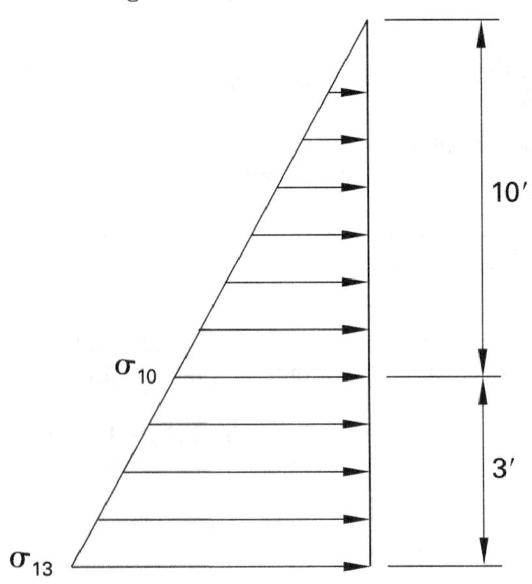

The specific weight of water is

$$\gamma = 62.4 \text{ lbf/ft}^3$$

The depth, h_1 and h_2, are given.

$$h_1 = 10 \text{ ft}$$
$$h_2 = 10 \text{ ft} + 3 \text{ ft} = 13 \text{ ft}$$

Find the pressure at depth h_1 and h_2.

$$\sigma_{10} = \gamma h_1 = \left(62.4 \ \frac{\text{lbf}}{\text{ft}^3}\right)(10 \text{ ft}) = 624 \text{ psf}$$

$$\sigma_{13} = \gamma h_2 = \left(62.4 \ \frac{\text{lbf}}{\text{ft}^3}\right)(13 \text{ ft}) = 811 \text{ psf}$$

The force is

$$F = (\text{pressure intensity})(\text{area})$$

The force on new steel plate is equal to the volume of the pressure block behind the plate.

The pressure block has a trapezoidal shape with a height of 3 ft and a thickness of 3 in, as shown.

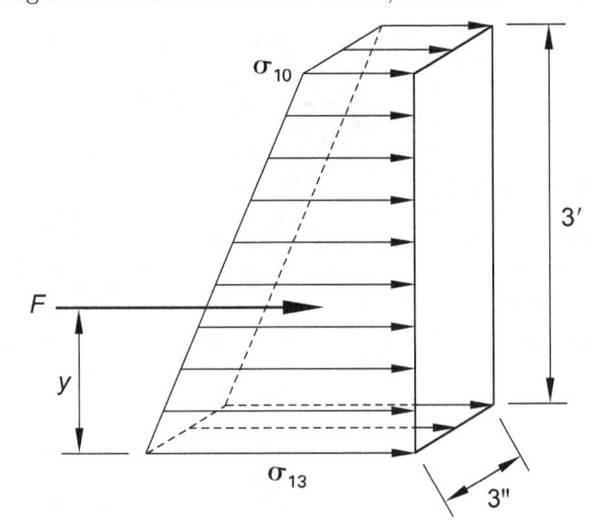

Find the total force on the weld.

$$F = \left(\frac{\sigma_{10} + \sigma_{13}}{2}\right)(\text{plate area})$$

$$= \left(\frac{624 \ \frac{\text{lbf}}{\text{ft}^2} + 811 \ \frac{\text{lbf}}{\text{ft}^2}}{2}\right)\left(\frac{3 \text{ in}}{12 \ \frac{\text{in}}{\text{ft}}}\right)(3 \text{ ft})$$

$$= 538 \text{ lbf}$$

The answer is (A).

162. The pressure block is trapezoidal in shape. The resultant force acts at the centroid of the stress block. Consider a unit width of the block, and use the properties table given in the Statics section of the *NCEES Handbook*, to determine the centroidal distance, y_c, from the base of the plate for a trapezoidal area.

$$y_c = \frac{h(2a + b)}{3(a + b)}$$

a and b are the top and base widths of the pressure block.

$$a = \sigma_{10} = \left(62.4 \ \frac{\text{lbf}}{\text{ft}^3}\right)(10 \text{ ft}) = 624 \text{ psf}$$

$$b = \sigma_{13} = \left(62.4 \ \frac{\text{lbf}}{\text{ft}^3}\right)(13 \text{ ft}) = 811 \text{ psf}$$

$$h = \text{trapezoidal height} = 3 \text{ ft} = 36 \text{ in}$$

$$y_c = \frac{(36 \text{ in})\big((2)(624 \text{ psf}) + 811 \text{ psf}\big)}{(3)(624 \text{ psf} + 811 \text{ psf})} = 17.21 \text{ in} \quad (17 \text{ in})$$

The answer is (A).

163. For a model to simulate the conditions of a prototype, the model must be geometrically, kinematically, and dynamically similar to the prototype system. For a prototype in which gravitational and inertial forces are the only important forces, the Froude number, Fr, is kept constant between the prototype and its model. Mathematically,

$$\frac{\text{inertial force of prototype}}{\text{gravity force of prototype}} = \frac{\text{inertial force of model}}{\text{gravity force of model}}$$

The answer is (B).

164. An equilateral triangle has three sides of equal length. This implies that each internal angle is 60°. The illustration shows the relationship between the side and the altitude.

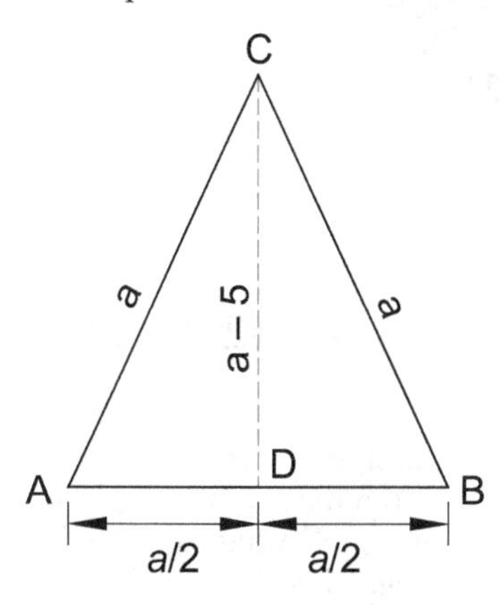

Since all angles are equal, use the equation

$$\sin A = \sin B = \frac{\overline{\text{CD}}}{\overline{\text{AC}}} = \frac{a-5}{a} = \sin 60° = 0.866$$

Rearrange the equation and find the side length, a.

$$\begin{aligned} a - 5\text{ ft} &= (0.866)a \\ a - (0.866)a &= 0.134a = 5\text{ ft} \\ a &= \frac{5\text{ ft}}{0.134} = 37.31\text{ ft} \quad (37.3\text{ ft}) \end{aligned}$$

The answer is 37.3 ft.

165. Use the area equation for the circular segment given in the Mathematics section of the *NCEES Handbook* to compute the area.

$$A = \frac{1}{2}r^2(\theta - \sin\theta)$$

Alternatively, consider the shaded area as the difference between the two areas A_1 and A_2.

$$\begin{aligned} \text{shaded area} &= \text{area under curve OAB } (A_1) \\ &\quad -\text{area of triangles OAC and OBC } (A_2) \end{aligned}$$

Find the area A_1.

$$A_1 = \pi r^2\left(\frac{60°}{360°}\right) = \pi(250\text{ m})^2\left(\frac{1}{6}\right) = 32{,}725\text{ m}^2$$

Since the two triangles AOC and OBC are identical, the area A_2 is

$$\begin{aligned} A_2 &= 2\left(\frac{(\overline{AC})(\overline{OC})}{2}\right) \\ &= \frac{(2)\left((250\text{ m})\sin 30°\right)\left((250\text{ m})\cos 30°\right)}{2} \\ &= 27{,}063\text{ m}^2 \end{aligned}$$

The shaded area of the curve is

$$\begin{aligned} \text{shaded area} &= A_1 - A_2 = 32{,}725\text{ m}^2 - 27{,}063\text{ m}^2 \\ &= 5662\text{ m}^2 \end{aligned}$$

The answer is (A).

166. The volume is calculated using the average end area method from the Earthwork Formulas section of the *NCEES Handbook*. The volume is

$$V = \frac{L(A_1 + A_2)}{2}$$

The length, L, is the distance between stations.

$$L = (3+50) - (1+65) = 350\text{ ft} - 165\text{ ft} = 185\text{ ft}$$

From the problem statement,

$$\begin{aligned} A_1 &= 200\text{ ft}^2 \quad \text{[cut]} \\ A_2 &= 125\text{ ft}^2 \quad \text{[fill]} \end{aligned}$$

Find the volume.

$$\begin{aligned} V &= \frac{(185\text{ ft})(200\text{ ft}^2 - 125\text{ ft}^2)}{2} = 6{,}938\text{ ft}^3 \\ &= \frac{6938\text{ ft}^3}{27\ \frac{\text{ft}^3}{\text{yd}^3}} \\ &= 257\text{ yd}^3 \end{aligned}$$

The answer is (A).

167. A deflection angle is the difference in angle from the prolongation of the back line to the forward line along a traverse. Based on the given bearing of segment AB, its prolongation AB′ is directed to S12°34′56″ W. The difference in angle between AB′ and AB is the sum of the angles S12°34′56″ W and S12°34′56″ E (= 25°09′ 52″). Since line AC is located at the left-hand side of the prolongation line, it is deflected to the left.

The answer is (B).

168. An inverted sight (IS) reading is taken by placing the staff upside-down under an object, as shown in the illustration. Let station A be the instrument location, station B be the underside of the beam, and station C be the point on a slope. Find the elevation at station B and station C.

$$\begin{aligned}\text{elevation at sta B} &= \text{elevation at sta A} + \text{HI} + \text{IS}\\ &= 123.45\text{ ft} + 5.15\text{ ft} + 3.13\text{ ft}\\ &= 131.73\text{ ft}\\ \text{elevation at sta C} &= \text{elevation at sta A} + \text{HI} - \text{FS}\\ &= 123.45\text{ ft} + 5.15\text{ ft} - 4.32\text{ ft}\\ &= 124.28\text{ ft}\end{aligned}$$

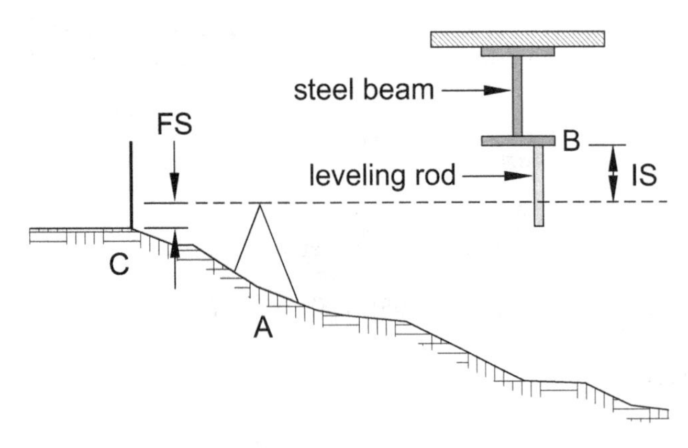

The answer is (B).

169. Given that $R = 250$ ft and angle $I = 66°$, use the horizontal curve formula to compute the curve length, L.

$$L = RI\left(\frac{\pi}{180}\right) = (250\text{ ft})(66°)\left(\frac{\pi}{180°}\right) = 288\text{ ft}$$

The answer is (B).

170. The rational formula is

$$Q = CIA$$

The runoff coefficient, C, in the rational formula depends on the type of surface, imperviousness, and slope of the area. The composite runoff coefficient is the weighted sum of the subarea coefficients based on the subareas, which are calculated using the following table.

surface type	area (acres)	runoff coefficient, C_j	C_jA_j (acres)
asphalt pavement	24	0.73	17.5
concrete pavement	16	0.85	13.6
lawns	35	0.22	7.7
roofs	24	0.90	21.6
total	99		60.4

$$C_{\text{composite}} = \frac{\sum C_jA_j}{\sum A} = \frac{60.4\text{ ac}}{99\text{ ac}} = 0.61 \quad (0.6)$$

The answer is (A).

171. Statements A, B, and C are correct. Statement D is incorrect because the hydraulic radius is 1/4 of the hydraulic diameter.

$$R_H = \frac{\text{hydraulic diameter}}{4}$$

The hydraulic radius of a pipe is the channel property which controls water discharge. The radius used helps to determine how much water and sediment can flow through the channel. The higher the radius of a pipe, the larger the volume of fluid the line carries.

The answer is (D).

172. The relation to estimate a pump horsepower is

$$\dot{W} = \frac{Q\gamma h}{\eta}$$

$$h = 100\text{ ft} + 48\text{ ft} = 148\text{ ft}$$

$$\dot{W} = \frac{\left(5\ \frac{\text{ft}^3}{\text{sec}}\right)\left(62.42\ \frac{\text{lbf}}{\text{ft}^3}\right)(148\text{ ft})}{0.85} = 54{,}342\text{ ft-lbf/sec}$$

Use the conversion factor to find the horsepower needed to pump the water.

$$1\text{ hp} = 550\text{ ft-lbf/sec}$$

$$\text{power} = \frac{54{,}342\ \frac{\text{ft-lbf}}{\text{sec}}}{550\ \frac{\frac{\text{ft-lbf}}{\text{sec}}}{\text{hp}}} = 98.8\text{ hp} \quad (100\text{ hp})$$

The answer is (D).

173. The pipe fitting loss is expressed as

$$h_{f,\ \text{fitting}} = C\frac{v^2}{2g}$$

For a round entrance, $C = 0.1$, and for a sharp exit, $C = 1.0$.

The pipe inside diameter, D, is

$$\begin{aligned} D &= \text{outside diameter} - (2)(\text{wall thickness}) \\ &= 12\ \text{in} - (2)(1\ \text{in}) \\ &= 10\ \text{in} \\ &= 0.83\ \text{ft} \end{aligned}$$

The pipe cross-sectional area is

$$A = \frac{\pi D^2}{4} = \frac{(\pi)(0.83\ \text{ft})^2}{4} = 0.55\ \text{ft}^2$$

The discharge rate, Q, is

$$Q = vA = 2\ \text{ft}^3/\text{sec}$$

Rearrange the equation and find the water velocity.

$$v = \frac{Q}{A} = \frac{2\ \frac{\text{ft}^3}{\text{sec}}}{0.55\ \text{ft}^2} = 3.64\ \text{ft/sec}$$

Find the pipe total fitting loss.

$$\begin{aligned} h_{f,\ \text{fitting}} &= \sum C\frac{v^2}{2g} = (C_{\text{entry}} + C_{\text{exit}})\left(\frac{v^2}{2g}\right) \\ &= \frac{(0.1 + 1.0)\left(3.64\ \frac{\text{ft}}{\text{sec}}\right)^2}{(2)\left(32.2\ \frac{\text{ft}}{\text{sec}^2}\right)} \\ &= 0.23\ \text{ft} \end{aligned}$$

The answer is (C).

174. Since in this case the water level is maintained at constant height, this is a constant head problem. The orifice flow is

$$Q = CA_0\sqrt{2gh}$$

The nozzle area is

$$A_0 = \frac{\pi D^2}{4} = \frac{(3.14)(2\ \text{in})^2}{(4)\left(12\ \frac{\text{in}}{\text{ft}}\right)^2} = 0.022\ \text{ft}^2$$

From the *NCEES Handbook* Orifices and Their Nominal Coefficients table, the coefficient for a sharp-edged orifice, C, is 0.61.

The head, h, above the orifice is 10 ft. Find the orifice flow.

$$Q = (0.61)(0.022\ \text{ft}^2)\sqrt{(2)\left(32.2\ \frac{\text{ft}}{\text{sec}^2}\right)(10\ \text{ft})} = 0.34\ \text{cfs}$$

Convert the flow from cubic feet per second to gallons per second.

$$\begin{aligned} Q &= \left(0.34\ \frac{\text{ft}^3}{\text{sec}}\right)\left(7.481\ \frac{\text{gallons}}{\text{ft}^3}\right) \\ &= 2.55\ \text{gps} \end{aligned}$$

The answer is (B).

175. The type of lining that is provided in the channel should be determined on the basis of the maximum allowable velocities that the lining can bear. The lining can be flexible or rigid, depending on the terrain condition. Channel slope is one of the major parameters affecting the selection of the channel linings. Examples of rigid linings include plain and reinforced concrete and soil cement. Examples of flexible linings include rock riprap and vegetation. The preferred method to design linings is based on the shear force during the service life of the lining; if the fluid shear force exceeds the shear capacity of the liner, erosion occurs.

The answer is (B).

176. Darcy velocity, v, and specific discharge, q, are synonymous terms. They are defined as discharge per unit cross-sectional area. This is also expressed as

$$\begin{aligned} q = v &= (\text{hydraulic conductivity})(\text{hydraulic gradient}) \\ &= (K)(\text{hydraulic gradient}) \end{aligned}$$

The hydraulic gradient is defined as the change in the hydraulic head per unit length. In other words, it is the slope of the pressure head line. The equation in option A is given in the Darcy's Law section of the *NCEES Handbook*. The equations in options B and C are variations of the equation in option A, which defines the hydraulic gradient in different mathematical formulations. The equation in option D, which defines the hydraulic gradient as the average head pressure, is incorrect. Also, a negative sign in the equation is needed to show that the flow is proportional to the head drop $(h_1 - h_2)$, which has a negative slope or negative hydraulic gradient.

The answer is (D).

177. There are two types of aquifers: confined and unconfined. The illustration shows an unconfined aquifer. To determine the well flow in unconfined aquifers, Q, use Dupuit's formula.

$$Q = \frac{\pi K(h_2^2 - h_1^2)}{ln\left(\frac{r_2}{r_1}\right)}$$

$$K = 0.001 \text{ ft/sec}$$
$$h_1 = 50 \text{ ft}$$
$$h_2 = 150 \text{ ft}$$
$$r_1 = 1 \text{ ft}$$
$$r_2 = 100 \text{ ft}$$

Find the well discharge rate.

$$Q = \frac{\pi\left(0.001 \ \frac{\text{ft}}{\text{sec}}\right)\left((150 \text{ ft})^2 - (50 \text{ ft})^2\right)}{\ln\left(\frac{100 \text{ ft}}{1 \text{ ft}}\right)} = 13.6 \text{ ft}^3/\text{sec}$$

The answer is (B).

178. The concentration of dissolved oxygen in water is called DO. The bio-oxygen demand in water is called BOD. The BOD test is a measure of organic matter in water. The more BOD the water has, the more waste it has. Therefore, it is less healthy for plants and animals. The term BOD_5 pertains to a sample tested after five days of decay. DO is an indicator of the general health of the water. A clean stream has a low BOD and a high DO. Wastewater has a high BOD and a low DO. At a location where wastewater is discharged into the stream, the BOD of the stream increases, and the DO concentration decreases.

The answer is (B).

179. Use the formula for hydraulic residence time.

$$\theta = \frac{V}{Q} = \frac{(100 \text{ ft})(6 \text{ ft})(8 \text{ ft})}{\left(20 \ \frac{\text{ft}^3}{\text{sec}}\right)\left(60 \ \frac{\text{sec}}{\text{min}}\right)} = 4 \text{ min}$$

The answer is (D).

180. In general terms, hard water denotes the water that does not lather well and leaves a hard, white, crusty deposit on pans and in hot water heaters. Monovalent cations do not cause hardness, but all polyvalent cations contribute to hardness. Chemically, hardness is defined as the sum of all polyvalent cations. The primary contributors to hardness are calcium and magnesium. Both are divalent. The hardness is expressed in terms of milligram per liter of $CaCO_3$. Water containing more than 150 mg/L as $CaCO_3$ is considered hard.

The answer is (C).

181. This is a three-hinge arch with two supports at points A and B and a crown hinge at C. The moment is zero at points A, B, and C. The resultants of respective vertical and horizontal loads V and H are shown.

$$V = \left(2 \ \frac{\text{kN}}{\text{m}}\right)(40 \text{ m}) = 80 \text{ kN}$$

$$H = \left(3 \ \frac{\text{kN}}{\text{m}}\right)(20 \text{ m}) = 60 \text{ kN}$$

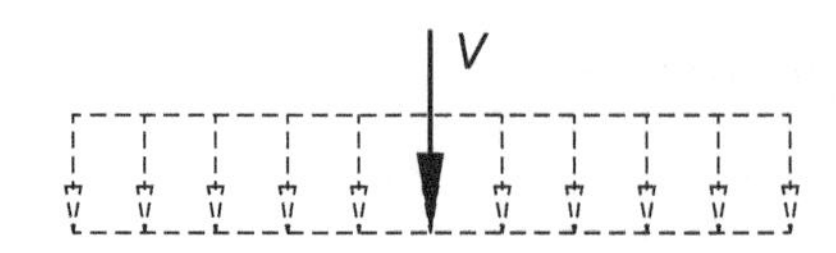

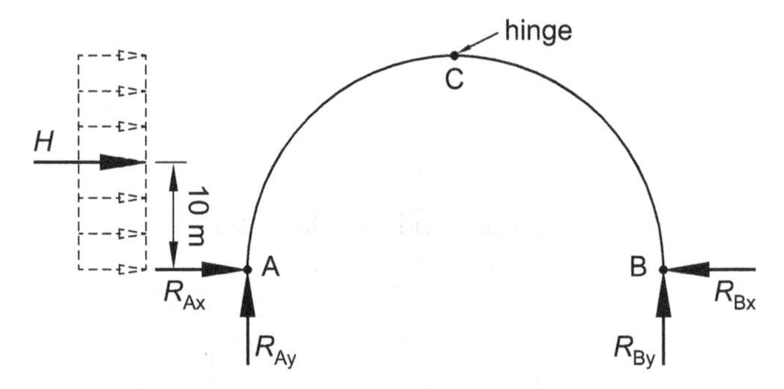

To determine the vertical reaction at A, R_{Ay}, let clockwise moments be positive. Taking the moment about support B,

$$\sum M_B = 0$$
$$(R_{Ay})(40 \text{ m}) + (60 \text{ kN})(10 \text{ m}) - (80 \text{ kN})(20 \text{ m}) = 0$$
$$R_{Ay} = 25 \text{ kN}$$

The positive sign of R_{Ay} shows that the assumed direction of reaction shown in the illustration is correct.

The answer is (A).

182. The factored uniformly distributed load, w_u, is

$$w_u = 1.2D + 1.6L = \frac{(1.2)\left(1.0 \ \frac{\text{kip}}{\text{ft}}\right) + (1.6)\left(0.8 \ \frac{\text{kip}}{\text{ft}}\right)}{12 \ \frac{\text{in}}{\text{ft}}}$$
$$= 0.21 \text{ kip/in}$$

The span length is

$$L = 16 \text{ ft} = (16 \text{ ft})\left(12 \ \frac{\text{in}}{\text{ft}}\right) = 192 \text{ in}$$

See the *NCEES Handbook*, Civil Engineering section, Table 1-1: W Shapes Dimensions and Properties. For W16 × 57,

$$I_{xx} = 758 \text{ in}^4$$

From the table of Typical Mechanical Properties, steel modulus of elasticity, $E = 29{,}000$ ksi.

The maximum deflection is at the cantilever tip.

$$\begin{aligned}\delta_{max} &= \frac{w_u L^4}{8EI} = \frac{(0.21)(192 \text{ in})^4}{(8)(29{,}000 \text{ ksi})(758 \text{ in}^4)} \\ &= 1.62 \text{ in} \quad (1.6 \text{ in})\end{aligned}$$

The answer is (D).

183. The radii of gyration are

$$r_x = \sqrt{\frac{I_x}{A_x}}$$

$$r_y = \sqrt{\frac{I_y}{A_y}}$$

For a rectangular shape, the area is

$$A = bh$$

For column bucking about the major axis, the moment of inertia is

$$I_x = \frac{bh^3}{12}$$

$$\frac{I_x}{A} = \frac{bh^3}{12bh} = \frac{h^2}{12}$$

Find the radii of gyration.

$$r_x = \sqrt{\frac{h^2}{12}} = 0.288h$$

$$r_y = \sqrt{\frac{b^3h}{12bh}} = 0.288b$$

The smaller radius of gyration controls buckling when the column is pinned at both ends. A 24 in × 36 in column will buckle about its weak axis, so the 24 in side is perpendicular to the buckling axis.

Therefore, the controlling radius of gyration is

$$r_y = 0.288b = (0.288)(24 \text{ in}) = 6.91 \text{ in} \quad (7 \text{ in})$$

The answer is (A).

184. The truss has 11 joints and 19 members. Each member has an unknown axial force. The support at F is a roller. Therefore, it is capable of developing only a vertical reaction. The support at A is a hinge, so it can develop both horizontal and vertical reactions. Therefore, the truss has three unknown reactions.

See the Stability, Determinacy, and Classification of Structures section of the *NCEES Handbook*. To determine whether a plane truss is stable and/or determinate, use the following criteria.

1. If the number of members + number of reactions < 2 times the number of joints, then it is unstable.

2. If the number of members + number of reactions = 2 times the number of joints, then it is stable and statically determinate.

3. If the number of members + number of reactions > 2 times the number of joints, then it is stable and statically indeterminate.

For the truss, the relationship between bars, reactions, and joints is

$$\begin{aligned}r &= \text{number of reactions} = 3 \\ b &= \text{number of bars} = 19 \\ j &= \text{number of joints} = 11\end{aligned}$$

Substitute the values in the criteria equation.

$$\text{LHS: } r + b = 3 + 19 = 22$$
$$\text{RHS: } 2j = (2)(11) = 22$$

Since both sides are equal, the truss is stable and statically determinate.

The answer is (B).

185. The fixed-end moments (FEM) for a beam carrying a point load are

$$\text{FEM}_{AB} = \frac{Pab^2}{L^2}$$

$$\text{FEM}_{BA} = \frac{Pa^2b}{L^2}$$

Find the moments at ends A and B.

$$P = 50 \text{ kips}$$
$$a = 15 \text{ ft}$$
$$b = 30 \text{ ft}$$
$$L = 45 \text{ ft}$$

$$\text{FEM}_{\text{AB}} = \frac{(50 \text{ kips})(15 \text{ ft})(30 \text{ ft})^2}{(45 \text{ ft})^2} = 333 \text{ ft-kips}$$

$$\text{FEM}_{\text{BA}} = \frac{(50 \text{ kips})(30 \text{ ft})(15 \text{ ft})^2}{(45 \text{ ft})^2} = 166 \text{ ft-kips}$$

A fixed-end moment is considered positive if it is acting clockwise on the support. The FEM at A is counterclockwise and therefore negative. The FEM at B is clockwise and therefore positive. Using this sign convention, the FEMs are

$$\text{FEM}_{\text{AB}} = -333 \text{ ft-kips}$$
$$\text{FEM}_{\text{BA}} = 167 \text{ ft-kips}$$

The answer is (B).

186. Use the method of superposition. Treat the uniformly distributed load (UDL) and the point load as loads that act alone, and then combine their effects. For a UDL w, the fixed-end moment is

$$\text{FEM}_{\text{AB}} = \frac{wL^2}{12}$$

For point load P at midspan, the distance is

$$a = b = 0.5L$$

The moment for the point load is

$$\text{FEM}_{\text{AB}} = \frac{Pab^2}{L^2} = \frac{P(0.5L)(0.5L)^2}{L^2} = \frac{PL}{8}$$

For a combined UDL and point load, the bending moment at support A is

$$\begin{aligned}\text{FEM}_{\text{AB}} &= \frac{wL^2}{12} + \frac{PL}{8} \\ &= \frac{\left(1 \ \frac{\text{kip}}{\text{ft}}\right)(30 \text{ ft})^2}{12} + \frac{(10 \text{ kips})(30 \text{ ft})}{8} \\ &= 112.5 \text{ ft-kips}\end{aligned}$$

The FEM at A is counterclockwise and positive.

The answer is (D).

187. The kinetic energy (KE) of a moving object with mass m and velocity v is

$$\text{KE} = \frac{1}{2}m\text{v}^2$$

Upon impact, the entire KE is converted into work done that is absorbed in crushing the vehicle. The reduction in length of the car, δ, due to crushing is called the *car crush.*

The impact force, F, on a vehicle barrier can be determined by equating its KE with the work done or energy absorbed by the car body at impact.

$$F\delta = \tfrac{1}{2}m\text{v}^2 = \frac{1}{2}\left(\frac{W}{g}\right)\text{v}^2$$

$$F = \frac{1}{2\delta}\left(\frac{W}{g}\right)\text{v}^2$$

$$W = 6000 \text{ lbf}$$
$$g = 32.2 \text{ ft/sec}^2$$
$$\text{v} = 35 \text{ mph} = \left(35 \ \frac{\text{mi}}{\text{hr}}\right)\left(1.47 \ \frac{\frac{\text{ft}}{\text{sec}}}{\frac{\text{mi}}{\text{hr}}}\right) = 51.45 \text{ fps}$$
$$\delta = 18 \text{ ft} - 16 \text{ ft} = 2 \text{ ft}$$

The impact force on the barrier is

$$\begin{aligned}F &= \frac{6000 \text{ lbf}}{(2)(2 \text{ ft})\left(32.2 \ \frac{\text{ft}}{\text{sec}^2}\right)}\left(51.45 \ \frac{\text{ft}}{\text{sec}}\right)^2 \\ &= 123{,}312 \text{ lbf} \quad (120{,}000 \text{ lbf})\end{aligned}$$

The answer is (D).

188. See the *NCEES Handbook*, Civil Engineering section, Table 1-1 W Shapes Dimensions and Properties, AISC Table 4-14 Available Critical Stress for Compression Members. The brace is in compression. Once it buckles, the frame will become unstable. Therefore, check the buckling capacity of the brace.

Use Table 1-1, W Shapes Dimensions and Properties, for W10 × 54.

$$A = 15.8 \text{ in}^2$$
$$r_y = 2.56 \text{ in}$$

Find the brace length.

$$\begin{aligned}el &= \sqrt{(12 \text{ ft})^2 + (30 \text{ ft})^2} \\ &= (32.31 \text{ ft})\left(12 \ \frac{\text{in}}{\text{ft}}\right) \\ &= 388 \text{ in}\end{aligned}$$

For the pinned-pinned case, the value of K is 1.

The slenderness ratio for the pinned-pinned end condition is

$$\frac{Kl}{r_y} = \frac{(1)(388\ \text{in})}{2.56\ \text{in}} = 1.52$$

It is known that the steel yield stress is 50 ksi. Therefore, use AISC Table 4-14 to determine the available critical stress of the brace for the slenderness ratio.

$$\phi F_{\text{cr}} = 9.78\ \text{ksi}$$

$$P_{\text{cr}} = \phi F_{\text{cr}} A = (9.78\ \text{ksi})\left(15.8\ \text{in}^2\right) = 155\ \text{kips}$$

Loads P and P_{cr} are related.

$$\begin{aligned} \theta &= \arctan(12\ \text{ft}/30\ \text{ft}) = 21.8^\circ \\ P &= P_{\text{cr}} \cos\theta = (155\ \text{kips})(\cos 21.8^\circ) \\ &= (155\ \text{kips})(0.93) \\ &= 144\ \text{kips} \end{aligned}$$

The answer is (B).

189. The deflection of a member depends on its stiffness. The stiffer a member, the less it deflects. The stiffness is inversely proportional to the product of the modulus of elasticity of the material, E, and the moment of inertia of the beam, I. Therefore, the deflection would decrease as EI is increased. Since E is the same for all sections, select the section with the largest moment of inertia. Section D has the largest moment of inertia.

The answer is (D).

190. In LRFD, the applied loads are factored. The LRFD denotes load and resistance factor design.

The answer is (C).

191. The cantilever is subjected to a negative bending moment. The deflected shape is also shown in the Mechanics of Materials section of the *NCEES Handbook*. The bending moment induces tension in the top fibers. Therefore, the bars should be placed near the top face. If the beam width is insufficient to accommodate all the bars in one row with minimum spacing, then they can be placed in two or more layers.

The answer is (A).

192. The column interaction diagram is based on column's size, concrete strength, amount, and the yield strength of the steel. The diagram is used to evaluate columns with different combinations of column axial load, P, and eccentricity, e. The y-axis represents the axial loading, P. The x-axis represents the bending moment, which is the product of the axial load and its eccentricity. The column with loadings enclosed within the diagram are considered adequate for the column. Point A, which corresponds to the maximum bending moment, is called the balanced point. The column failure mode is compressive above point A and tensile below point A. See the illustration given below.

The answer is

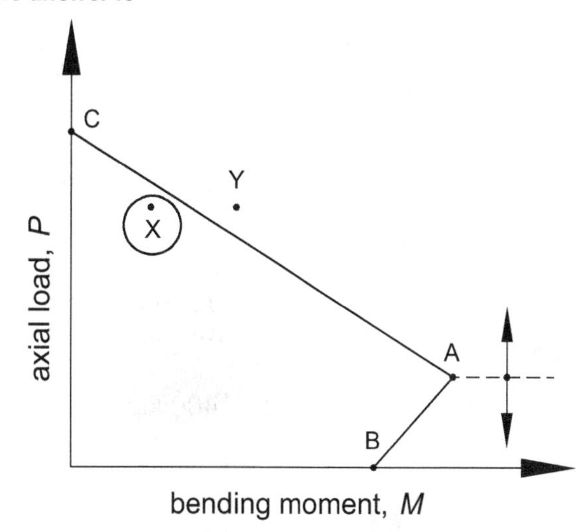

193. AASHTO soil classification is used mainly for roadwork. It groups soils into eight groups, namely A-1 through A-7-6. For groups A4 through A7, it is also necessary to report its group index (GI) with the classification. The higher the GI, the less suitable the soil. The plasticity index is

$$\begin{aligned} \text{PI} &= \text{liquid limit, LL} - \text{plastic limit, PL} \\ &= 52 - 35 \\ &= 17 \end{aligned}$$

The percentage of fine passing of sieve No. 200 is

$$F = 50$$

Per AASHTO, soil with more than 35% passing No. 200 sieve is classified as silty clay (A-4 to A-7).

For LL = 52, the choice is limited to three types: A-5, A-7-5 and A-7-6.

The soil group A-5 can be eliminated because it requires PI $\leq$ 10. In this case, PI = 17.

By definition, plasticity index of A-7-5 subgroup is equal to or less than LL – 30. In this case,

$$\text{PI} = 17 < \text{LL} - 30 = 52 - 30 = 22$$

The condition is satisfied, so the soil group is A-7-5.

Find the group index.

$$\begin{aligned} \text{GI} &= (F-35)\big(0.2+(0.005)(\text{LL}-40)\big) \\ &\quad +(0.01)(F-15)\ (\text{PI}-10) \\ \text{GI} &= (50-35)\big(0.2+(0.005)(52-40)\big) \\ &\quad +(0.01)\ (50-15)(17-10) \\ &= 6.35 \quad (6) \end{aligned}$$

The soil group is A-7-5 (6).

The answer is (A).

194. The soil pressure under the footing is a near-maximum allowable pressure of 6 ksf. For the soil pressure to remain within the allowable limit requires that the footing be placed so that centroid of column loads is in line with soil-resisting pressure. For a uniform soil pressure, the load is uniformly distributed over the 15 ft length, as shown in the illustration.

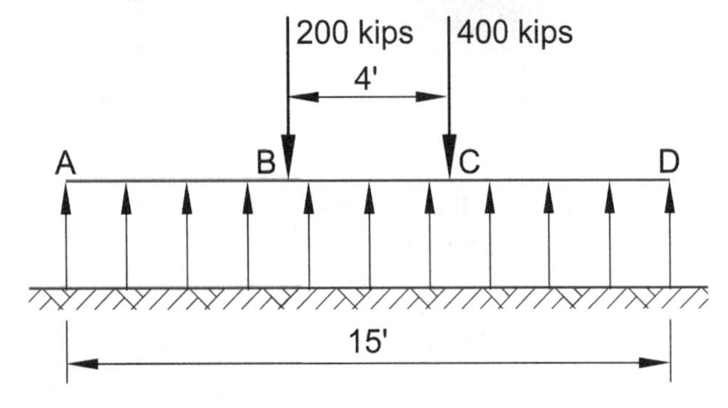

The total soil-resisting force is

$$F = 400 \text{ kips} + 200 \text{ kips} = 600 \text{ kips}$$

Let distance AB be x.

Take the moment about point A. For this, two column loads and their points of application are known. On resistance side, the total applied load of 600 kips is uniformly distributed under the footing, so that its center of resistance coincides with the center of the footing, which is at a distance of 7.5 ft from point A. Using equilibrium,

$$\begin{aligned} \sum M_A &= 0 \\ (200 \text{ kips})x + (400 \text{ kips})(x+4 \text{ ft}) & \\ -(600 \text{ kips})\left(\frac{15 \text{ ft}}{2}\right) &= 0 \end{aligned}$$

$$\begin{aligned} 600x + 1600 - 4500 &= 0 \\ 600x &= 2900 \\ x &= \frac{2900}{600} \\ &= 4.83 \text{ ft} \quad (4 \text{ ft } 10 \text{ in}) \end{aligned}$$

The answer is (A).

195. There are two types of unit weights: wet unit weight and dry unit weight. Unless specified in the problem as the dry unit weight, assume that the given unit weight pertains to wet soil.

The soil sample weight, W, is 222 g. The sample volume, V, is

$$\begin{aligned} V &= \text{container capacity} - \begin{array}{c}\text{water needed to fill the} \\ \text{container with sample inside}\end{array} \\ &= 500 \text{ cm}^3 - 390 \text{ cm}^3 \\ &= 110 \text{ cm}^3 \end{aligned}$$

By oven drying, the water evaporates, leaving the pores filled with air. The soil volume does not decrease. The total unit weight is

$$\gamma = \frac{W}{V} = \frac{222 \text{ g}}{110 \text{ cm}^3} = 2.02 \text{ g/cm}^3$$

The answer is (A).

196. The problem involves using the given relationships and formulas to determine the unknown. The relatively density is defined as

$$D_r = \left(\frac{(\gamma_{D\text{ field}} - \gamma_{D\text{ min}})}{(\gamma_{D\text{ max}} - \gamma_{D\text{ min}})}\right)\left(\frac{\gamma_{D\text{ max}}}{\gamma_{D\text{ field}}}\right)(100\%)$$

$$\begin{aligned} \gamma_{D\text{ min}} &= 110 \text{ pcf} \\ \gamma_{D\text{ max}} &= 140 \text{ pcf} \end{aligned}$$

Substitute the values into the equation. Rearrange the equation and find the minimum acceptable dry density of the compacted soil.

$$D_r = \left[\frac{\left(\gamma_{D\text{ field}} - 110 \ \frac{\text{lbf}}{\text{ft}^3}\right)}{\left(140 \ \frac{\text{lbf}}{\text{ft}^3} - 110 \ \frac{\text{lbf}}{\text{ft}^3}\right)}\right]\left[\frac{140 \ \frac{\text{lbf}}{\text{ft}^3}}{\gamma_{D\text{ field}}}\right](100\%) = 70\%$$

$$\gamma_{D\text{ field}} = 129.4 \text{ pcf} \quad (130 \text{ pcf})$$

The answer is (C).

197. The concrete block is fully submerged in water, as shown.

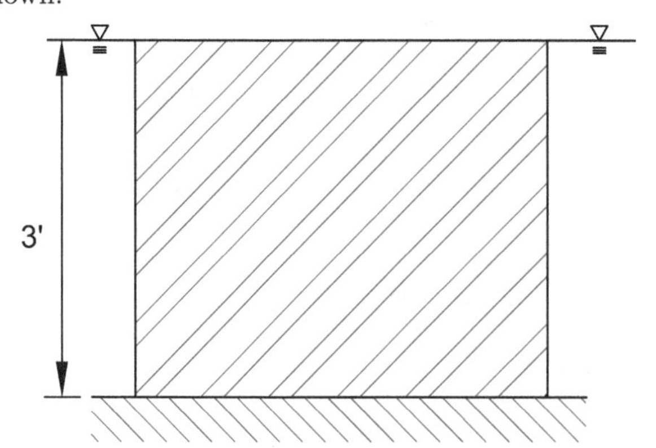

The concrete density in the submerged condition is

$$\begin{aligned}\text{concrete density} &= \gamma - \gamma_w \\ &= 120\ \frac{\text{lbf}}{\text{ft}^3} - 62.4\ \frac{\text{lbf}}{\text{ft}^3} \\ &= 57.6\ \text{pcf}\end{aligned}$$

The concrete submerged weight is

$$W = \left(57.6\ \frac{\text{lbf}}{\text{ft}^3}\right)(3\ \text{ft})(3\ \text{ft})(3\ \text{ft}) = 1555\ \text{lbf}$$

The contact area is

$$A = (3\ \text{ft})(3\ \text{ft}) = 9\ \text{ft}^2$$

Find the contact pressure.

$$\text{contact pressure} = \frac{1555\ \text{lbf}}{9\ \text{ft}^2} = 172.77\ \text{psf} \quad (172\ \text{psf})$$

The answer is (B).

198. The overturning moment is the product of the resultant pressure and the block's centroid distance above the base. In this case, it can be the sum of the products for each pressure block. For the triangular pressure block, the centroid distance is located one-third the distance above the base. For the rectangular pressure block, the centroid distance is located at a distance of 3.5 ft above the base, as noted in the illustration.

Find the overturning moments caused by the pressure blocks A_1, A_2, and A_3.

$$M_{A1} = P_{A1}\left(\frac{5}{3}\ \text{ft} + 7\ \text{ft}\right) = (500\ \text{lbf})(8.67\ \text{ft}) = 4333\ \text{ft-lbf}$$

$$M_{A2} = P_{A2}\left(\frac{7}{2}\ \text{ft}\right) = (1400\ \text{lbf})(3.5\ \text{ft}) = 4900\ \text{ft-lbf}$$

$$M_{A3} = P_{A3}\left(\frac{7}{3}\ \text{ft}\right) = (470\ \text{lbf})(2.33\ \text{ft}) = 1095\ \text{ft-lbf}$$

Find the overturning moment caused by the pore pressure.

$$M_W = P_w\left(\frac{7}{3}\ \text{ft}\right) = (1529\ \text{lbf})(2.33\ \text{ft}) = 3568\ \text{ft-lbf}$$

The base moment per unit length of the wall is

$$\begin{aligned}\sum M_i &= 4333\ \text{ft-lbf} + 4900\ \text{ft-lbf} \\ &\quad + 1095\ \text{ft-lbf} + 3568\ \text{ft-lbf} \\ &= 13{,}896\ \text{ft-lbf} \quad (14{,}000\ \text{ft-lbf})\end{aligned}$$

The answer is (D).

199. No information about the water table is given. Assume unsaturated soil (i.e., no pore pressure).

Shear stress at failure is

$$\tau_F = c + \sigma_N \tan\phi$$

Since the soil is sandy, its cohesion, c, is 0.

The illustration shows the relationship between shear stress and normal stress at failure.

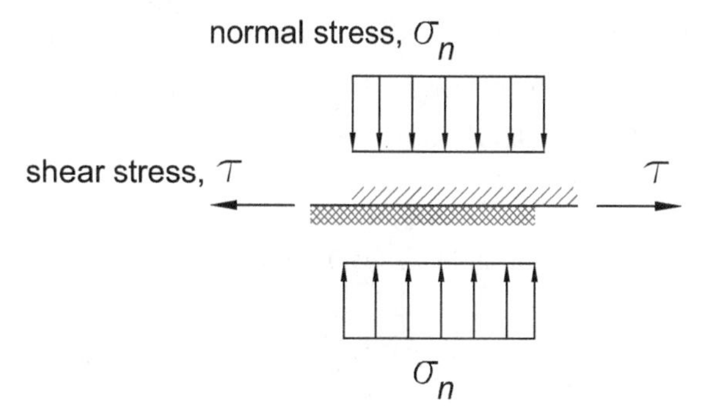

The normal stress at a depth of 20 ft is

$$\sigma_n = (20\ \text{ft})\left(115\ \frac{\text{lbf}}{\text{ft}^3}\right) = 2300\ \text{psf}$$

The corresponding shear stress is

$$\tau_F = \frac{0 + \left(2300\ \frac{\text{lbf}}{\text{ft}^2}\right)\tan 35^\circ}{\left(12\ \frac{\text{in}}{\text{ft}}\right)^2} = 11.2\ \text{psi}\quad (10\ \text{psi})$$

The answer is (B).

200. For a circular drilled pier of D feet in diameter, the area of the cross section is

$$A = \frac{\pi D^2}{4}$$

The load-carrying capacity is

$$P = A\ q_{\text{bearing}}$$

$$P = 600\ \text{kips} = \left(\frac{\pi D^2}{4}\right)(12\ \text{ksf}) = 3\pi D^2$$

Rearrange the equation and find the diameter.

$$\begin{aligned} D^2 &= \frac{600}{3\pi}\ \text{ft}^2 = 63.66\ \text{ft}^2 \\ D &= 7.93\ \text{ft}\quad (8\ \text{ft}) \end{aligned}$$

The answer is (B).

201. The problem states that the soil is a normally compacted clay. This assumption precludes overconsolidated clay and limits the choice to the virgin compression range on the e-log p curve. The compressive index, C_C, is defined as the slope in the virgin compression portion of consolidation curve.

$$\begin{aligned} C_C &= 0.009(\text{LL} - 10) = (0.009)(40 - 10) \\ &= (0.009)(30) \\ &= 0.27 \end{aligned}$$

The compression index is given by the equation

$$C_C = \frac{\Delta e}{\Delta \log_{10} p}$$

Rearrange the equation and find the change in void, Δe.

$$\Delta e = C_C(\Delta \log p)$$

$$\begin{aligned} \Delta e &= (0.27)(\log_{10} p_2 - \log_{10} p_1) \\ &= (0.27)\ \log_{10}\frac{p_2}{p_1} \\ &= (0.27)\ \log_{10}\frac{2500\ \frac{\text{lbf}}{\text{in}^2}}{1500\ \frac{\text{lbf}}{\text{in}^2}} \\ &= (0.27)\ \log_{10} 1.666 \\ &= 0.06 \end{aligned}$$

The answer is (A).

202. Since the clay layer is between two pervious layers, water drains out from both faces adjoining the sandy soils, allowing the clay layer to settle. The ultimate consolidation settlement is

$$S_{\text{ULT}} = (\text{unit strain})(\text{length}) = \varepsilon_V H_S$$

The average ultimate strain in the soil layer is

$$\varepsilon_V = \frac{\Delta e_{\text{TOT}}}{(1 + e_0)}$$

$$\Delta e_{\text{TOT}} = \begin{array}{l}\text{total change in void ratio due to} \\ \text{recompression and virgin compression}\end{array}$$

In this case, the soil is not overconsolidated. Therefore, the entire void ratio change is due to virgin compression. The initial void ratio, e_0, pertains to the virgin compression.

$$\begin{aligned} \Delta e_{\text{TOT}} &= C_C \log_{10}\frac{p_2}{p_1} = (0.18)\log_{10}\frac{2500\ \frac{\text{lbf}}{\text{in}^2}}{1500\ \frac{\text{lbf}}{\text{in}^2}} \\ &= (0.18)(0.22) \\ &= 0.04 \end{aligned}$$

The average ultimate strain is

$$\varepsilon_V = \frac{\Delta e_{\text{TOT}}}{(1 + e_0)} = \frac{0.04}{1 + 0.7} = 0.024$$

Find the total settlement of the clay layer.

$$\begin{aligned} S_{\text{ULT}} &= \varepsilon_V H_S = (0.024)(10\ \text{ft}) = 0.24\ \text{ft} \\ &= 2.9\ \text{in}\quad (3.0\ \text{in}) \end{aligned}$$

The answer is (C).

203. To be effective in erosion control, the control method should be able to reduce high velocities of runoff water and to dissipate its energy so that the surface flow cannot erode the soil surface it flows on. There are two methods of soil erosion control: natural and structural. In natural methods, the terrain of the area is not disturbed at all or is minimally disturbed. In structural methods, structural elements are introduced, and the soil slope is engineered to control erosion. This problem is about the natural methods.

Seeding and fertilizing of the area is typically the most effective natural method of erosion control. Options A and C are also quite effective. Option B is the least effective since the vegetative filler strip is planted parallel to the runoff flow, and thus it offers no resistance to the runoff water. To be effective, the vegetative filler strip should be planted perpendicular to the runoff flow.

The answer is (B).

204. The design speed for a superelevated horizontal curve can be determined by the following formula

$$0.01e + f = \frac{V^2}{15R}$$

$$e = 8$$
$$R = 950 \text{ ft}$$
$$f = 0.12$$

Find the design speed.

$$\begin{aligned} V^2 &= 15R(0.01e + f) \\ &= (15)(950 \text{ ft})\big((0.01)(8) + 0.12\big) \\ &= 2850 \\ V &= 53.3 \text{ mph} \end{aligned}$$

The posted speed should be lower than the calculated speed, or 50 mph.

The answer is (C).

205. Use the stopping sight distance (SSD) formula given in the Stopping Sight Distance section of the *NCEES Handbook*.

$$\text{SSD} = \text{distance travelled during reaction time} + \text{braking distance}$$

$$\text{SSD} = 1.47Vt + \left(\frac{V^2}{30\left(\left(\frac{a}{32.2}\right) \pm G\right)}\right)$$

The first term in the above equation pertains to the driver's reaction distance and the second term pertains to the braking distance. The skid marks pertain to the braking distance. The braking distance is

$$S_b = \frac{V^2}{30\left(\left(\frac{a}{32.2}\right) \pm G\right)} = 80 \text{ ft}$$

The term a denotes deceleration.

From the basic friction formula, the force is

$$F = \mu N$$

The friction coefficient, f, is 0.6.

$$\mu = \frac{F}{N} = \frac{ma}{mg} = \frac{a}{g}$$

$$\frac{a}{32.2 \ \frac{\text{ft}}{\text{sec}^2}} = f = 0.6$$

G = the percentage grade divided by 100 = 0.03.

Find the speed by rearranging the braking distance equation.

$$\begin{aligned} S_b &= \frac{V^2}{30((0.6) + 0.03)} = \frac{V^2}{18.9} = 80 \text{ ft} \\ V^2 &= (80 \text{ ft})(18.9) \\ &= 1512 \\ V &= 38.9 \text{ mph} < 40 \text{ mph posted limit} \end{aligned}$$

The car was traveling within the posted speed limit when the brakes were applied.

The answer is (B).

206. Centripetal force is what pulls a rotating object towards the center of rotation. For example, if one swings a ball on a rope around horizontally, the centripetal force is the pull on the rope so the ball doesn't fly away. High superelevation rates can be difficult to attain in urban settings due to closely spaced intersections, numerous driveways, and limited right of ways. Therefore, option D is incorrect. Options A, B, and C are correct.

The answer is (D).

207. Options B, C, and D are correct. The SN method is used for ESALs greater than 50,000 for the design period. The roads with lower ESALs are considered low-volume roads. The use of SN in designing pavements is given in the Highway Pavement Design section of the *NCEES Handbook*. Option A is incorrect.

The answer is (A).

208. The speed at which a pedestrian crosses the walkway is not given. So, the default speed of 3.5 ft/sec is assumed, as given in the Traffic Signal Timing section in the *NCEES Handbook*. The minimum green light time, in seconds, for pedestrians crossing is given by

$$G_p = 3.2 + \frac{L}{S_p} + 0.27N_{\text{ped}}$$

$$L = \text{crosswalk length} = 105 \text{ ft}$$

$$S_p = \text{average ped. speed} = 3.5 \text{ ft/sec}$$

$$N_{\text{ped}} = \text{number of pedestrians in interval} = 10$$

Find the minimum green light time interval.

$$G_p = 3.2 + \frac{105 \text{ ft}}{3.5 \ \frac{\text{ft}}{\text{sec}}} + (0.27)(10) = 35.9 \text{ sec} \quad (36 \text{ sec})$$

The answer is (C).

209. The purpose of the yellow (or amber) light is two-fold as follows

1. Allow vehicles, which cannot stop at the stopping line using comfortable deceleration or which are in the intersection, to clear the intersection, such as vehicle #1 shown in the illustration.
2. Alert the motorist approaching the intersection to the fact that the light is about to change to red and they should decelerate and stop before entering the intersection, such as vehicle #2 shown in the illustration.

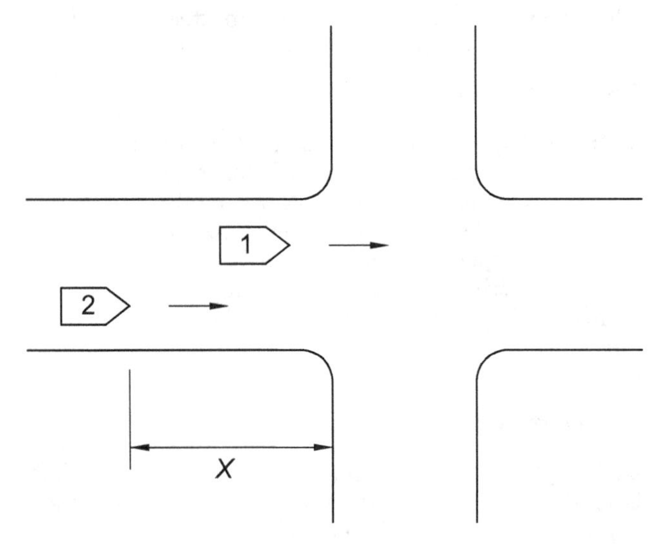

The problem concerns item 2 above. In this case, the length of the yellow light time interval is, therefore, the time needed to stop a vehicle without excessive deceleration. The time interval, y, is expressed as

$$y = \text{driver's perception time} + \text{braking time}$$

$$y = t + \frac{\text{v}}{2a \pm 64.4G}$$

Since no grade is mentioned, assume $G = 0$. Find the time interval of the yellow light.

$$y = t + \frac{\text{v}}{2a}$$

$$t = 2.3 \text{ sec}$$

$$a = 11 \text{ ft/sec}^2$$

$$\text{v} = 40 \text{ mph} = \left(40 \ \frac{\text{mi}}{\text{hr}}\right)\left(1.47 \ \frac{\frac{\text{ft}}{\text{sec}}}{\frac{\text{mi}}{\text{hr}}}\right) = 58.8 \text{ ft/sec}$$

$$y = 2.3 \text{ sec} + \frac{58.8 \ \frac{\text{ft}}{\text{sec}}}{(2)\left(11 \ \frac{\text{ft}}{\text{sec}^2}\right)}$$

$$= 4.97 \text{ sec} \quad (5.0 \text{ sec})$$

The answer is (C).

210. There are two types of mean speeds: time mean speed and space mean speed. The time mean speed is the arithmetic mean, and the space mean speed is the harmonic mean of the speeds of vehicles passing at a point on a highway during an interval of time. The time mean speed can be determined by

$$\overline{\text{v}}_t = \frac{1}{n}\sum_{i=1}^{n} \text{v}_i$$

In this case, $n = 6$ and the vehicles speeds are shown in the illustration.

$$\overline{\text{v}}_t = \frac{30 \ \frac{\text{mi}}{\text{hr}} + 30 \ \frac{\text{mi}}{\text{hr}} + 35 \ \frac{\text{mi}}{\text{hr}} + 36 \ \frac{\text{mi}}{\text{hr}} + 40 \ \frac{\text{mi}}{\text{hr}} + 45 \ \frac{\text{mi}}{\text{hr}}}{6} = 36 \text{ mph}$$

The answer is (B).

211. See the Traffic Flow Relationships diagrams in the Transportation section of the *NCEES Handbook*. The density-speed relationship is linear. As the density increases, speed decreases. When density reaches its maximum, the vehicles are lined up or packed bumper to bumper, and the flow comes to a halt. This is called jam density, k_j. In other words, $\text{v} = 0$ at k_j. The value of

k_j can be determined by inserting v = 0 in the given equation.

$$2\left(0\ \frac{\text{mi}}{\text{hr}}\right) + k_j - 120 = 0$$
$$k_j = 120 \text{ vehicles/mi}$$

The equation is plotted as shown.

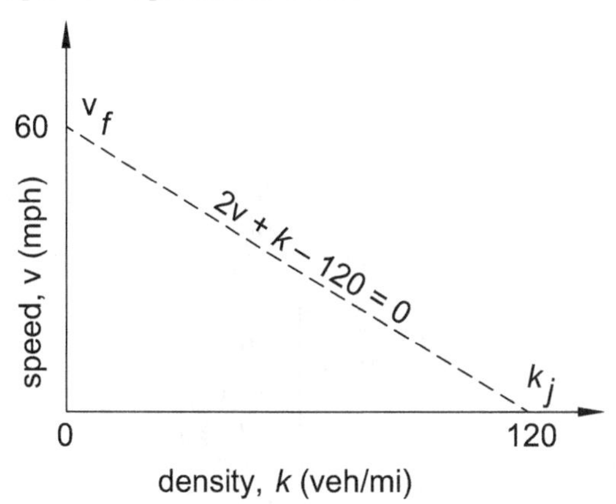

The answer is (C).

212. The rate per million of entering vehicles (RMEV) is defined as the number of crashes per million vehicles entering the intersection.

$$\text{RMEV} = \frac{A(1{,}000{,}000)}{V}$$

Find the RMEV.

$$\text{study period} = 1 \text{ yr} = 365 \text{ days}$$
$$\begin{array}{c}\text{crashes per yr}\\ \text{at intersection}\end{array} = A = 20$$
$$\begin{aligned} V &= (\text{average daily traffic})(365 \text{ days}) \\ &= \left(10{,}000\ \frac{\text{vehicles}}{\text{day}}\right)(365 \text{ days}) \\ &= 3{,}650{,}000 \text{ vehicles/yr} \end{aligned}$$
$$\begin{aligned} \text{RMEV} &= \frac{(20)(1{,}000{,}000)}{3{,}650{,}000\ \frac{\text{vehicles}}{\text{yr}}} \\ &= 5.48 \end{aligned}$$

The answer is 5.48.

213. Compare the total cost for bids A and B.

$$\begin{aligned} \text{bid A} &= \left(10{,}000 \text{ yd}^3\right)(\$8) + \left(1000 \text{ yd}^3\right)(\$15) \\ &= \$95{,}000 \\ \text{bid B} &= \left(10{,}000 \text{ yd}^3\right)(\$4.40) + \left(1000 \text{ yd}^3\right)(\$50) \\ &= \$94{,}000 \end{aligned}$$

Bid B is a low-cost bid. However, if the work quantities were to change so that the soil is 100 yd^3 less and rock is 100 yd^3 more than the estimate, the new cost would be

$$\begin{aligned} \text{bid A} &= \left(9900 \text{ yd}^3\right)(\$8) + \left(1100 \text{ yd}^3\right)(\$15) \\ &= \$95{,}700 \\ \text{bid B} &= \left(9900 \text{ yd}^3\right)(\$4.40) + \left(1100 \text{ yd}^3\right)(\$50) \\ &= \$98{,}560 \end{aligned}$$

In this case, bid B is no longer the low-cost bid.

In bid B, the contractor lowered cost on the soil excavation and disproportionally raised the cost on rock excavation. This makes bid B attractive initially, but it is an unbalanced bid. Such bids are undesirable and, when detected, they should be rejected. Options B and C call for subdividing the excavation work, which is impractical.

The answer is (D).

214. The contract price was based on the unit cost and actual crack lineage. The unit cost includes the contractor's 10% net profit.

$$\begin{aligned} \text{contract price} &= (\text{actual crack lineage})(\text{unit repair cost}) \\ &= (30{,}000 \text{ ft})\left(\frac{\$5}{\text{ft}}\right) \\ &= \$150{,}000 \end{aligned}$$

Because of the terms of the contract, the owner's estimate and budget are not relevant.

The answer is (C).

215. Since concrete is fully liquid when placed between the wall forms, the Rankine active coefficient, K_A, for the newly placed concrete against the formwork is 1.0. The lateral pressure against the wall at distance h below the top is shown in the illustration.

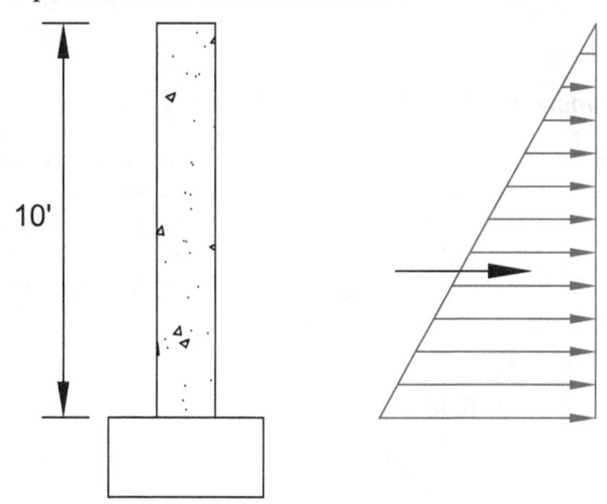

The pressure is

$$p = wh$$

The concrete density, w, is 145 pcf, and the formwork height, h, from base to the top is 10 ft.

$$p = \left(145\ \frac{\text{lbf}}{\text{ft}^3}\right)(10\ \text{ft}) = 1450\ \text{psf}$$

Find the total force against the formwork per unit length.

$$\begin{aligned} P &= \left(1450\ \frac{\text{lbf}}{\text{ft}^2}\right)\left(\frac{10\ \text{ft}}{2}\right)(1\ \text{ft}) \\ &= 7250\ \text{lbf} \quad (7000\ \text{lbf}) \end{aligned}$$

The answer is (C).

216. Options A, B, and C are correct, and option D is incorrect. For compressed air used for cleaning, reduce the pressure to 30 psi, and only use it with the proper guards and other protective equipment. Never clean your clothing with compressed air; you could be injured by particles driven into your eyes or skin by the force.

The answer is (D).

217. Calculate the budgeted cost of the entire work.

$$\text{budgeted cost} = \frac{\$1{,}100{,}000}{(1 + 10\%)} = \$1{,}000{,}000$$

Total material to be hauled and compacted is 100,000 yd^3. Find the budgeted cost per cubic yard.

$$\text{budgeted cost per cu. yd} = \frac{\$1{,}000{,}000}{100{,}000\ \text{yd}^3} = \$10\ \text{per yd}^3$$

Material hauled is 15,000 yd^3. Find the budgeted cost of work performed (BCWP).

$$\text{BCWP} = \left(\frac{\$10}{\text{yd}^3}\right)(15{,}000\ \text{yd}^3) = \$150{,}000$$

The answer is $150,000.

218. A CPM arrow diagram is also called an activity-on-arrow network. In this diagram, the nodes are events, and the arrows represent activity. To complete an event, all activities culminating at the node must be completed. The minimum time required to finish the project is 36 days, with activities A, B, F, D, G, and E on the critical path.

The answer is (C).

219. See the illustration showing the wall dimensions. Building the wall requires formwork on both faces and on both vertical ends of the wall. No formwork is required at the base and at the top.

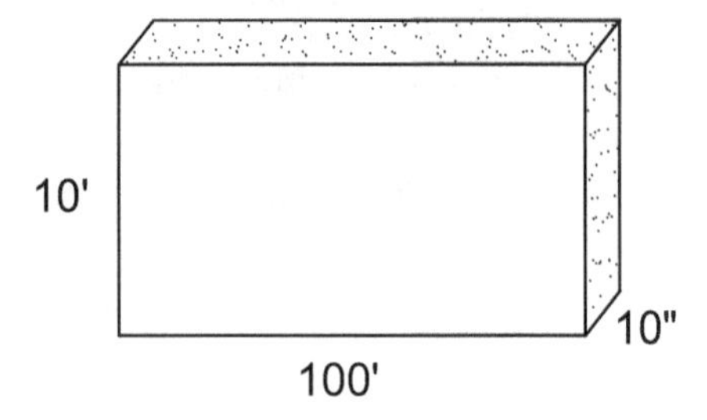

Find the total footage area of the forms.

$$\begin{aligned} \text{area} &= (100\ \text{ft})(10\ \text{ft})(2\ \text{faces}) \\ &\quad + (10\ \text{ft})\left(\frac{10\ \text{in}}{12\ \frac{\text{in}}{\text{ft}}}\right)(2\ \text{ends}) \\ &= 2017\ \text{ft}^2 \end{aligned}$$

Find the cost of leasing forms and the labor cost.

$$\text{leasing cost} = \left(\frac{\$2}{\text{ft}^2}\right)(2017\ \text{ft}^2) = \$4034$$

$$\begin{aligned} \text{labor cost for erecting forms} &= \left(\frac{\$5}{\text{ft}^2}\right)(2017\ \text{ft}^2) \\ &= \$10{,}085 \\ \text{labor cost for stripping forms} &= \left(\frac{\$2}{\text{ft}^2}\right)(2017\ \text{ft}^2) \\ &= \$4034 \end{aligned}$$

Find the total cost of the formwork.

$$\begin{aligned} \text{total cost} &= \$4034 + \$4034 + \$10{,}085 \\ &= \$18{,}153 \quad (\$18{,}000) \end{aligned}$$

The answer is (D).

220. The illustration shows the column reinforcement and section ties.

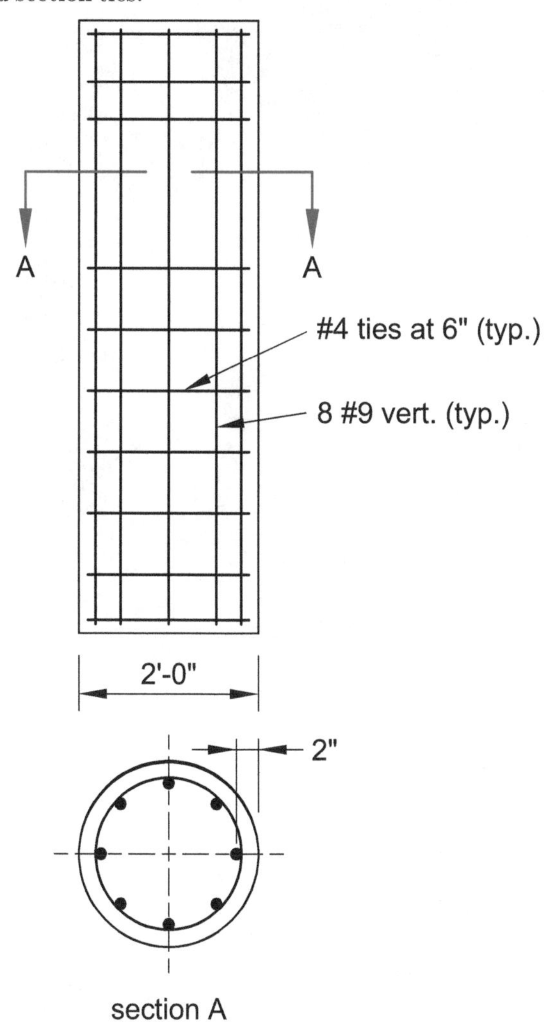

$$\text{weight of \#3 bar} = 0.376 \text{ lbf/ft}$$

$$\text{tie diameter} = 24 \text{ in} - 2 \text{ in} - 2 \text{ in} = 20 \text{ in}$$

$$\begin{aligned}\text{length of a tie bar} &= \pi D + 12 \text{ in} \\ &= \pi(20 \text{ in}) + 12 \text{ in} \\ &= 75 \text{ in}\end{aligned}$$

The number of ties per column is

$$\begin{aligned}\text{no. of ties per column} &= \left(\frac{12.75 \text{ ft}}{12 \text{ in}}\right)\left(12 \ \frac{\text{in}}{\text{ft}}\right) + 1 \\ &= 13.75 \cong 14\end{aligned}$$

Find the weight of column ties.

$$\begin{aligned}\text{weight} &= \left(\frac{14 \text{ ties}}{1 \text{ column}}\right)(100 \text{ columns}) \\ &\quad \times \left(\frac{75 \text{ in}}{12 \ \frac{\text{in}}{\text{ft}}}\right)\left(\frac{0.376 \ \frac{\text{lbf}}{\text{ft}}}{2000 \ \frac{\text{lbf}}{\text{ton}}}\right) \\ &= 1.65 \text{ tons}\end{aligned}$$

The answer is (B).

Solutions
Exam 3

221. Rewrite the given line equation in terms of y.

$$3x - 5y = 7$$
$$y = \frac{3}{5}x - \frac{7}{5}$$

The slope of the given line is

$$m_1 = 3/5$$

The slope of the perpendicular line is

$$m_2 = \frac{-1}{m_1} = \frac{-5}{3}$$

The equation of the perpendicular line is

$$y = \frac{-5}{3}x + b$$

The perpendicular line passes through (2, 5). Use these values of x and y to find the y-intercept, b.

$$b = \frac{5}{3}x + y = \left(\frac{5}{3}\right)(2) + 5$$
$$= 25/3$$

The equation of the perpendicular line is

$$y = \frac{-5}{3}x + \frac{25}{3}$$
$$3y = -5x + 25$$
$$5x + 3y = 25$$

Alternatively, the problem can be solved graphically as shown.

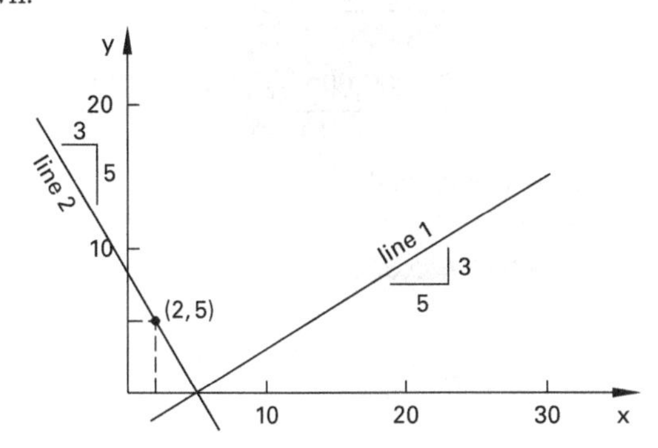

The answer is (D).

222. The intensity ratio is the ratio of the exponential values of the two earthquakes' Richter scale measurements.

$$\text{intensity ratio} = \frac{10^{8.4}}{10^{8.3}} = 1.26$$

The answer is 1.26.

223. The equation represents a parabola, as shown. The rotation of the parabola about its y-axis generates a two-dimensional paraboloid (bowl-like) shape.

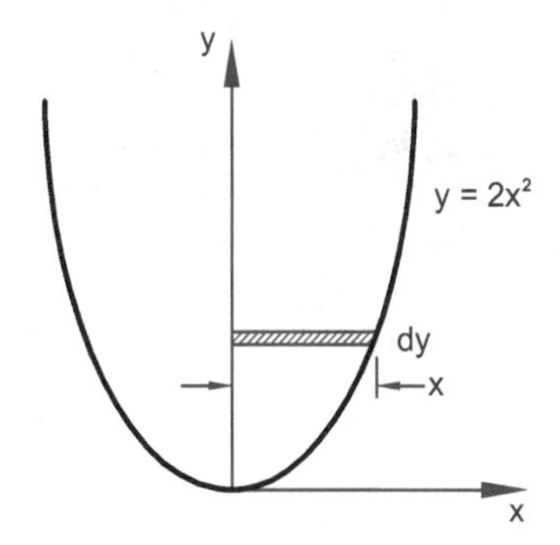

The area of a slice of width x rotated about the y-axis is found from the equation for the area of a circle.

$$A = \pi x^2$$

The equation for the volume of the slice is

$$dV = \pi x^2 dy = \frac{\pi(2x^2)}{2}dy$$
$$= \frac{\pi y}{2}dy$$

The volume between $y = 0$ and $y = 10$ is

$$V = \int dV = \int_0^{10} \frac{\pi y}{2}dy = \frac{\pi}{2}\int_0^{10} y\,dy$$
$$= \frac{\pi}{2}\frac{y^2}{2} = \frac{\pi}{4}(10^2 - 0^2) = \frac{100\pi}{4}$$
$$= 25\pi$$

The answer is 25π.

224. The angle between two vectors can be determined by using dot product multiplication. The dot product is scalar and is defined by

$$\begin{aligned}\mathbf{A}\cdot\mathbf{B} &= AB\cos\theta \\ &= (2)(6)\mathbf{i}^2 + (2)(-3)\mathbf{j}^2 + (1)(-2)\mathbf{k}^2 \\ &= 12 - 6 - 2 = 4\end{aligned}$$

The scalar magnitude of vector **A** is

$$A = \sqrt{2^2 + 2^2 + 1^2} = \sqrt{9} = 3$$

The scalar magnitude of vector **B** is

$$B = \sqrt{6^2 + (-3)^2 + (-2)^2} = \sqrt{49} = 7$$

Find the angle between the vectors using the dot product identity and substituting the scalar values of **A** and **B**.

$$\begin{aligned}\cos\theta &= \frac{\mathbf{A}\cdot\mathbf{B}}{AB} = \frac{4}{(3)(7)} = \frac{4}{21} = 0.1905 \\ \theta &= \cos^{-1}0.1905 = 79.02^\circ\end{aligned}$$

The answer is 79.02°.

225. Since the productivity increase is always 5% of the productivity from the previous period, it is a geometric progression series. If the first term in the series is 100, the common ratio is

$$r = \frac{i\text{th term}}{(i-1)\text{th term}} = 105 \div 100 = 1.05$$

In total, 11 concrete placements are made: one before the training started, then one after each of the ten training sessions. For the geometric progression series, the value of the last term is

$$l = ar^{n-1} = (100\text{ yd}^3)(1.05)^{11-1} = 163\text{ yd}^3$$

The answer is 163 yd³.

226. The curvature K is defined as

$$K = \frac{y''}{[1 + (y')^2]^{\frac{3}{2}}}$$

Given the parabola described, y' and y'' are

$$\begin{aligned}y' &= 1 - \frac{x}{3} \\ y'' &= -\frac{1}{3}\end{aligned}$$

For $x = 3$,

$$\begin{aligned}y' &= 1 - \frac{3}{3} = 0 \\ y'' &= -\frac{1}{3}\end{aligned}$$

The curvature is

$$K = \frac{-\frac{1}{3}}{[1 + (0)^2]^{\frac{3}{2}}} = -\frac{1}{3}$$

The answer is –1/3.

227. For two events A and B in which neither $P(A)$ nor $P(B)$ is zero, the compound probability of both events is

$$P(A, B) = P(A)P(B|A) = P(B)P(A|B)$$

This probability property can be extended to more than two events. In this case, three events are considered. Since the balls drawn are not being replaced, the events are dependent. The first drawing has 10 balls to draw from, the second has 9 balls, and the third has 8 balls.

$$P = \left(\frac{1}{10}\right)\left(\frac{1}{9}\right)\left(\frac{1}{8}\right) = \frac{1}{720}$$

The answer is 1/720.

228. The weighted arithmetic mean of a set of values can be determined using the expression from the Engineering Probability and Statistics section of the *NCEES Handbook.*

$$\overline{X}_w = \frac{\sum w_i X_i}{\sum w_i}$$

Since one report is given twice the weight, $w_{\text{B}} = 1$ and $w_{\text{A}} = 2$. The expected cost is

$$\begin{aligned}\overline{X}_w &= \frac{w_{\text{A}}X_{\text{A}} + w_{\text{B}}X_{\text{B}}}{w_{\text{A}} + w_{\text{B}}} \\ &= \frac{(2)(\$125{,}000{,}000) + (1)(\$95{,}000{,}000)}{(2+1)} \\ &= \$115{,}000{,}000\end{aligned}$$

The answer is $115,000,000.

229. The coefficient of variation is generally expressed as a percentage rather than in terms of the units of the particular data. The sample coefficient of variation, CV, is

$$\begin{aligned} \text{CV} &= \frac{s}{\overline{X}} = \frac{6.6}{30} = 0.22 \\ &= 22\% \end{aligned}$$

The answer is 22.

230. Use the formula from the *NCEES Handbook* Least Squares section. To evaluate the parameters in the regression equations, tabulate the data.

	x	y	xy	x^2
	2	9	18	4
	3	11	33	9
	5	15	75	25
	9	22	198	81
Σ	19	57	324	119

Use the summed values in the formulas.

$$\begin{aligned} S_{xy} &= \sum_{i=1}^{n} x_i y_i - \left(\frac{1}{n}\right)\left(\sum x_i\right)\left(\sum y_i\right) \\ &= 324 - \frac{(19)(57)}{4} \\ &= 53.25 \end{aligned}$$

$$\begin{aligned} S_{xx} &= \sum_{i=1}^{n} (x_i)^2 - \left(\frac{1}{n}\right)\left(\sum x_i\right)^2 \\ &= 119 - \left(\frac{1}{4}\right)(19)^2 \\ &= 28.75 \end{aligned}$$

The slope is

$$\widehat{b} = \frac{S_{xy}}{S_{xx}} = \frac{53.25}{28.75} = 1.85$$

The equation for the y-intercept is

$$\widehat{a} = \overline{y} - \widehat{b}\,\overline{x} = \frac{\sum y}{4} - 1.85\left(\frac{\sum x}{4}\right)$$

Using the calculated values from the table,

$$\widehat{a} = \frac{57}{4} - 1.85\left(\frac{19}{4}\right) = 5.46$$

The line equation is

$$\begin{aligned} y &= \widehat{a} + \widehat{b}\,x \\ y &= 5.46 + 1.85x \end{aligned}$$

The answer is y = 5.46 + 1.85x.

231. Statements B, C, and D are correct. Statement A is a definition of wage properly earned and is not considered bribery. According to the Model Rules Sec. 240.15, a licensee shall not make any political contribution with the intent to influence the award of a contract by public authority, so statement E is incorrect.

The answer is (A) and (E).

232. Fraud, incompetence, negligence, gross incompetence, and gross negligence are examples of professional misconduct. Setting up and running a website that denies climate change is permitted as free speech under the U.S. Constitution. As such, it is not considered professional misconduct.

The answer is (D).

233. The building codes often lag behind the state-of-the-art, so the engineer in responsible charge should use the state-of-the-art methods, which include the governing code and other guidelines used by the industry. Option A is incorrect because a standard of perfection cannot be met. Option B is incorrect because meeting the minimum code requirement may not result in meeting the standard of care. Option D is incorrect because the engineer is not responsible for job site safety unless the engineer assumes such responsibility by contract or conduct.

The answer is (C).

234. A licensee who is an employee should inform the licensee's employer. A self-employed licensee should inform the licensee's client. In addition, a licensee should inform other authorities as appropriate. See Model Rules in the *NCEES Handbook*.

The answer is (D).

235. The contractor is not obligated to do the work for less than the price they offered, which is \$1.1M. However, the contractor can't demand that the owner accept their \$1.1M bid either. The contractor may accept the counteroffer, reduce the price and do the work for \$1M, but is not required to do so. A reasonable period to respond to an offer is within three months, so the contractor calling the bid expired would be unreasonable.

The answer is (A).

236. The monthly return is 0.5%. The number of monthly deposits in 5 years would be 60. From the economic factor tables in the *NCEES Handbook*, the minimum monthly deposit is

$$\begin{aligned} A &= \$500{,}000\left(\frac{A}{F}, 0.5\%, 60\right) \\ &= \$500{,}000(0.0143) \\ &= \$7150 \end{aligned}$$

The answer is $7150.

237. See the Progression and Series section in the Mathematics chapter of the *NCEES Handbook*. The cash flow increase is not linear, but it follows geometric progression. The growth rate shows a 50% increase every year from the previous year. Since the growth is given in inflated dollars, the dollar figures need to be deflated to account for general inflation using the interest rate per interest period. For a nominal growth rate, r, of 50% and an inflation rate, i, of 10%, the real growth rate is

$$\begin{aligned} \text{real growth rate} &= \frac{1+r}{1+i} - 1 \\ &= \frac{1+0.5}{1+0.1} - 1 \\ &= 1.36 - 1 \\ &= 0.36 \quad (36\%) \end{aligned}$$

The answer is 36.

238. The present worth, P, for Equipment 1 with a life expectancy of 10 years is

$$\begin{aligned} P_1 &= \text{initial cost} + (\text{annual maintenance})\left(\frac{P}{A}, 8\%, 10\right) \\ &\quad -(\text{salvage value})\left(\frac{P}{F}, 8\%, 10\right) \\ &= \$50{,}000 + (\$15{,}000)(6.7101) - (\$5000)(0.4632) \\ &= \$148{,}335.50 \end{aligned}$$

The present worth for Equipment 2 with a life expectancy of 15 years is

$$\begin{aligned} P_2 &= \text{initial cost} + (\text{annual maintenance})\left(\frac{P}{A}, 8\%, 15\right) \\ &\quad -(\text{salvage value})\left(\frac{P}{F}, 8\%, 15\right) \\ &= \$75{,}000 + (\$10{,}000)(8.5595) - (\$12{,}000)(0.3152) \\ &= \$156{,}812.60 \end{aligned}$$

Equipment 1 costs less than Equipment 2. Therefore, it is a better choice. The cost savings of choosing Equipment 1 is

$$\$156{,}812.60 - \$148{,}335.50 = \$8477.10 \quad (\$8500)$$

The answer is (A) and (E).

239. For a number of years after construction, t, the life-cycle cost equations for alternatives A and B are

$$A = \$1{,}000{,}000 + \$100{,}000t$$

$$B = \$1{,}800{,}000 + \$65{,}000t$$

The break-even point for the two options is

$$\begin{aligned} \$1{,}000{,}000 + \$100{,}000t &= \$1{,}800{,}000 + \$65{,}000t \\ t &= 22.9 \text{ years} \end{aligned}$$

This can be shown graphically as

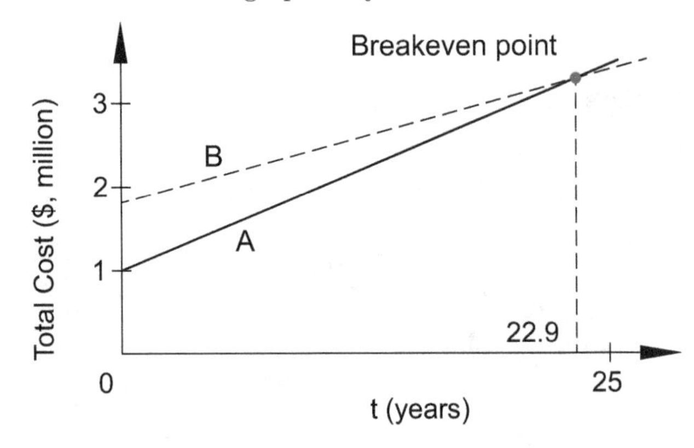

After 23 years of service life, alternative B becomes more economical. The service life of the project is 25 years, so the engineer should recommend alternative B.

The answer is (B).

240. The initial cost and annual cost are both expenses. The rate of return (ROR) on the investment is unknown. No factor table is available to determine ROR precisely. However, it can be determined by interpolating between two applicable factor tables. Equating the expenses and returns,

$$\begin{aligned} \$40{,}000 + (\$3000)\left(\frac{P}{A}, i, 5\right) &= (\$15{,}000)\left(\frac{P}{A}, i, 5\right) \\ \left(\frac{P}{A}, i, 5\right) &= 3.3333 \end{aligned}$$

The factor tables for interest rates of 12% and 18% provide P/A factors from which the required ROR can be interpolated.

$$\left(\frac{P}{A},12\%,5\right) = 3.6018$$

$$\left(\frac{P}{A},18\%,5\right) = 3.1272$$

Use linear interpolation between the 12% and 18% interest rates and their ROR. Use the line equation from the Straight Line section of the Mathematics chapter of the *NCEES Handbook*. The slope is

$$m = \frac{3.1272 - 3.6018}{0.18 - 0.12} = -7.91$$

By definition,

$$y_2 - y_1 = m(x_2 - x_1)$$

The ROR is

$$\begin{aligned} i - 0.12 &= \frac{(3.3333 - 3.6018)}{-7.91} = 0.034 \\ i &= 0.034 + 0.12 = 0.154 \\ &= 15.4\% \quad (15\%) \end{aligned}$$

The answer is 15%.

241. The decision to drill has two possible outcomes: finding oil and not finding oil. The probability of finding oil is $p_1 = 0.3$. The probability of not finding oil is

$$\begin{aligned} p_2 &= 1 - p_1 = 1 - 0.3 \\ &= 0.7 \end{aligned}$$

1. The decision to drill has two consequences:

(a) If oil is struck, the oil would be worth \$13M. With drilling cost being \$3M, the net payoff by drilling and striking oil is \$10M. It has a probability, p_1, of 0.3.

(b) If no oil is struck, the drilling cost is \$3M, with a probability, p_2, of 0.7.

The expected payoff of deciding to drill is

$$\begin{aligned} E_1 &= (\$10{,}000{,}000)(0.3) + (-3{,}000{,}000)(0.7) \\ &= \$900{,}000 \end{aligned}$$

2. The decision not to drill has two consequences: Sell the property to company B for \$1M with a probability, p_2, of 0.7, plus receive another \$1M contingent upon presence of oil (with a probability, p_1, of 0.3).

(a) If no oil is struck, the drilling rights are worth \$1M, with a probability, p_2, of 0.7.

(b) If oil is struck, the drilling rights are worth \$2M, with a probability, p_1, of 0.3.

The expected payoff of deciding not to drill and selling to Company B is

$$\begin{aligned} E_2 &= (\$1{,}000{,}000)(0.7) + (\$2{,}000{,}000)(0.3) \\ &= \$1{,}300{,}000 \end{aligned}$$

The maximum expected payoff is \$1,300,000 if Company A decides not to drill.

The answer is (B).

242. The tower is a plane determinate truss. See the Statics section of the *NCEES Handbook*. Two equilibrium conditions must be met.

$$\sum F_x = 0$$

$$\sum F_y = 0$$

For the pole AD to remain plumb, the horizontal components of the tension forces in each cable must be equal. Using the free-body diagram at joint A, resolve the forces in the x- and y-directions.

For the first equilibrium condition,

$$\begin{aligned} T_{AC}\cos 45^\circ - T_{AB}\cos 30^\circ &= 0 \\ T_{AC}\cos 45^\circ &= T_{AB}\cos 30^\circ \\ T_{AC}(0.707) &= (100\ \text{N})(0.866) \\ T_{AC} &= 122.5\ \text{N} \quad (122\ \text{N}) \end{aligned}$$

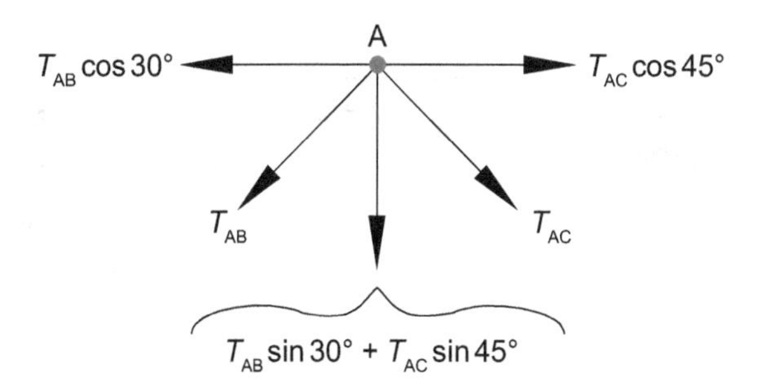

The answer is (A) and (E).

243. In the initial condition, cable AB is in tension under the vertical load, W. As the cable is pulled to point B, its inclination with respect to vertical is

$$\alpha = \tan^{-1}\frac{3\ \text{m}}{10\ \text{m}} = \tan^{-1}(0.3) = 16.7^\circ$$

Draw the free-body diagram.

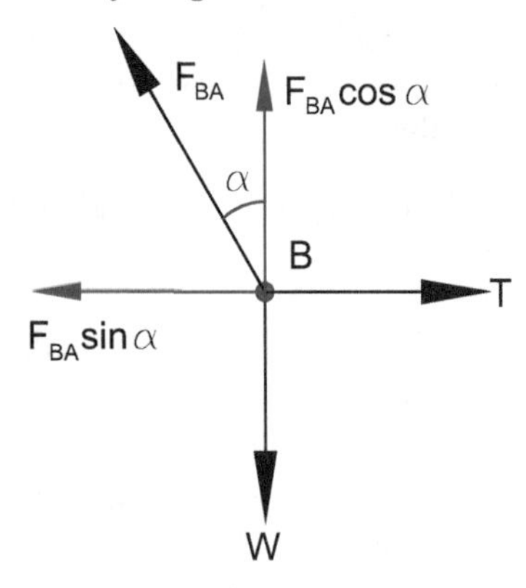

Find the equilibrium conditions at point B.

$$\sum F_y = F_{BA}\cos\alpha - W = 0$$
$$F_{BA} = \frac{W}{\cos\alpha}$$
$$\sum F_x = T - F_{BA}\sin\alpha = 0$$

The required pull is

$$\begin{aligned} T &= F_{BA}\sin\alpha = \frac{W\sin\alpha}{\cos\alpha} = W\tan\alpha \\ &= (1200\ \text{lbf})(\tan 16.7°) \\ &= 360\ \text{lbf} \end{aligned}$$

The answer is (B).

244. Substitute the uniformly distributed load (UDL) with a single load that represents the sum of the UDL acting at its centroid, as shown.

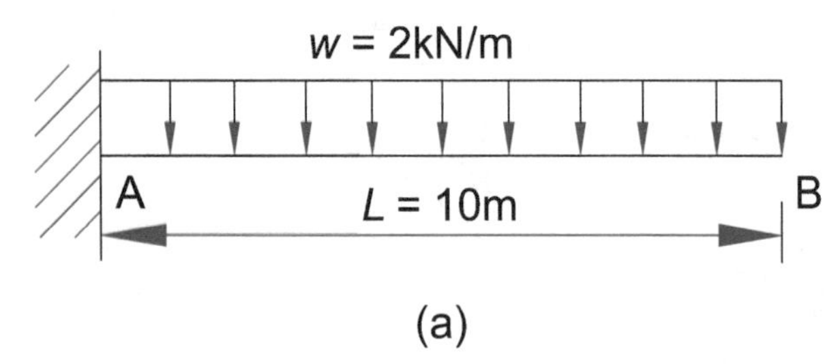

(a)

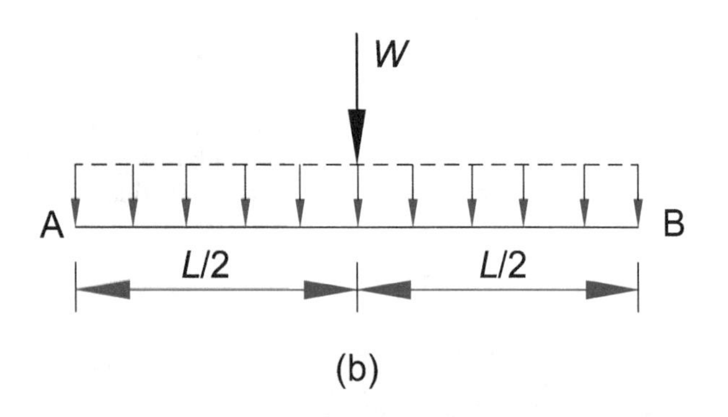

(b)

The sum of the UDL is

$$W = wL = \left(2\ \frac{\text{kN}}{\text{m}}\right)(10\ \text{m}) = 20\ \text{kN}$$

The distance from W to support A is

$$d_{W,A} = 0.5L = 5\ \text{m}$$

The moment at support A is

$$\begin{aligned} M_A &= (\text{force})(\text{distance}) = (20\ \text{kN})(5\ \text{m}) \\ &= 100\ \text{kN-m} \end{aligned}$$

The load causes clockwise moment at the support.

The answer is (A) and (D).

245. The inclination of cable AB with respect to vertical is

$$\alpha = \tan^{-1}\frac{3\ \text{m}}{10\ \text{m}} = \tan^{-1}(0.3) = 16.7°$$

Convert the mass to weight.

$$\begin{aligned} W &= (1000\ \text{kg})\left(9.81\ \frac{\text{m}}{\text{s}^2}\right) = 9810\ \text{N} \\ &= 9.81\ \text{kN} \end{aligned}$$

The required tension is

$$T = W\tan\alpha = (9.81\ \text{kN})(\tan 16.7°) = 2.94\ \text{kN} \quad (3\ \text{kN})$$

The answer is (A).

246. Assume that (1) an upward reaction is positive, and (2) the reaction at A is acting upward.

To determine the reaction at A, take the moment about B. Let counterclockwise moments be positive.

$$\begin{aligned} \sum M_B &= -R_A(10\ \text{m}) + (2\ \text{kN}) \\ &\quad \times(7\ \text{m}) + (3\ \text{kN})(3\ \text{m}) \\ &\quad -(4\ \text{kN})(2\ \text{m}) - (5\ \text{kN})(6\ \text{m}) \\ &= 0 \\ 10R_A &= 14 + 9 - 8 - 30 \\ &= -15\ \text{kN} - m \\ R_A &= -1.5\ \text{kN} \end{aligned}$$

R_A is negative, which shows that the reaction is acting downward to resist an uplift force.

The answer is (B) and (C).

247. The distance from any point to the centroid of the composite section can be determined using the relationships given in the Statics section of the *NCEES Handbook.*

$$\bar{x} = \frac{\sum A_i x_i}{\sum A_i}$$

$$\bar{y} = \frac{\sum A_i y_i}{\sum A_i}$$

Only the distance $\bar{y}$ needs to be computed. Divide the T-section into two rectangular areas marked 1 and 2, and select the bottom of the section as the datum, as shown.

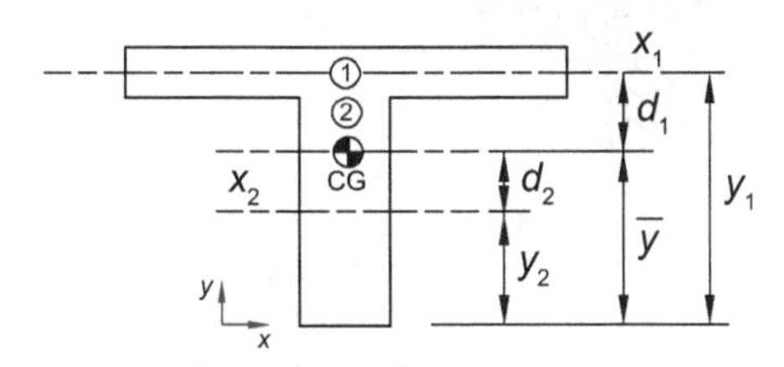

The values are tabulated.

area number	b_i	h_i	A_i	y_i	$A_i y_i$
1	60 in	5 in	300 in^2	32.5 in	9750 in^3
2	12 in	30 in	360 in^2	15 in	5400 in^3
		total	660 in^2		15,150 in^3

The distance from the bottom of the beam to its centroid is

$$\bar{y} = \sum \frac{A_i y_i}{A_i} = \frac{15{,}150 \text{ in}^3}{660 \text{ in}^2} = 22.95 \text{ in}$$

The answer is (C).

248. First, determine the distance to the centroid of the T-section in the y-direction.

The values are tabulated.

area number	b_i	h_i	A_i	y_i	$A_i y_i$
1	60 in	5 in	300 in^2	26.5in	7950 in^3
2	12 in	24 in	288 in^2	12 in	3456 in^3
		total	588 in^2		11,406 in^3

The distance from the bottom of the beam to its centroid is

$$\bar{y} = \sum \frac{A_i y_i}{A_i} = \frac{11{,}406 \text{ in}^3}{588 \text{ in}^2} = 19.4 \text{ in}$$

For each rectangular shape, the moment of inertia about its centroidal axis can be determined by using the expression shown.

$$I_x = \frac{bd^3}{12}$$

Use the parallel axis theorem.

$$I'_x = I_x + Ad_x^2$$

The centroid of the T-section and distances d_1 and d_2 from the centroidal axis to the centroids of individual rectangles are shown.

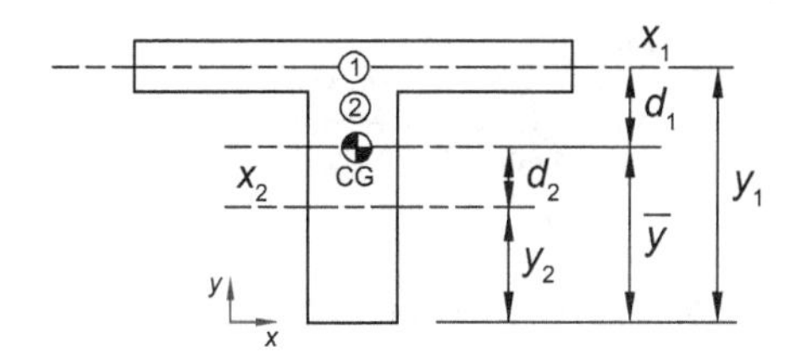

Divide the T-section into two rectangular areas and tabulate the parameters.

no.	b_i	h_i	A_i	I_{xi}	$\bar{y}_i$	$d_i = \lvert y_i - \bar{y} \rvert$	$A_i d_i^2$
1	60 in	5 in	300 in^2	625 in^4	26.5 in	26.5 in – 19.4 in = 7.1 in	15,123 in^4
2	12 in	24 in	288 in^2	13,824 in^4	12 in	12 in – 19.4 in = 7.4 in	15,771 in^4
	total			14,449 in^4			30,894 in^4

$$I_{total} = \sum I_{xi} + A_i d_i^2 = 14{,}449 \text{ in}^4 + 30{,}894 \text{ in}^4 = 45{,}343 \text{ in}^4 \quad (45{,}000 \text{ in}^4)$$

The answer is (C).

249. For the weight, W, consider equilibrium along the x- and y-axes as impending motion between the block and the plane.

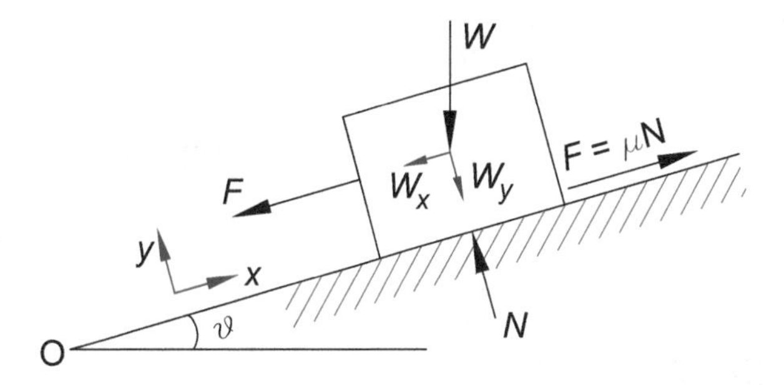

In the y-direction, which is perpendicular to the plane, there are two forces acting.

$$\sum F_y = 0$$

$$W = mg = \left(9.81 \ \frac{\text{m}}{\text{s}^2}\right)(100 \text{ kg}) = 981 \text{ N}$$

The force normal to the plane is

$$N = W\cos 30° = (981 \text{ N})(0.866) = 850 \text{ N}$$

The friction force available to resist sliding of the block is

$$F_{\text{friction}} = \mu_s N = (0.3)(850 \text{ N}) = 255 \text{ N}$$

Summing up the forces in the x-direction, along the plane surface,

$$\begin{aligned}\sum F_x &= F_{\text{applied}} + F_{\text{friction}} - W_x \\ &= F_{\text{applied}} + F_{\text{friction}} - W\sin\theta \\ &= 0\end{aligned}$$

$$F_{\text{applied}} + 255 \text{ N} - (981 \text{ N})\sin 30° = 0$$

$$F_{\text{applied}} = 235.5 \text{ N} \quad (240 \text{ N})$$

The answer is (A).

250. Use the projectile equations.

$$\text{Eq. 1}: \quad y = -\frac{1}{2}gt^2 + v_0\sin\theta \times t + y_0$$

$$\text{Eq. 2}: \quad x = v_0\cos\theta t$$

The initial velocity in ft/sec is

$$v_0 = \frac{\left(12 \ \frac{\text{mi}}{\text{hr}}\right)\left(5280 \ \frac{\text{ft}}{\text{mi}}\right)}{3600 \ \frac{\text{sec}}{\text{hr}}} = 17.64 \text{ ft/sec}$$

Solve Eq. 1 for the time, t. Use $y = 0$ ft at the ground and $y = 12$ ft at the cliff, as shown.

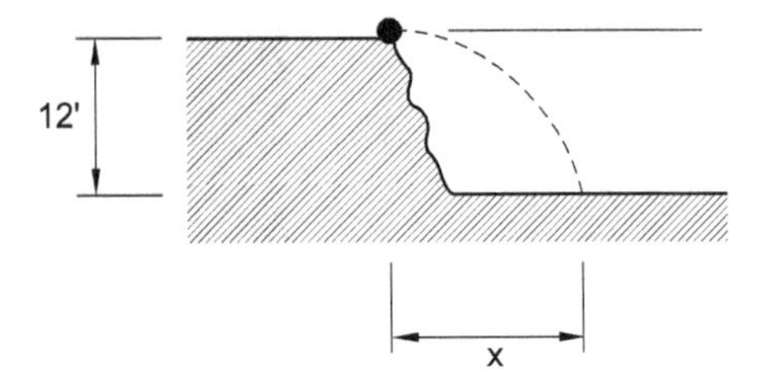

The angle at which the car is projected from the top the cliff is 0° to the horizontal.

$$y = -\frac{1}{2}gt^2 + v_0\sin\theta \times t + y_0$$

$$0 = \left(-\frac{1}{2}\right)(32.2)t^2 + 12 \text{ ft}$$

$$t = 0.86 \text{ sec}$$

Substitute the value of t into Eq. 2.

$$\begin{aligned}x &= v_0\cos\theta t \\ &= \left(17.64 \ \frac{\text{ft}}{\text{sec}}\right)(1)(0.86 \text{ sec}) \\ &= 15.2 \text{ ft} \quad (15 \text{ ft})\end{aligned}$$

The answer is (C).

251. The ring has its entire mass far away from its center and has a greater moment of inertia than the solid cylinder. For a solid cylinder with a radius of r_1, the mass radius of gyration about its center is

$$r_{\text{solid}} = \sqrt{\frac{r_1^2}{2}}$$

For a ring with an outer radius of r_1 and inner radius of r_2, the mass radius of gyration about its center is

$$r_{\text{ring}} = \sqrt{\frac{r_1^2 + r_2^2}{2}}$$

Therefore, the mass radius of gyration of the ring is greater than the mass radius of gyration of the solid cylinder. The ring would offer more resistance to the angular motion and would reach the bottom of the ramp later than the solid cylinder.

This may seem counterintuitive, but it should be noted that since both the hollow and solid cylinders have equal mass, their densities are not equal.

The answer is (C).

252. The equation for the kinetic energy of rotation of the wheel is

$$\text{KE} = \frac{1}{2}I\omega^2$$

The angular velocity is

$$\omega = 2\pi\left(\frac{60 \ \frac{\text{rev}}{\text{min}}}{60 \ \frac{\text{sec}}{\text{min}}}\right) = 2\pi \text{ rad/s}$$

The equation for the mass moment of inertia of a flywheel is given the Dynamics section of the *NCEES Handbook*.

$$\begin{aligned} I &= Mr^2 = (1000\ \text{kg})(1\ \text{m})^2 \\ &= 1000\ \text{kg·m}^2 \end{aligned}$$

The kinetic energy of rotation is

$$\begin{aligned} \text{KE} &= \left(\frac{1}{2}\right)(1000\ \text{kg·m}^2)\left(2\pi\ \frac{\text{rad}}{\text{s}}\right)^2 \\ &= 19{,}738\ \text{N·m} \quad (20{,}000\ \text{N·m}) \end{aligned}$$

The answer is (D).

253. It is given that the angular deceleration of the flywheel is uniform. Therefore, use the formula given in the *NCEES Handbook*, Dynamics section.

$$\alpha = \frac{\omega_2 - \omega_1}{t}$$

Convert the initial angular velocity to units of radians per second.

$$\omega_1 = \frac{\left(240\ \frac{\text{rev}}{\text{min}}\right)(2\pi)}{60} = 8\pi\ \text{rad/sec}$$

Convert the reduced angular velocity to units of radians per second.

$$\omega_2 = \frac{\left(180\ \frac{\text{rev}}{\text{min}}\right)(2\pi)}{60} = 6\pi\ \text{rad/sec}$$

The angular deceleration is

$$\alpha = \frac{6\pi\ \frac{\text{rad}}{\text{sec}} - 8\pi\ \frac{\text{rad}}{\text{sec}}}{20\ \text{sec}} = -0.1\pi\ \text{rad/sec}^2$$

The time needed to come to a complete stop is

$$\begin{aligned} \omega_2 &= \omega_1 + \alpha t = 0 \\ \omega_2 &= 8\pi\ \frac{\text{rad}}{\text{sec}} - \left(0.1\pi\ \frac{\text{rad}}{\text{sec}^2}\right)t = 0 \\ t &= 80\ \text{sec} \end{aligned}$$

To determine the number of revolutions the flywheel makes before coming to a full stop, find the angular distance.

$$\begin{aligned} \theta &= \omega_1 t + \frac{1}{2}\alpha t^2 = \left(8\pi\ \frac{\text{rad}}{\text{sec}}\right)(80) \\ &\quad + \frac{1}{2}\left(-0.1\pi\ \frac{\text{rad}}{\text{sec}^2}\right)(80\ \text{sec})^2 \\ &= 320\pi\ \text{rad} \end{aligned}$$

The number of revolutions is

$$N = \frac{\theta}{2\pi} = \frac{320\pi\ \text{rad}}{2\pi} = 160$$

The answer is (A).

254. The kinetic energy is a function of mass and velocity.

$$\text{KE} = \frac{1}{2}mv^2$$

Convert the weight to mass.

$$m = \frac{30{,}000\ \text{N}}{9.81\ \frac{\text{m}}{\text{s}^2}} = 3060\ \text{kg}$$

Convert the velocity to meters per second.

$$v = 50\ \frac{\text{km}}{\text{h}} = \frac{50\ \frac{\text{km}}{\text{h}}\left(1000\ \frac{\text{m}}{\text{km}}\right)}{3600\ \frac{\text{s}}{\text{h}}} = 13.89\ \text{m/s}$$

The kinetic energy is

$$\begin{aligned} \text{KE} &= \left(\frac{1}{2}\right)(3060\ \text{kg})\left(13.89\ \frac{\text{m}}{\text{s}}\right)^2 \\ &= 295\ \text{kJ} \end{aligned}$$

The answer is (A).

255. By symmetry, the two-span beam is equivalent to a single-span propped cantilever, as shown, because the middle support in the two-span loading does not rotate and can be considered fixed.

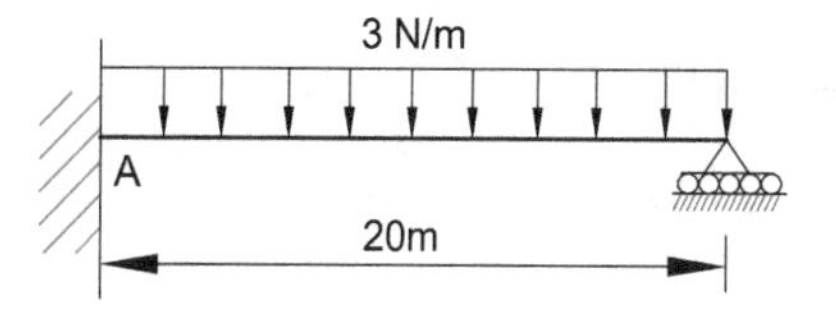

The maximum moment at the fixed end of a propped cantilever is

$$M = \frac{wL^2}{8} = \frac{\left(3\ \frac{\text{N}}{\text{m}}\right)(20\ \text{m})^2}{8}$$
$$= 150\ \text{N·m}$$

The answer is (C).

256. *Segment CD*: The shear varies linearly in the interval CD. The variation indicates a uniformly distributed load over the interval. The load applied in the interval CD is

$$\begin{aligned} W &= w_1 L_{\text{CD}} \\ &= -18.93\ \text{kips} - (-28.93\ \text{kips}) \\ &= 10\ \text{kips} \end{aligned}$$

The uniformly distributed load across the 10 ft span is

$$w_1 = \frac{10\ \text{kips}}{10\ \text{ft}} = 1\ \text{kip/ft}$$

Segment DE: The portion is a cantilever with the tip at point E. The shear force in the beam increases linearly from zero at point E to 6 kips at support D. A linear variation denotes a uniformly distributed load. The total load is

$$\begin{aligned} w_2 L &= w_2(6\ \text{ft}) = 6\ \text{kips} \\ w_2 &= \frac{6\ \text{kips}}{6\ \text{ft}} = 1\ \text{kip/ft} \end{aligned}$$

The answer is 1 kip/ft and 1 kip/ft.

257. Starting from point A, the beam shear is 26.07 kips. Therefore, the reaction at support A is 26.07 kips.

Segment AB: Since the shear load is constant in segment AB, there is no load applied in the AB interval. At point B, the shear changes from positive to negative. The change in shear equals the applied load at B. Therefore,

$$P_1 = 26.07\ \text{kips} - (-3.93\ \text{kips}) = 30\ \text{kips}$$

Segment BC: The shear load is constant between points B and C. Therefore, there is no load in this interval. The applied load at C is

$$P_2 = -3.93\ \text{kips} - (-18.93\ \text{kips}) = 15\ \text{kips}$$

The answer is 30 kips and 15 kips.

258. For a circular section of radius r, its moment of inertia, I, and polar moment of inertia, J, are determined using the relations shown.

$$J = \frac{\pi r^4}{2}$$

$$I = \frac{\pi r^4}{4}$$

Therefore,

$$\frac{I}{J} = \frac{1}{2}$$

The maximum shear stress occurs at the outermost fiber and is calculated using the relation shown.

$$\tau_{\text{max}} = \frac{Tr}{J}$$

The maximum bending stress occurs at the outermost fiber.

$$\sigma_{\text{max}} = \frac{Mr}{I}$$

The ratio of the maximum shear stress to the maximum bending stress is

$$\begin{aligned} \frac{\text{maximum shear stress}}{\text{maximum bending stress}} &= \frac{Tr}{J}\left(\frac{I}{Mr}\right) = \frac{T}{M}\left(\frac{I}{J}\right) \\ &= \frac{T}{2M} \end{aligned}$$

The answer is (C).

259. In an elastic-plastic material, the fracture stress equals its yield stress. The punched area is

$$\begin{aligned} A &= (\text{plate thickness})(\text{hole perimeter}) \\ &= t_{\text{plate}}(\pi D) \\ &= (0.5\ \text{in})\pi(1\ \text{in}) \\ &= 1.57\ \text{in}^2 \end{aligned}$$

The required punching force is

$$\begin{aligned} F &= F_y A \\ &= \left(50\ \frac{\text{kips}}{\text{in}^2}\right)(1.57\ \text{in}^2) \\ &= 78.5\ \text{kips}\quad (80\ \text{kips}) \end{aligned}$$

The answer is (D).

260. As the temperature rises, bar A will tend to expand more than bar B. In an unrestrained situation, neither bar A nor bar B would have any induced stress. If they are restrained so that they expand or contract equally, stresses would be induced in both bars. Bar A is not able to fully expand due to restraint caused by bar B. As such, bar A is in compression, bar B is in tension, and both bars expand equally.

The answer is

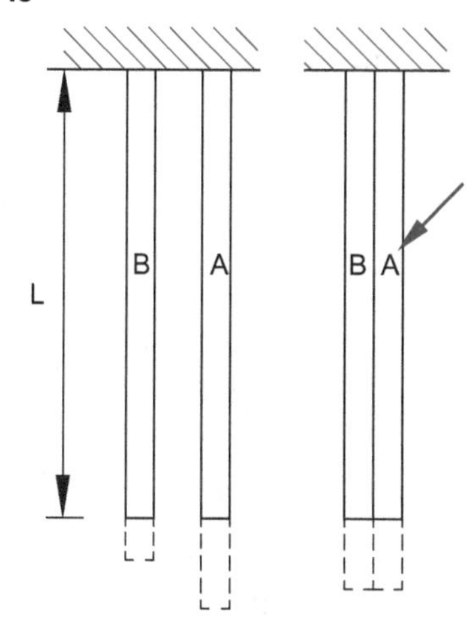

261. The maximum shear stress is given by the relation shown.

$$\tau_{\text{max}} = \frac{\sigma_1 - \sigma_2}{2}$$

In this case, $\sigma_1 = 100$ ksi and $\sigma_2 = -100$ ksi. The maximum shear stress is

$$\tau_{max} = \frac{100\ \frac{\text{kips}}{\text{in}^2} - \left(-100\ \frac{\text{kips}}{\text{in}^2}\right)}{2} = 100\ \text{ksi}$$

The maximum shear stress equals the radius of the Mohr circle, as shown.

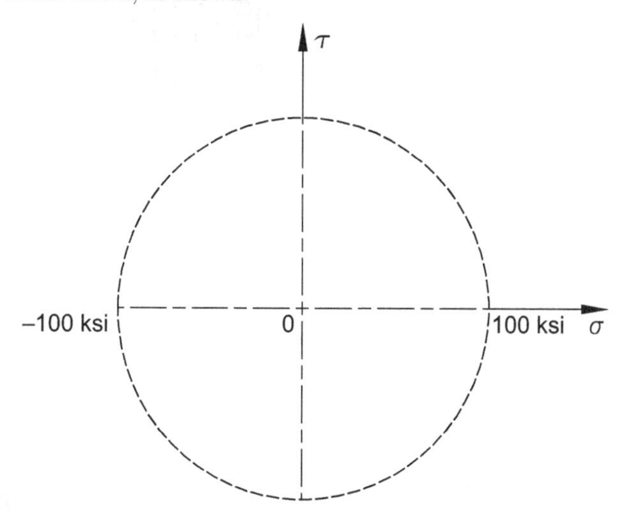

The answer is (C).

262. For steel, the modulus of elasticity is 29,000 ksi. For a 36 ksi yield stress, the yield strain is

$$\epsilon_y = \frac{\sigma_y}{E} = \frac{36\ \frac{\text{kips}}{\text{in}^2}}{29{,}000\ \frac{\text{kips}}{\text{in}^2}} = 0.001241$$

The elongation at yield is

$$\varepsilon = \frac{\Delta L}{L_0}$$

$$\begin{aligned}\Delta L_y &= \varepsilon L_0 = (0.001241)(3\ \text{ft})\left(12\ \frac{\text{in}}{\text{ft}}\right)\\ &= 0.0447\ \text{in}\end{aligned}$$

Since the specimen elongation is greater than the yield elongation, the specimen went into its plastic range, causing permanent elongation, as shown.

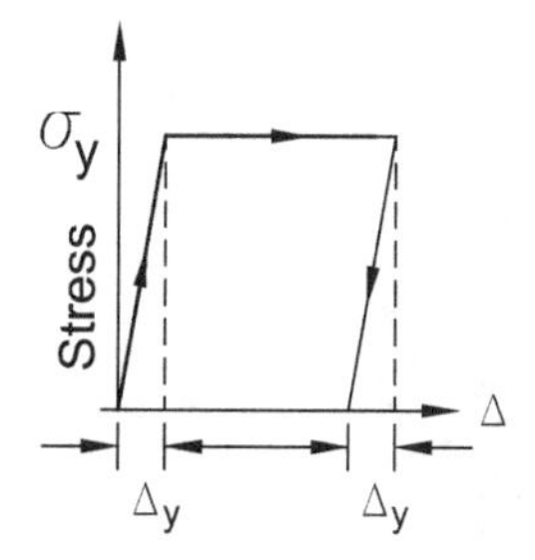

The permanent elongation is

$$\Delta L_{\text{permanent}} = 0.20\ \text{in} - 0.0447\ \text{in} = 0.155\ \text{in}$$

The answer is (C).

263. The mix ratio is by weight, not by volume, so option A is incorrect. The mix designation stands for one part cement, two parts fine aggregate, and four parts coarse aggregate, so option B is incorrect. Synthetic aggregates such as blast-furnace slag, expanded clay, shale, pumice, or scoria are used in concrete mixes. The aggregates are generally in the range of 60% to 75%. The weight of aggregates in the mix is six times the weight of the Portland cement. Statements C, D, and E are correct.

The answer is (A) and (B).

264. Steel reinforcement rusts when it is exposed to air, water, and corrosive chemicals. Epoxy coating is the most common method to protect steel reinforcement from rusting and corrosion.

The answer is (C).

265. For the mix ratio of 1:2:3, each pound of cement in the mix requires a total of 5 lbf aggregate (2 lbf fine aggregate and 3 lbf coarse aggregate). Therefore, the weight of aggregate needed to maintain the mix proportion is five times the cement required to produce one cubic yard of mix.

$$\text{aggregate required} = (611.7 \text{ lbf})(5) = 3058.5 \text{ lbf} \quad (3060 \text{ lbf})$$

The answer is (D).

266. The development of a high-strength mix is not an objective of the specifications, so option D is incorrect.

The answer is (D).

267. *Hardness* is defined as the resistance of a material to localized plastic deformation, so option A is incorrect. *Ductility* is defined as a material's amenability to drawing without losing strength, so option C is incorrect. *Toughness* is defined as the ability of a material to absorb or store energy without fracturing, so option D is incorrect.

The ability of a material to absorb or store energy without permanent deformation is called *resilience*. The corresponding energy is the area under its stress-strain diagram in the elastic range, as shown.

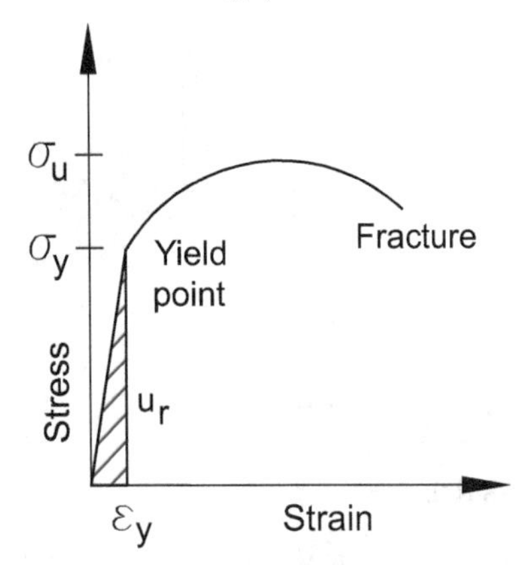

The answer is (B).

268. The pressure difference between points A and B is equal to the pressure difference between points C and D. Since the pressure in the U-tube is at the atmospheric level, the density of the air can be neglected. See the *NCEES Handbook*, Units and Conversion Factors section, for conversion equations.

$$1 \text{ unit of atmospheric pressure} = 33.9 \text{ ft of water column} = 14.7 \text{ psi}$$

The height of water column and the pressure (in psi) are related. Solve the equations simultaneously to express water column height in psi units. The pressure difference for a 15 in water column is

$$\Delta P = 15 \text{ in}\left(\frac{14.7 \text{ psi}}{33.9 \text{ ft } H_2O}\right)\left(\frac{1 \text{ ft}}{12 \text{ in}}\right) = 0.54 \text{ psi}$$

The answer is (A).

269. The absolute pressure at the bottom of the tank is equal to the sum of the atmospheric pressure at the surface of the tank and the pressure created by 3 m of oil.

The unit weight of water is 9.8 kN/m^3. The unit weight of mercury is

$$\gamma_{Hg} = (SG)(\gamma_w) = (13.6)\left(9.8 \ \frac{\text{kN}}{\text{m}^3}\right) = 133.28 \text{ kN/m}^3$$

The atmospheric pressure at the top of the tank is

$$P_{atm} = \left(133.28 \ \frac{\text{kN}}{\text{m}^3}\right)(0.76 \text{ m}) = 101.29 \text{ kPa}$$

The unit weight of the oil is

$$\gamma_{oil} = (0.8)\left(9.8 \ \frac{\text{kN}}{\text{m}^3}\right) = 7.84 \text{ kN/m}^3$$

The pressure created by the oil is

$$P_{oil} = \left(7.84 \ \frac{\text{kN}}{\text{m}^3}\right)(3 \text{ m}) = 23.52 \text{ kPa}$$

The absolute pressure at the base is

$$P_{abs} = 101.29 \text{ kPa} + 23.52 \text{ kPa} = 124.81 \text{ kPa} \quad (125 \text{ kPa})$$

The vertical pressure distribution is shown.

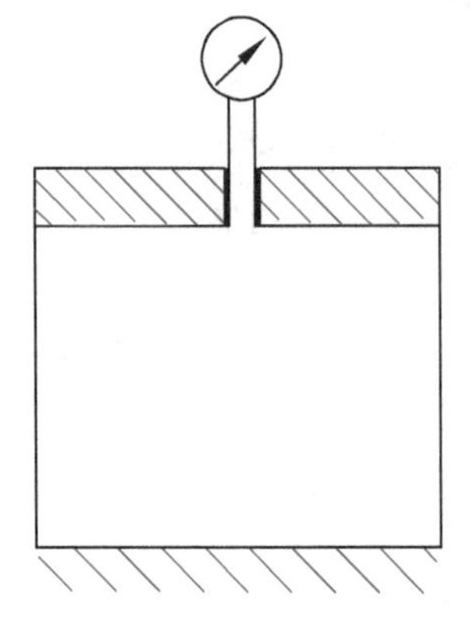

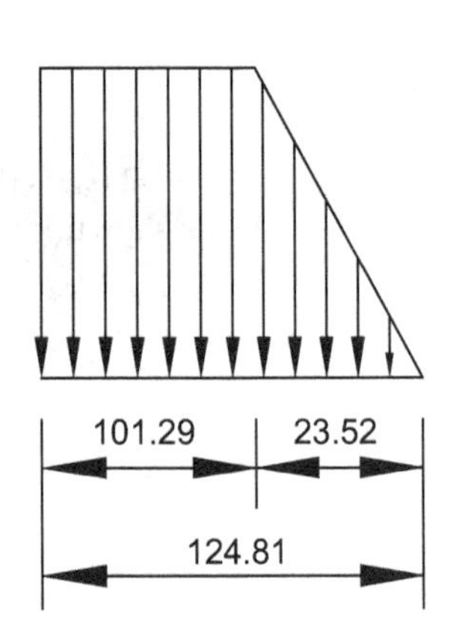

The answer is (D).

270. Since oil is lighter than water, it floats above the 6 ft deep layer of water. The pressure at the base is the sum of both the oil and water pressures, as shown.

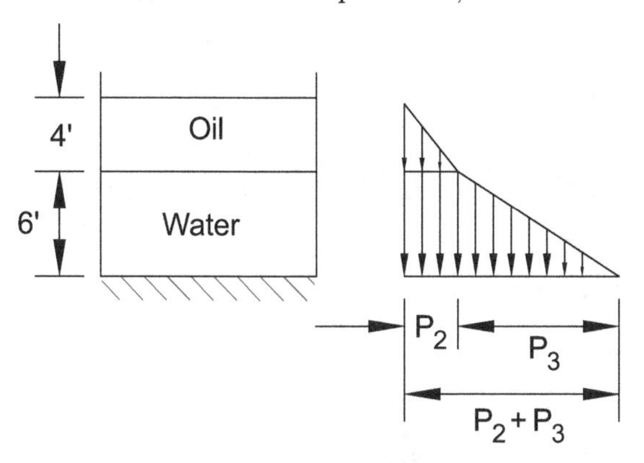

The pressure at the oil-water interface is the sum of the atmospheric pressure and the pressure due to the 4 ft of oil.

$$\begin{aligned} p_2 &= p_1 + \gamma_{\text{oil}}(4\ \text{ft}) \\ &= 0\ \frac{\text{lbf}}{\text{ft}^2} + (0.75)\left(62.4\ \frac{\text{lbf}}{\text{ft}^3}\right)(4\ \text{ft}) \\ &= 187.2\ \text{psf} \end{aligned}$$

The pressure at the base is equal to the pressure at the oil-water interface and the pressure due to the water.

$$\begin{aligned} p_3 &= p_2 + \gamma_{\text{water}}(6\ \text{ft}) \\ &= 187.2\ \frac{\text{lbf}}{\text{ft}^2} + \left(62.4\ \frac{\text{lbf}}{\text{ft}^3}\right)(6\ \text{ft}) \\ &= 561.6\ \text{psf} \quad (560\ \text{psf}) \end{aligned}$$

The answer is (C).

271. Use the impulse-momentum principle.

$$F = Q_2\rho_2 v_2 - Q_1\rho_1 v_1 = \dot{m}(v_2 - v_1)$$

The discharge is

$$Q = Av = \left(\frac{\pi(1\ \text{in})^2}{4\left(144\ \frac{\text{in}^2}{\text{ft}^2}\right)}\right)\left(190\ \frac{\text{ft}}{\text{sec}}\right) = 1.036\ \text{ft}^3/\text{sec}$$

The mass flow rate is

$$Q(\rho) = (1.036\ \text{ft}^3/\text{sec})\left(\frac{62.4\ \frac{\text{lbm}}{\text{ft}^3}}{32.2\ \frac{\text{lbm}}{\text{slug}}}\right) = 2\ \text{slugs/sec}$$

At impact, the plate does not move, and the jet spreads out radially over the plate, as shown.

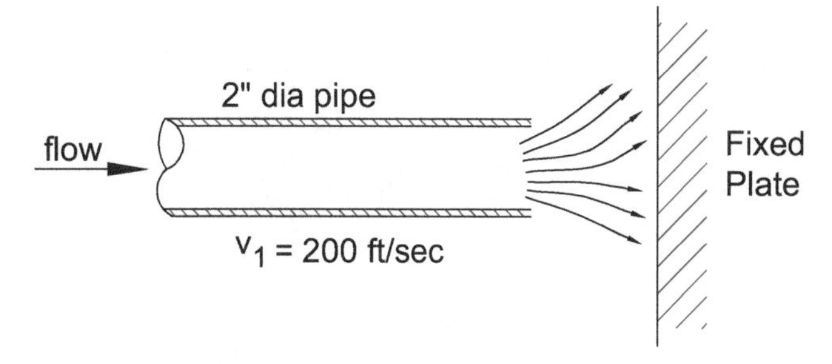

Thus, the jet velocity perpendicular to the plate, v_2, is destroyed, and the reaction force is

$$F = \dot{m}(\Delta v) = \left(2\ \frac{\text{slug}}{\text{sec}}\right)\left(190\ \frac{\text{ft}}{\text{sec}}\right) = 382\ \text{lbf} \quad (380\ \text{lbf})$$

The answer is (B).

272. Use the impulse turbine formula for maximum turbine output given in the Fluid Mechanics section of the *NCEES Handbook*.

$$\begin{aligned} \dot{W}_{\text{max}} &= \frac{Q\rho(v_1)^2}{2} = \frac{\dot{m}(v_1)^2}{2} \\ &= \frac{\left(8.5\ \frac{\text{slugs}}{\text{sec}}\right)\left(200\ \frac{\text{ft}}{\text{sec}}\right)^2}{(2)\left(550\ \frac{\frac{\text{ft-lbf}}{\text{sec}}}{\text{hp}}\right)} \\ &= 309\ \text{hp} \quad (300\ \text{hp}) \end{aligned}$$

The answer is (D).

273. It is known that water flow is laminar and not turbulent. The cavitation occurs when water pressure equals its vapor pressure. The vapor pressure is also defined as the pressure at which liquid and vapor are in equilibrium. From the table in the *NCEES Handbook*: Properties of Water (English Units), for water at 100°F, the corresponding vapor pressure is 0.95 psi.

The answer is 0.95 psi.

274. A horizontal curve is a circular curve, not a parabolic curve, so option B is incorrect. A horizontal curve does not need to maintain a constant elevation along the curve path, so option E is incorrect.

The answer is (B) and (E).

275. Draw the polygon shape and number the vertices in order, going either clockwise or counterclockwise, as shown.

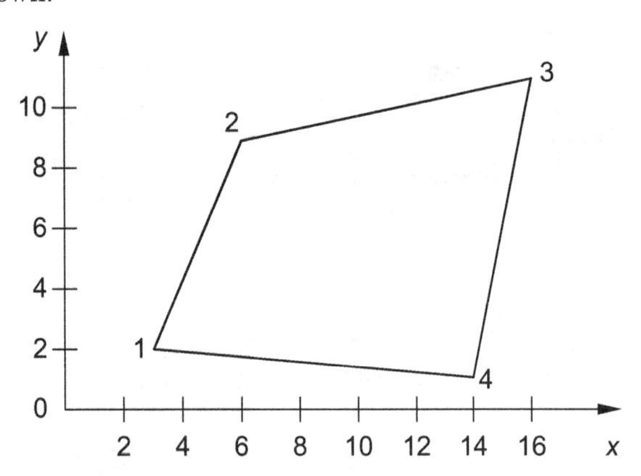

Use the coordinates method to determine the area.

$$A = \frac{1}{2}\left|\begin{array}{l}(x_1y_2 - x_2y_1) \\ \quad +(x_2y_3 - x_3y_2) \ . \ . \ . + (x_ny_1 - x_1y_n)\end{array}\right|$$

List the coordinates and wrap around to the first vertex again, closing the polygon. Calculate the area as follows.

vertex, i	x (ft)	y (ft)	x_iy_i (ft²)	$x_{i+1}y_i$ (ft²)	difference (ft²)
1	3	2			15
2	6	× 9	3 × 9 = 27	6 × 2 = 12	
3	16	× 11	6 × 11 = 66	16 × 9 = 144	−78
4	14	× 1	16 × 1 = 16	14 × 11 = 154	−138
1	3	× 2	14 × 2 = 28	3 × 1 = 3	25
sum					−176

Use the coordinates method to determine the area.

$$\begin{aligned} A &= \frac{1}{2}\left|\begin{array}{l}(x_1y_2 - x_2y_1) \\ \quad +(x_2y_3 - x_3y_2) \ . \ . \ . + (x_ny_1 - x_1y_n)\end{array}\right| \\ &= \left(\frac{1}{2}\right)\left|-176 \text{ ft}^2\right| \\ &= 88 \text{ ft}^2 \end{aligned}$$

The answer is (B).

276. It is known that the stockpile is conical in shape. As given in the *NCEES Handbook*, Mathematics section and Earthwork Formulas section, the volume of a cone, V, is

$$V = \frac{(\text{base area})(\text{height})}{3}$$

The area of the base is

$$\text{base area} = \frac{\pi D^2}{4} = \frac{\pi(20 \text{ m})^2}{4} = 314.15 \text{ m}^2$$

Therefore,

$$V = \frac{(314.15 \text{ m}^2)\ (15 \text{ m})}{3} = 1571 \text{ m}^3 \quad (1600 \text{ m}^3)$$

The answer is (B).

277. Draw the figure based on the facts given in the problem.

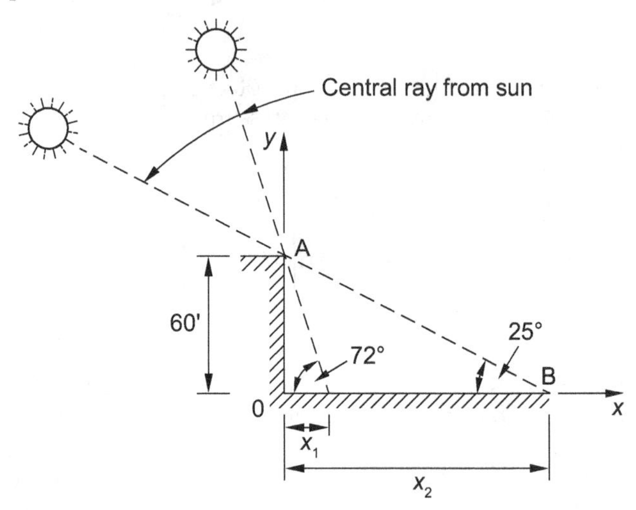

The shadow on the winter solstice, x_2, would to be the longest. Therefore, it would be the determining factor.

$$\frac{60 \text{ ft}}{x_2} = \tan 25° = 0.466$$

$$x_2 = \frac{60 \text{ ft}}{0.466} = 128.87 \text{ ft} \quad (130 \text{ ft})$$

The answer is (C).

278. Vertical curves are generally parabolic and not circular.

$$y = ax^2$$

Option A is incorrect. Options B and C state the properties of a parabolic curve and are correct. A vertical curve may have grades g_1 and g_2 that are either both positive or both negative, so option D is correct.

The answer is (A).

279. Draw a figure of the situation described.

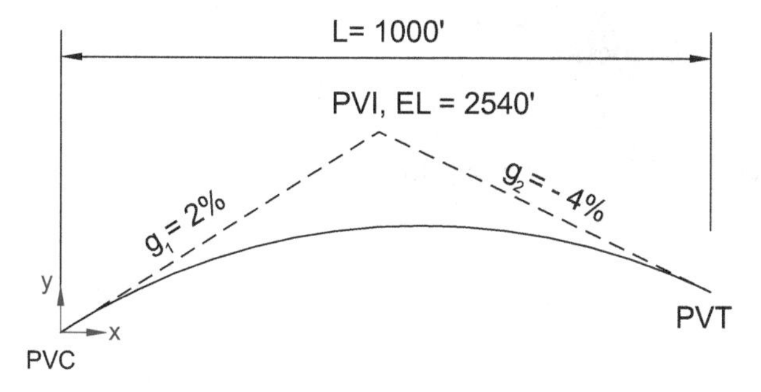

See the Vertical Curves section of the *NCEES Handbook*. The given curve is a crest curve, since the back slope is positive and the forward slope is negative. Since the back tangent is a sloping line, the equation for the elevation of any point on the tangent, Y_{PVx}, is linear.

$$Y_{PVx} = Y_{PVC} + g_1 x$$

Therefore, the elevation at the point of intersection (PVI) is

$$Y_{PVI} = Y_{PVC} + g_1\left(\frac{L}{2}\right)$$

Therefore, the elevation at PVC is

$$\begin{aligned} Y_{PVC} &= Y_{PVI} - g_1\left(\frac{L}{2}\right) \\ &= 2540 \text{ ft} - 0.02\left(\frac{1000 \text{ ft}}{2}\right) \\ &= 2530 \text{ ft} \end{aligned}$$

The answer is (B).

280. The rational formula for discharge is

$$Q = CIA$$

The area, A, is expressed in acres; the intensity, i, is expressed in in/hr; and the resulting Q is in ft^3/sec. In general, when an area consists of several subareas of different imperviousness, the rational formula is

$$Q = i\sum_{j=1}^{n} A_j C_j$$

Convert the rainfall intensity to in/hr, leaving the area units in acres, so that the formula yields the discharge in ft^3/sec.

$$i = \left(0.1 \frac{\text{in}}{\text{min}}\right)\left(60 \frac{\text{min}}{\text{hr}}\right) = 6 \text{ in/hr}$$

Multiply the areas by the coefficients and tabulate the results.

surface type	area (ac)	C	A_jC_j
buildings	16	0.6	9.6
open unpaved space	14	0.2	2.8
light industrial	20	0.65	13.0
paved roads and parking lots	10	0.9	9.0
total			34.4

The peak discharge is

$$\begin{aligned} Q &= i\sum_{j=1}^{n} A_j C_j = \left(6 \frac{\text{in}}{\text{hr}}\right)(34.4 \text{ ac}) \\ &= 206 \text{ ac-in/hr} \end{aligned}$$

The answer is (C).

281. The equation for the hydraulic radius is

$$R = \frac{\text{cross-sectional area}}{\text{wetted perimeter}}$$

The wetted perimeter consists of the base and the two sloping sides of length x.

$$\text{wetted perimeter} = 8 \text{ ft} + (2)(4 \text{ ft})\csc 60^\circ = 17.24 \text{ ft}$$

Divide the cross section into three parts to calculate the channel cross-sectional area, as shown.

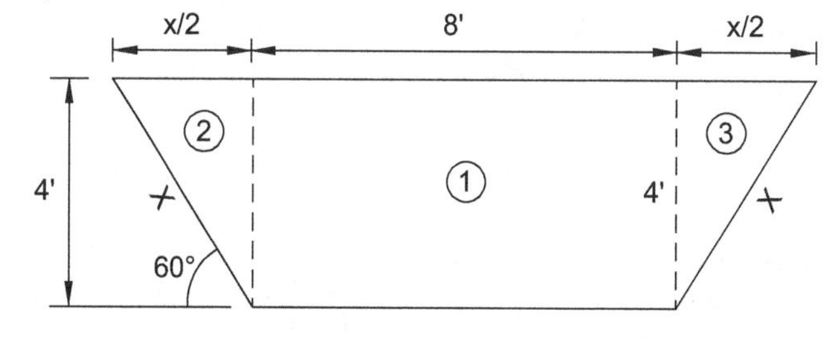

$$A_1 = (8 \text{ ft})(4 \text{ ft}) = 32 \text{ ft}^2$$

$$\begin{aligned} A_2 = A_3 &= \left(\frac{1}{2}\right)(4 \text{ ft})\left(\frac{x}{2}\right) \\ &= 4.62 \text{ ft}^2 \end{aligned}$$

The total area is

$$\begin{aligned} A &= A_1 + A_2 + A_3 = 32 + x + x \\ &= 32 \text{ ft}^2 + (2)(4.62 \text{ ft}^2) \\ &= 41.24 \text{ ft}^2 \end{aligned}$$

The hydraulic radius is

$$R = \frac{A}{P} = \frac{41.24\ \text{ft}^2}{17.24\ \text{ft}} = 2.39\ \text{ft} \quad (2.4\ \text{ft})$$

The answer is (A).

282. As stated in the Fluid Mechanics section of the *NCEES Handbook*, turbines produce power by extracting energy from a working fluid. The energy loss in the fluid results in a decrease in the fluid pressure (head). Therefore, option A is incorrect.

The equation for the net positive suction head available (NPSH_A) is

$$\text{NPSH}_\text{A} = \frac{P_\text{atom}}{\rho g} + H_s - H_f - \frac{V^2}{2g} - \frac{P_\text{vapor}}{\rho g}$$

The net positive suction head available for a pump decreases as the fluid vapor pressure at the pump inlet increases. Therefore, option B is correct and C is not.

The centrifugal pump characteristics diagram in the Fluid Mechanics section of the *NCEES Handbook* shows that pump flow rate increases and then drops as the head is increased. The pump shuts off when it reaches its maximum operating head. At that point, its discharge is zero. Therefore, option D is incorrect. For a porous fixed bed of solid particles, the Ergun equation can be used to estimate pressure loss for both laminar and turbulent flow conditions. Therefore, option E is correct.

The answer is (B) and (E).

283. Per the Fluid Mechanics section of the *NCEES Handbook*, the friction loss in the 6 in diameter pipe will be equal to the friction loss in the 8 in diameter pipe, and the equation for the total flow is

$$Q_{10} = Q_6 + Q_8$$

The friction loss in a pipe is given by the Darcy-Weisbach equation. The loss is equal for both branches, so the equation can be rewritten as shown.

$$h_f = f\left(\frac{L}{D}\right)\frac{v^2}{2g} = f\left(\frac{L_8}{D_8}\right)\frac{v_8^2}{2g} = f\left(\frac{L_6}{D_6}\right)\frac{v_6^2}{2g}$$

Because the friction coefficient is the same throughout, the Moody diagram is not needed. Since $L_6 = L_8$, the Darcy-Weisbach equation can be simplified as shown.

$$\frac{v_8^2}{D_8} = \frac{v_6^2}{D_6} = \frac{v_8^2}{8\ \text{in}} = \frac{v_6^2}{6\ \text{in}}$$

$$v_8 = \sqrt{\frac{8\ \text{in}}{6\ \text{in}}}\ v_6 = 1.15v_6$$

The flow rate in the 10 in pipe is known. The ratio-of-discharge rate is also known, so determine the flow rate in both pipes.

$$A_6v_6 + A_8v_8 = 10\ \text{ft}^3/\text{sec}$$

Determine the flow rate in both the 8 in and 6 in pipes.

$$\left(\frac{\pi}{4}\right)\left(\frac{6\ \text{in}}{12\ \frac{\text{in}}{\text{ft}}}\right)^2 v_6 + \left(\frac{\pi}{4}\right)\left(\frac{8\ \text{in}}{12\ \frac{\text{in}}{\text{ft}}}\right)^2 v_8 = 10\ \text{ft}^3/\text{sec}$$

$$0.2v_6 + (0.35\ \text{ft}^2)(1.15)v_6 = 10\ \text{ft}^3/\text{sec}$$

$$0.6v_6 = 10\ \text{ft}^3/\text{sec}$$

$$v_6 = 16.7\ \text{ft/sec}$$

The flow in the 6 in pipe is

$$Q_6 = v_6A_6 = \left(16.7\ \frac{\text{ft}}{\text{sec}}\right)(0.2\ \text{ft}^2) = 3.34\ \text{ft}^3/\text{sec} \quad (3\ \text{ft}^3/\text{sec})$$

The answer is (B).

284. The terms *spillway* and *weir* are synonymous. The weir discharge formulas and coefficients are given in the Weir Formulas section of the *NCEES Handbook*. The equation for the discharge over a rectangular weir is

$$Q = CLH^{1.5}$$

The coefficient C is 3.33 for a rectangular weir (USCS units). Therefore,

$$Q = CLH^{1.5} = (3.33)(150\ \text{ft})(3\ \text{ft})^{1.5} = 2595\ \text{ft}^3/\text{sec} \quad (2600\ \text{ft}^3/\text{sec})$$

The answer is (C).

285. Darcy's equation can be expressed as follows.

$$Q = -KA\left(\frac{dh}{dx}\right)$$

The head loss, dh, is measured as the drop in the water table in the direction of the flow. A drop in the water table over a length of bed denotes a negative gradient. The term in parentheses denotes the hydraulic gradient, so the formula can also be written as

$$Q = -\frac{KA(h_1 - h_2)}{L} = \frac{KA(h_2 - h_1)}{L}$$

The equation shows that the discharge rate is proportional to the hydraulic conductivity, the head drop in the direction of the flow, and the cross-sectional area of the ground filter. However, it is inversely proportional to the length of the filter.

The answer is (B).

286. All contaminated water is considered polluted, but not all polluted water is considered contaminated. Therefore, statement B is correct and statement A is not.

For water to be considered contaminated, it must contain pathogenic bacteria. Polluted water may not have pathogenic bacteria, but it is still unsafe and unfit for drinking and domestic consumption. Therefore, option C is incorrect but D is correct.

Potable water can be consumed in any desired amount without concern for adverse health effects, but it may not always taste good. Option E is incorrect.

The answer is (B) and (D).

287. The 20 day BOD is the ultimate BOD, and it is denoted by L. The amount of BOD exerted at any time, t, is expressed as

$$y_t = L(1 - e^{-k_1 t})$$

The ultimate BOD is 320 mg/L, and for $t = 5$, the BOD is 240 mg/L. Therefore, the rate of logarithmic decay is

$$240\ \frac{\text{mg}}{\text{L}} = \left(320\ \frac{\text{mg}}{\text{L}}\right)(1 - e^{-5k_1})$$
$$k_1 = 0.28\ \text{days}^{-1}$$

The answer is (C).

288. Using per capita demand, the total daily wastewater discharge is

$$Q_{\text{total}} = (100{,}000\ \text{people})(100\ \text{GPCD}) = 10\ \text{MG/day}$$

The hourly flow rate of wastewater handled by each clarifier in is

$$Q = \frac{Q_{\text{total}}}{5\ \text{clarifiers}} = \left(\frac{10\ \frac{\text{MG}}{\text{day}}}{5}\right) = 2\ \frac{\text{MG}}{\text{day}}$$
$$= \left(2{,}000{,}000\ \frac{\text{gal}}{\text{day}}\right)\left(0.134\ \frac{\text{ft}^3}{\text{gal}}\right) = 268{,}000\ \frac{\text{ft}^3}{\text{day}}$$
$$= \frac{268{,}000\ \frac{\text{ft}^3}{\text{day}}}{\left(24\ \frac{\text{hr}}{\text{day}}\right)} = 11{,}167\ \text{ft}^3/\text{hr}$$

A diagram of a clarifier is shown. No recycling is involved. The residence time is 2 hr.

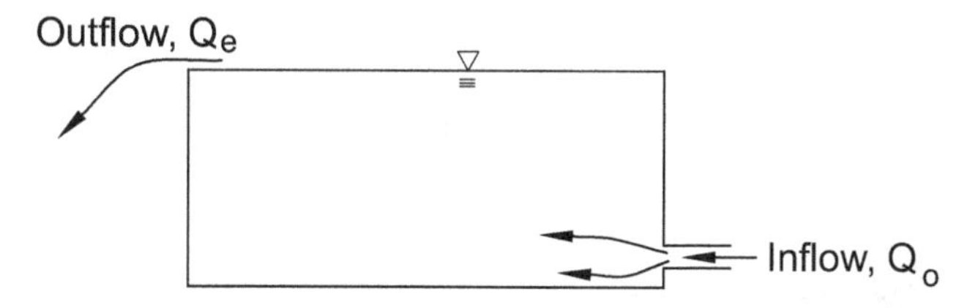

The volume of wastewater handled by each clarifier during the residence time is

$$\text{volume of wastewater} = (\text{flow rate})(\text{residence time})$$
$$= \left(11{,}167\ \frac{\text{ft}^3}{\text{hr}}\right)(2\ \text{hr})$$
$$= 22{,}333\ \text{ft}^3$$

The equation for the volume of each clarifier is therefore

$$V = AD = \frac{\pi d^2 D}{4} = 22{,}333\ \text{ft}^3$$

The depth of each clarifier is

$$D = \frac{(22{,}333\ \text{ft}^3)(4)}{\pi(50\ \text{ft})^2} = 11.37\ \text{ft} \quad (12\ \text{ft})$$

The answer is (B).

289. The molecular weight of HCl is 1 g/mol + 35.5 g/mol = 36.5 g/mol.

The molecular weight of $Ca(OH)_2$, or lime, is 40 g/mol + (2 g/mol)(16 g/mol + 1g/mol) = 74 g/mol.

The number of moles of acid in the spill is

$$N_{\text{HCl}} = \frac{(10{,}000\ \text{kg})\left(1000\ \frac{\text{kg}}{\text{g}}\right)}{36.5\ \frac{\text{g}}{\text{mol}}} = 274{,}000\ \text{mol HCl}$$

The number of moles of lime required is

$$N_{\text{lime}} = (274{,}000\ \text{mol})\left(\frac{1\ \text{mol Ca(OH}_2)}{2\ \text{mol HCl}}\right)$$
$$= 137{,}000\ \text{mol lime}$$

The mass of lime required is

$$m_{\text{lime}} = (137{,}000\ \text{mol})\left(74\ \frac{\text{g}}{\text{mol}}\right)$$
$$= 10{,}138\ \text{kg} \quad (10{,}100\ \text{kg})$$

The answer is (C).

290. A diagram of a mixed flow and its ingredients is shown.

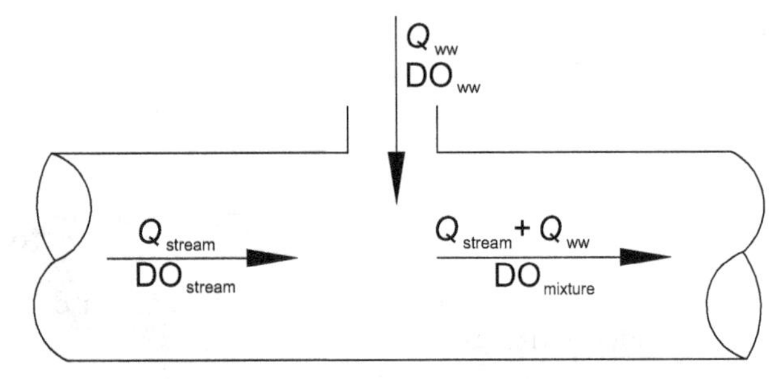

The DO of the mixture is determined using the mass-balance equation. Use the weighted-average concept in the equation, which uses the DOs of the stream and the wastewater (ww).

$$\text{Total flow of the mixture} = Q_{\text{stream}} + Q_{\text{ww}}$$

For DO in the stream and in wastewater, write the above equation as

$$(Q_{\text{stream}} + Q_{\text{ww}})(\text{DO}_{\text{mixture}}) = (Q_{\text{stream}})(\text{DO}_{\text{stream}}) + (Q_{\text{ww}})(\text{DO}_{\text{ww}})$$

Rearrange the equation to determine the DO of the mixture.

$$\text{DO}_{\text{mixture}} = \frac{(Q_{\text{stream}})(\text{DO}_{\text{stream}}) + (Q_{\text{ww}})(\text{DO}_{\text{ww}})}{Q_{\text{stream}} + Q_{\text{ww}}}$$

Convert the weighted-average wastewater flow into units of cubic meters per second.

$$Q_{\text{ww}} = \left(\frac{\left(20{,}000\ \frac{\text{m}^3}{\text{day}}\right)(1\ \text{day})}{\left(24\ \frac{\text{hr}}{\text{day}}\right)\left(3600\ \frac{\text{s}}{\text{hr}}\right)}\right) = 0.23\ \text{m}^3/\text{s}$$

The DO after mixing is

$$\text{DO after mixing} = \frac{\left(1\ \frac{\text{m}^3}{\text{s}}\right)\left(7.5\ \frac{\text{mg}}{\text{L}}\right) + \left(0.23\ \frac{\text{m}^3}{\text{s}}\right)\left(1.5\ \frac{\text{mg}}{\text{L}}\right)}{\left(1\ \frac{\text{m}^3}{\text{s}} + 0.23\ \frac{\text{m}^3}{\text{s}}\right)} = 6.38\ \text{mg/L} \quad (6\ \text{mg/L})$$

The answer is (C).

291. This structure is a three-hinge arch with two supports at point A and point B, and a crown hinge at point C. The moment is zero at point A, point B, and point C. To determine the horizontal reaction at point A, take the moment about point C for the arch segment AC, as shown.

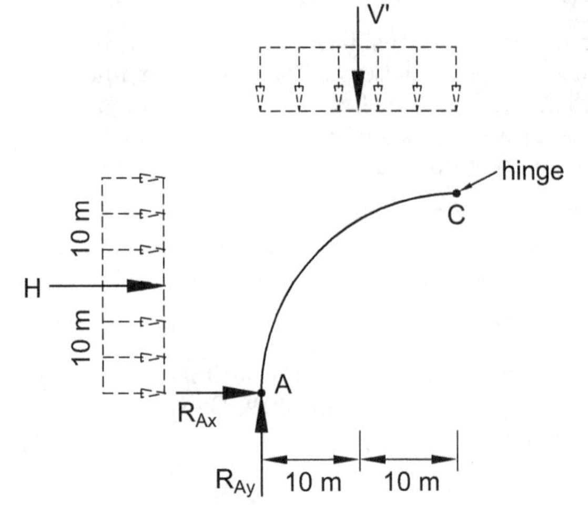

The horizontal reaction at point A is

$$\sum M_C = 0$$

$$R_{Ay} = 25\ \text{kN}$$

$$H = \left(3\ \frac{\text{kN}}{\text{m}}\right)(20\ \text{m}) = 60\ \text{kN}$$

Assume the horizontal reaction at A is acting toward the center, as shown. Let counterclockwise moments be positive.

$$\begin{aligned}\sum M_C &= (R_{Ax})(20) - (R_{Ay})(20) + (V')(10) + (H)(10) = 0\\ &(R_{Ax})(20\ \text{m}) - (25\ \text{kN})(20\ \text{m}) + (40\ \text{kN})(10\ \text{m}) + (60\ \text{kN})(10\ \text{m}) = 0\\ R_{Ax} &= -25\ \text{kN}\end{aligned}$$

The negative sign indicates that the reaction is acting away from the center.

The answer is (A) and (E).

292. A single degree of freedom system is shown.

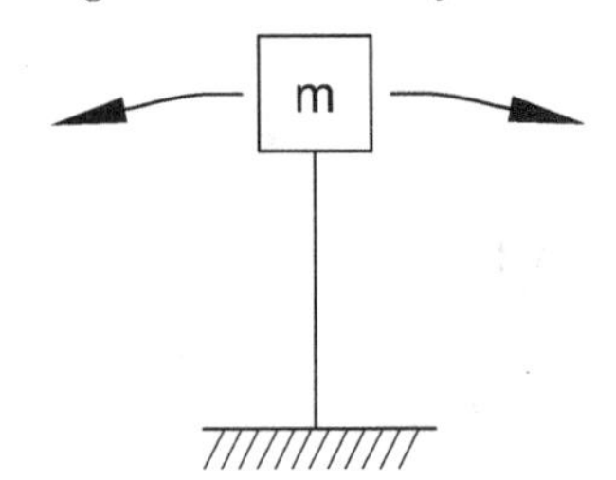

The equation for the natural frequency of the system is

$$\omega_n = \sqrt{\frac{K}{m}}$$

Substituting into the equation for the fundamental period,

$$T_n = \frac{2\pi}{\omega_n} = 2\pi\sqrt{\frac{m}{K}} = (2)(3.14)\sqrt{\frac{8000\ \text{kg}}{2000\ \frac{\text{kN}}{\text{m}}}}$$

$$= 12.56\ \text{s} \quad (12\ \text{s})$$

The answer is (C).

293. The column base is fixed against both rotation and translation. The top end is fixed against rotation but free to translate. The recommended design value of the effective length factor, k, is 1.2.

$$kL = (1.2)(15\ \text{ft}) = 18\ \text{ft}$$

Use AISC Table 4-1, given in the *NCEES Handbook*. The available strength in axial compression given in the table for a W12 × 50 section with $kL = 18$ is 270 kips.

The answer is (B).

294. The frame has six joints and six members. One truss support is a roller, which is capable of developing only a vertical reaction. The other support is a hinge, which can develop both a horizontal and a vertical reaction. Therefore, the frame has three unknown reactions. The frame also has an additional four hinges at the internal truss joints. Therefore, the frame has the following properties.

- 6 joints
- 6 members
- 3 independent reaction components
- 4 condition equations based on known internal moments or forces

As described in the Structural Analysis section of the *NCEES Handbook*, the stability and determinacy of an arch depends on whether the value of $3m + r$ is equal to, less than, or greater than the value $3j + c$. Find these values.

$$3m + r = (3)(6) + 3 = 21$$

$$3j + c = (3)(6) + 4 = 22$$

Because $3m + r$ is less than $3j + c$, the frame is unstable.

The answer is (A).

295. The rotation of a propped cantilever at its pinned end can be determined by the following equation, given in the *NCEES Handbook*, Structural Analysis section.

$$\theta = \frac{ML}{4EI}$$

For a W24 × 68 beam, the moment of inertia is 1830 in^4. The modulus of elasticity of structural steel is 29,000 ksi. Convert the rotation angle from degrees to radians.

$$\theta = \frac{(2^\circ)\pi}{180^\circ} = 0.035\ \text{rad}$$

Solve the two equation simultaneously.

$$\theta = \frac{ML}{4EI} = 0.035$$

Rearrange the equation for the rotation of a propped cantilever to find the moment.

$$\theta = \frac{ML}{4EI} = 0.035$$

$$M = \frac{(0.035)(4)EI}{L} = \frac{(0.035)(4)\left(29{,}000\ \frac{\text{kips}}{\text{in}^2}\right)(1830\ \text{in}^4)}{(25\ \text{ft})\left(12\ \frac{\text{in}}{\text{ft}}\right)\left(12\ \frac{\text{in}}{\text{ft}}\right)}$$

$$= 2064\ \text{ft-kips} \quad (2100\ \text{ft-kips})$$

The answer is (C).

296. The equation for the bending moment at each end of a beam carrying a uniformly distributed load is

$$FEM_{AB} = FEM_{BA} = \frac{wL^2}{12}$$

The moment at the midspan can be determined by using the midspan bending moment for a simply supported condition and superimposing the fixed-end moments on it, as shown. The midspan bending moment for a simply supported beam under a uniformly distributed load is

$$M_c = \frac{wL^2}{8}$$

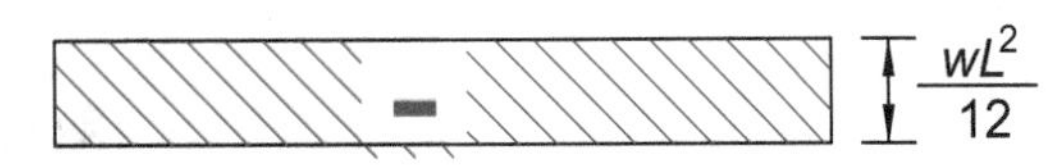

A. Fixed End Moment Diagram

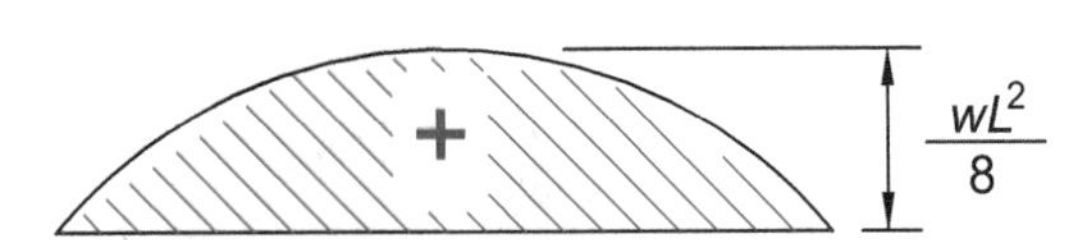

B. Simply-Supported B.M. Diagram

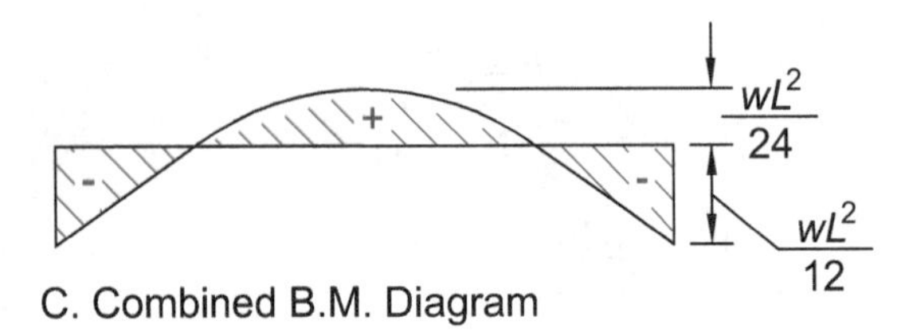

C. Combined B.M. Diagram

Since the bending moment due to the uniformly distributed load is a sagging moment, and the moments to keep the ends fixed are hogging moments, the bending moment along the beam span is the difference between the two.

$$\begin{aligned} M_c &= \frac{wL^2}{8} - \frac{wL^2}{12} = \frac{wL^2}{24} \\ &= \frac{\left(2\ \frac{\text{kips}}{\text{ft}}\right)(24\ \text{ft})^2}{24} \\ &= 48\ \text{ft-kips} \end{aligned}$$

The answer is (A).

297. When a vehicle accelerates or brakes, longitudinal forces are transmitted from its wheels to the bridge, as shown.

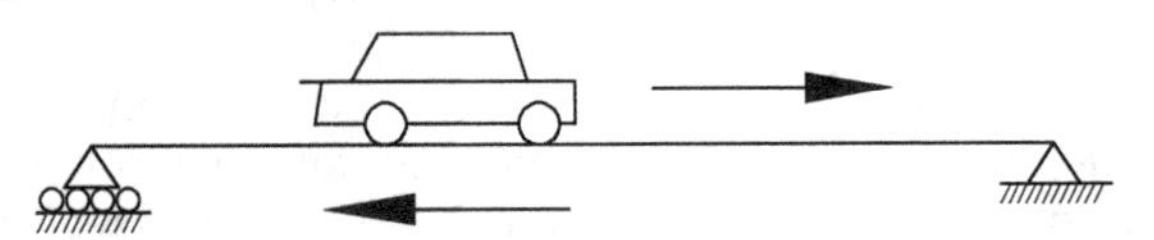

The magnitude of the force can be determined using Newton's law of motion.

$$F = ma = \frac{W}{g}\frac{dv}{dt}$$

Determine the rate of acceleration/deceleration in units of ft/sec^2.

$$\frac{dv}{dt} = \frac{\left(50\ \frac{\text{mi}}{\text{hr}}\right)\left(5280\ \frac{\text{ft}}{\text{mi}}\right)}{(15\ \text{sec})\left(3600\ \frac{\text{sec}}{\text{hr}}\right)} = 4.9\ \text{ft/sec}^2$$

Substitute all known values into the force equation.

$$\begin{aligned} F &= \left(\frac{72{,}000\ \text{lbf}}{32.2\ \frac{\text{ft}}{\text{sec}^2}}\right)\left(4.9\ \frac{\text{ft}}{\text{sec}^2}\right) \\ &= 10{,}957\ \text{lbf} \quad (11{,}000\ \text{lbf}) \end{aligned}$$

The answer is (C).

298. Allowable stress design does use safety factors. Load and resistance factor design uses resistance factors. Both design methods use the same methods of structural analysis. Performance-based design criteria involve non-linear analysis, which is not required in either method.

The answer is (D).

299. The design of a member in tension using load and resistance factor design requires a three-step process:

step 1: Determine the yield strength based on the gross area and the yield strength of the steel.

step 2: Determine the fracture strength based on the net area and the ultimate strength of the steel.

step 3: Determine the design strength by selecting the smaller strength of the first two steps.

See *NCEES Handbook*, Design of Steel Components for design formulas.

Find the gross area.

$$A_g = (6\ \text{in})\left(\frac{1}{2}\ \text{in}\right) = 3\ \text{in}^2$$

The bolt diameter is 3/4 in, so the hole diameter is 3/4 in + 1/16 in = 13/16 in.

Find the net area.

$$\begin{aligned} A_n &= \left(b_g - \sum d_h\right)t \\ &= \left(6\ \text{in} - \frac{13}{16}\ \text{in}\right)(0.5\ \text{in}) \\ &= 2.59\ \text{in}^2 \end{aligned}$$

For yielding,

$$\phi_y = 0.90$$

The yield strength is

$$\begin{aligned}\phi T_n &= \phi_y F_y A_g \\ &= (0.90)\left(50\ \frac{\text{kips}}{\text{in}^2}\right)(3\ \text{in}^2) \\ &= 135\ \text{kips}\end{aligned}$$

For fracture,

$$\phi_f = 0.75$$

The fracture strength is

$$\begin{aligned}\phi T_n &= \phi_f F_u A_n \\ &= (0.75)\left(65\ \frac{\text{kips}}{\text{in}^2}\right)(2.59\ \text{in}^2) \\ &= 126.3\ \text{kips} \quad (125\ \text{kips})\end{aligned}$$

The answer is (A).

300. See *NCEES Handbook*, Design of Steel Components for design formulas. The bolt is in double shear. In double shear, the shear force required to shear the bolt is twice the shear force required in single shear. Therefore, the available bolt strength is

$$\phi r_n = (2)(35.3\ \text{kips}) = 70.6\ \text{kips} \quad (70\ \text{kips})$$

The answer is (D).

301. See the Design of Reinforced Concrete Components section in the *NCEES Handbook*, which shows the interaction diagram of a rectangular column. A bending moment with $e/h = 1.2$ indicates that the column failure is close to the balanced point, and likely occurs due to tension yielding of the longitudinal reinforcement. Therefore, the reinforcement should be placed on the side where the tension is greater. Since the applied moment is subject to reversal, as shown, reinforcement is most effective if placed at faces C and D.

The answer is (C).

302. The resistance factors are tabulated in the Design of Reinforced Concrete Components section of the *NCEES Handbook*. The ACI 318 building code calls the factors; the strength reduction factors. The factors vary depending on the type of loading and are always less than 1. A resistance factor accounts for material quality control achievable for a given member, and for uncertainties in the strength of a given member. Higher uncertainties require more reduction in strength. So the greater the uncertainty, smaller would be the factor. Statement D is incorrect.

The answer is (D).

303. A soil sample that contains particles of similar sizes is called poorly graded soil, and a soil sample that contains particles of all sizes is called well-graded soil. Therefore, options A and B are incorrect. The symbol M in the USCS is used for silt. Therefore, option C is incorrect. Grains of clay soil cannot be seen with the naked eye because they have cohesive properties and are lumped together due to electromagnetic and chemical forces.

The answer is (D).

304. The coefficients represent the grading or textural characteristics of the soil. These are based on the geometric properties of the grading curve that describes soil texture. The coefficients are defined as shown.

$$\text{coefficient of uniformity},\ C_u = \frac{D_{60}}{D_{10}}$$

$$\text{coefficient of curvature},\ C_c = \frac{(D_{30})^2}{D_{10}D_{60}}$$

D_{10} is the effective particle size: 10% of the particles in the sample are finer than this size, and 90% are larger and will be retained on a sieve sized to catch particles of size D_{10}. Similarly, D_{60} is the effective particle size at which 40% of particles are caught in a properly sized sieve, and D_{30} is the effective particle size at which 70% of particles are caught in a properly sized sieve.

The coefficients represent textural or grading properties. Two types of soil classifications are possible: well-graded soils and poorly graded soils. For well-graded soils, the second suffix is "W"; for poorly graded soils, the second suffix is "P." Per USCS classifications, only sand (type "S") and gravel (type "G") have such suffixes. Therefore, out of 15 classifications, four could be used for this soil sample: SP, SW, GP, and GW.

Out of the four options, only GP (poorly graded gravel) has the required coefficients.

The answer is (B).

305. Since the dry unit weight, γ_D, is to be determined, the weight of the wet soil sample is not relevant.

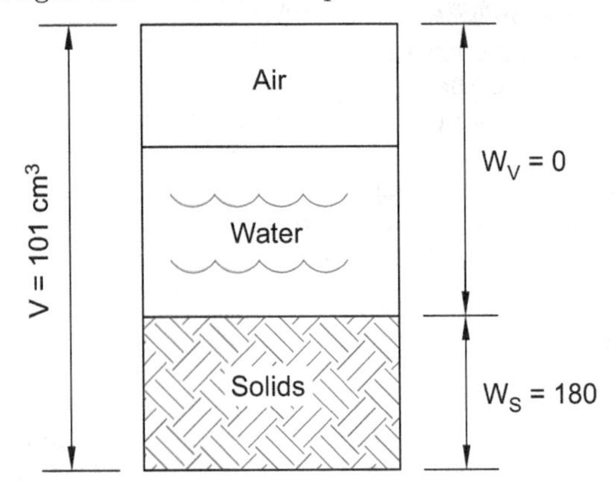

The dry unit weight is

$$\gamma_D = \frac{\text{weight of solids}}{\text{total volume}} = \frac{W_S}{V} = \frac{180 \text{ g}}{101 \text{ cm}^3} = 1.78 \text{ g/cm}^3 \quad (1.8 \text{ g/cm}^3)$$

The answer is (A).

306. The relative compaction is

$$\text{RC} = \frac{\gamma_{D,\text{field}}}{\gamma_{D,\text{max}}}(100\%) = \left(\frac{129.4 \ \frac{\text{lbf}}{\text{ft}^3}}{140 \ \frac{\text{lbf}}{\text{ft}^3}}\right)(100\%) = 92.4\% \quad (92\%)$$

The answer is (B).

307. Total overburden pressure is defined as the pressure intensity due to the weight of soil, its moisture, and surcharge on any horizontal plane below before construction starts. The unsaturated soil height is

$$z - h = 6 \text{ ft} - 3.5 \text{ ft} = 2.5 \text{ ft}$$

The total overburden at a depth of 2.5 ft is

$$p_1 = \gamma(z-h) = \left(120 \ \frac{\text{lbf}}{\text{ft}^3}\right)(2.5 \text{ ft}) = 300 \ \text{lbf/ft}^2$$

The total overburden at the footing base is

$$p = p_1 + \gamma_{\text{sat}}(h) = 300 \ \frac{\text{lbf}}{\text{ft}^2} + \left(128 \ \frac{\text{lbf}}{\text{ft}^3}\right)(3.5 \text{ ft}) = 748 \text{ lbf/ft}^2 \quad (750 \text{ lbf/ft}^2)$$

The answer is (C).

308. Use the horizontal stress profile and force equations given in the *NCEES Handbook*, Geotechnical section. The equation for active lateral pressure is

$$P_a = 0.5k_a\gamma H^2$$

The equation for passive lateral pressure is

$$P_p = 0.5k_p\gamma h^2$$

The angle of internal friction of the soil is given. Determine the coefficients.

$$k_a = \tan^2\left(45° - \frac{\phi}{2}\right) = \tan^2\left(45° - \frac{30}{2}\right) = 0.33$$

$$k_p = \tan^2\left(45° + \frac{\phi}{2}\right) = 3$$

Substitute the known values to determine the active and passive pressures.

$$P_a = (0.5)(0.33)\gamma H^2 = 0.167\gamma H^2$$

$$P_p = (0.5)(3)\gamma h^2 = 1.5\gamma h^2$$

The active lateral pressure tends to slide, and passive lateral pressure tends to resist the sliding. The factor of safety is

$$\text{FOS} = \frac{P_p}{P_a} = \frac{1.5\gamma h^2}{(0.167)\gamma H^2} = 9\left(\frac{h}{H}\right)^2 = (9)\left(\frac{15 \text{ ft}}{30 \text{ ft}}\right)^2 = 2.25$$

The answer is (C).

309. See *NCEES Handbook*, Civil Engineering section. The equation for the shear stress at failure of soil is

$$\tau_F = c + \sigma_N \tan\phi$$

For sands, the cohesion coefficient, c, is 0. Therefore, the equation can be simplified.

$$\tau_F = \sigma_N \tan\phi$$

The equation for the ultimate shear capacity is therefore

$$\tau_{\text{ult}} = \sigma_N \tan 30°$$

The shear stress at the working level is

$$\tau_{\text{working}} = \sigma_N \tan 25°$$

The factor of safety is

$$\text{FOS} = \frac{\tau_{\text{ult}}}{\tau_{\text{working}}} = \frac{\sigma_N \tan 30°}{\sigma_N \tan 25°} = \frac{\tan 30°}{\tan 25°} = 1.24$$

The answer is (C).

310. The ultimate load capacity of a drilled shaft depends on the soil bearing capacity at the base and the skin friction between the shaft and the soil around the drilled pier. It is expressed as

$$Q_u = (\text{FS})\,Q_{\text{base}} + Q_{\text{skin}}$$

The drilled pier is circular in cross section. The soil footing that supports the pier loading is also circular in shape, and its diameter equals the diameter of the pier. Therefore, the area is

$$A_{\text{base}} = \frac{\pi}{4}D^2 = \frac{\pi}{4}(3\ \text{ft})^2 = 7.07\ \text{ft}^2$$

The soil bearing capacity at the base is

$$\begin{aligned} Q_{\text{base}} &= A_{\text{base}}q_{\text{soil}} = (7.07\ \text{ft}^2)\left(10\ \frac{\text{kips}}{\text{ft}^2}\right) \\ &= 70.7\ \text{kips} \end{aligned}$$

The friction resistance is around the drilled pier's perimeter. Calculate the side (perimeter) area of the circular shaft.

$$A_{\text{skin}} = \pi DL = \pi(3\ \text{ft})(30\ \text{ft}) = 282.7\ \text{ft}^2$$

The skin friction is

$$\begin{aligned} Q_{\text{skin}} &= A_{\text{skin}}\,q_{\text{skin}} = (282.7\ \text{ft}^2)\left(0.1\ \frac{\text{kips}}{\text{ft}^2}\right) \\ &= 28.3\ \text{kips} \end{aligned}$$

The ultimate load capacity is

$$\begin{aligned} Q_u &= (\text{FS})\,Q_{\text{base}} + Q_{\text{skin}} \\ &= (3.0)(70.7\ \text{kips} + 28.3\ \text{kips}) \\ &= 297\ \text{kips} \quad (295\ \text{kips}) \end{aligned}$$

The answer is (C).

311. Consolidation is a void ratio reduction that takes place over a period of time. This includes primary consolidation. Statement B is incorrect.

A consolidation test is performed on an undisturbed sample. It is also called an odometer test. Therefore, statement A is correct.

Secondary consolidation is the additional void–ratio change after all pore pressure has been dissipated, meaning it takes place over a long period of time. Therefore, statement C is correct.

In computing the length of a drainage path, the longest path a particle can take to escape normal pressure is used. Therefore, statement D is correct.

The answer is (B).

312. Find the at-rest lateral pressure. The coefficient, k_0, is 0.58.

$$\begin{aligned} P_0 &= 0.5k_0\gamma H^2 = 0.5(0.58)\left(120\ \frac{\text{lbf}}{\text{ft}^3}\right)(12\ \text{ft})^2 \\ &= 5011\ \text{lbf/ft} \end{aligned}$$

The wall acts as a simply supported beam under a triangular load. The reaction at the base is 2/3 of the total wall pressure.

$$R_{\text{base}} = \left(\frac{2}{3}\right)5011\ \frac{\text{lbf}}{\text{ft}} = 3341\ \text{lbf/ft} \quad (3300\ \text{lbf/ft})$$

The answer is (C).

313. This area is called a bench. It is a slightly reverse sloping step and is an effective measure to slow down the runoff water. A sediment trap has a depression to slow down the water and allow the silt to settle. A sediment basin is designed to catch sediments. It has a dam and a pipe outlet and/or an emergency spillway.

The answer is (A).

314. This is a kinematics problem. Since the deceleration rate is given, the weight of the car, the road friction coefficient, and the grade incline are not needed to compute the stopping distance. Therefore, choices C and D are incorrect. To determine the stopping distance, use the formula given in the Constant Acceleration section of the *NCEES Handbook*. The initial acceleration, a_0, is $-16\ \text{ft/sec}^2$. The initial velocity, v_0, is 35 mph, and the final velocity, v, is 0 mph. The initial displacement, s_0, is 0. Calculate the final displacement.

$$v^2 = (v_0)^2 + 2a_0(s - s_0)$$

$$(0)^2 = \left(\frac{\left(35\ \frac{\text{mi}}{\text{hr}}\right)\left(5280\ \frac{\text{ft}}{\text{mi}}\right)}{3600\ \frac{\text{sec}}{\text{hr}}}\right)^2 + 2\left(-16\ \frac{\text{ft}}{\text{sec}^2}\right)(s-0)$$

$$s = 82.24\ \text{ft}$$

Since the required stopping distance is greater than 55 ft, the car would not be able to stop at the stopping line.

The answer is (B).

315. See the Vertical Curves table in the *NCEES Handbook.* First, determine if the curve is a sag curve or a crest curve. In this case, since its back tangent has a positive 2% grade and the forward grade is negative, it is a crest curve.

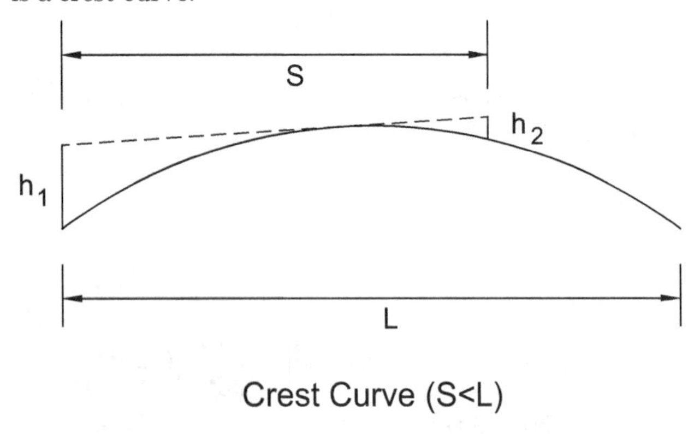

Crest Curve (S<L)

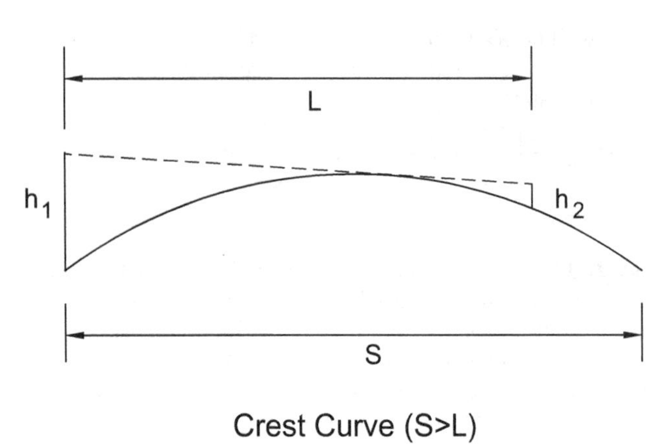

Crest Curve (S>L)

Both the height of driver's eyes and the object height meet the standard criteria. Therefore, the sight distance equation for the crest curve simplified for standard criteria can be used.

$$A = |g_1 - g_2| = |2\% - (-4\%)| = 6\%$$

The minimum curve length is

$$L_{\text{min}} = \frac{AS^2}{2158} = \frac{(6\%)(496\ \text{ft})^2}{2158\ \text{ft}} = 684\ \text{ft} \quad (700\ \text{ft})$$

The answer is (B).

316. First, determine if it is a sag curve or a crest curve. Both tangents have negative slopes, and the back tangent has a steeper downgrade than the forward tangent, so it is a sag curve. The sight distance is

$$A = |g_1 - g_2| = |-5\% - (-1\%)| = 4\%$$

The minimum length of the curve is

$$L_{\text{min}} = \frac{AV^2}{46.5} = \frac{(4\%)(45\ \text{ft})^2}{46.5\ \text{ft}} = 174\ \text{ft} \quad (170\ \text{ft})$$

The answer is (B).

317. The objective of the AASHTO method is to design a flexible pavement. For this, an adequate structural number (SN) is determined to carry the projected design load equivalency factor (LEF). Therefore, the SN is not selected at will. However, the selection of the number of layers is left to the engineer in charge. A flexible pavement is a layered system. The required SN above the subgrade is first determined, and then the required SNs above the base and sub-base layers are determined using the appropriate strength of each layer. The minimum allowable thickness of each layer is then determined.

The strength of each construction material, such as asphalt concrete, base, and sub-base materials, are given by AASHTO as the layer coefficient. Several values of depth D_1, D_2, etc. can be used to satisfy the required SN value. Layer thicknesses should not be less than the minimum value prescribed by AASHTO. There is no maximum limit on the thickness of a layer. Therefore, a single layer may be adequate to satisfy the required SN, though it may be impractical or uneconomical to go beyond the practical range of the thickness of a layer.

The answer is (B).

318. The heavy vehicle adjustment factor accounts for the presence of heavy vehicles in the traffic stream. It can be computed from the equation shown.

$$f_{HV} = \frac{1}{1 + P_T(E_T - 1)}$$

E_T is the passenger-car equivalent (PCE) of a single unit truck or tractor-trailer in a traffic stream, which is equal to 3 for rolling terrain.

$$\begin{aligned} f_{HV} &= \frac{1}{1 + P_T(E_T - 1)} = \frac{1}{1 + (0.15)(3 - 1)} \\ &= 0.77 \end{aligned}$$

The answer is (C).

319. In this case, since the driver decided not to stop, the reaction time is not applicable. Use the time-distance formula. The total distance is

$$\begin{aligned} s &= \text{distance to stopping line} \\ &\quad + \text{width of intersection} + \text{car length} \\ &= 80 \text{ ft} + 60 \text{ ft} + 15 \text{ ft} \\ &= 155 \text{ ft} \end{aligned}$$

The travel time needed is

$$t = \frac{\text{distance}}{\text{speed}} = \frac{155 \text{ ft}}{\left(35 \ \frac{\text{mi}}{\text{hr}}\right)\left(1.47 \ \frac{\text{ft-hr}}{\text{mi-sec}}\right)} = 3 \text{ sec} \quad (3.0 \text{ sec})$$

The answer is (B).

320. The rate per million of entering vehicles (RMEV) is the number of crashes per million vehicles entering the intersection. It is expressed as

$$\text{RMEV} = \frac{A(1{,}000{,}000)}{V}$$

A is the total number of crashes per year at the intersection. V is the average daily traffic at the intersection in vehicles per year. Rearrange the equation and solve for A.

$$\begin{aligned} \text{RMEV} &= \frac{A(1{,}000{,}000)}{V} \\ A &= \left(\frac{\text{RMEV}}{1{,}000{,}000}\right)V \\ &= \left(\frac{8}{1{,}000{,}000}\right)\left(8000 \ \frac{\text{veh}}{\text{day}}\right)\left(365 \ \frac{\text{day}}{\text{yr}}\right) \\ &= 23.36 \quad (24) \end{aligned}$$

The answer is (C).

321. See the Traffic Flow Relationships diagrams in the Transportation section of the *NCEES Handbook*. The speed-flow relationship is

$$q = kv$$

When the density is very low, each driver is free to travel at the maximum speed possible, depending on the physical characteristics of the roadway. In other words, when the density is approximately 0, the mean free speed and the speed are equivalent, and as the density, k, increases, the flow (or the number of vehicles crossing in a given time) increases. The flow is at its maximum when the following conditions are true.

- density, $k_0 = 0.5k_j$
- speed, $v_0 = 0.5v_f$

Therefore, the maximum flow is

$$q_{\max} = k_0 v_0$$

For the k-v relationship,

$$k_j = 120 \text{ veh/mi}$$

$$v_f = 60 \text{ mph}$$

$$q_{\max} = \left[(0.5)\left(120 \ \frac{\text{veh}}{\text{mi}}\right)\right]\left[(0.5)\left(60 \ \frac{\text{mi}}{\text{hr}}\right)\right] = 1800 \text{ veh/hr}$$

The answer is (B).

322. When density reaches its maximum, the vehicles are lined up or packed bumper to bumper, and the flow comes to a halt. When the density is very low, there is little interaction between the vehicles traveling on the roadway. This speed is also known as the mean free speed. In other words, when the density is approximately 0, the mean free speed and the speed are equivalent. The value of the mean free speed can thus be determined by inserting a density of 0 into the equation of the line in the graph.

$$\begin{aligned} 2v + k - 120 &= 0 \\ 2(v_f) + 0 - 120 &= 0 \\ v_f &= 60 \text{ mph} \end{aligned}$$

The answer is (C).

323. The responsibility is determined by the contract between the parties. The formwork system is part of the means and methods used by the contractor working on the job site. Therefore, the contractor is responsible for the formwork design.

The answer is (C).

324. Liquidated damages are a specified amount agreed upon in advance to represent damages for breach of contract, usually due to delay. Since quantifying future damages is time consuming and virtually impossible, the owner and contractor contractually agree that a delay in completion of the project will cost the contractor a fixed amount per day. As such, liquidated damages are not a penalty, but a contractually agreed upon amount. Option D is correct.

The answer is (D).

325. Shores are designed to carry the weight of the formwork, concrete, and construction loads above. Therefore, option A is correct. Reshores are shores placed snugly under a stripped concrete slab or other structural member after the original forms and shores have been removed from a large area, so option B is correct. Once the shores are removed, the floor carries itself in its entirety, so option

C is correct. Reshores require few bracing or horizontal elements, so option D is not correct.

The answer is (D).

326. According to 29 CFR 1926.501(b)(7) – Excavations, the employee has to be protected from a fall if the depth of the excavation, pit, or shaft exceeds 6 ft.

The answer is (C).

327. The contractor has completed hauling 15,000 yd^3 and actually spent \$210,000. The actual cost of the work performed is \$210,000.

The answer is (C).

328. The bid price is \$1.1 million. It includes 10% profit over the budgeted cost. Therefore, the budgeted cost of the entire work is

$$\frac{\$1,100,000}{1+0.1} = \$1,000,000$$

The monthly budgeted cost is therefore

$$\frac{\$1,000,000}{10 \text{ mo}} = \$100,000/\text{mo}$$

For two months, the budgeted cost of the work scheduled is

$$(2)(\$100,000) = \$200,000$$

The answer is (B).

329. The illustration shows the vertical joints between the segments in the wall.

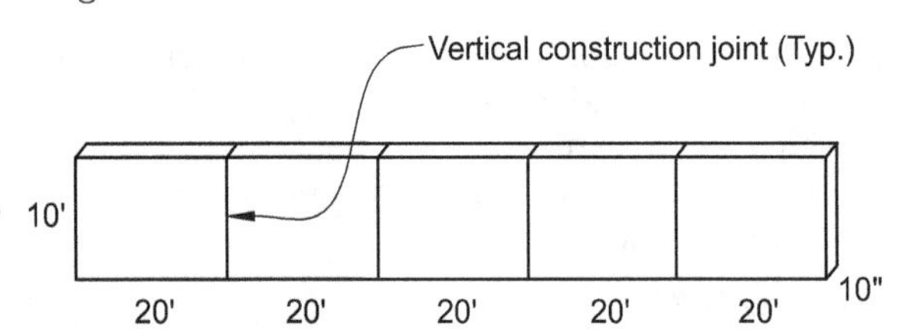

The construction of a free-standing wall requires formwork on both faces. The wall is built in five segments of 20 ft each. The forms for the wall face will be used five times. In addition, the first segment requires formwork for both ends. The remaining four segments need formwork for one end only, since the other side of the wall segment would butt against the adjacent, previously-built segment, so no formwork is needed at that end.

The footage of the forms required for the first segment is

$$\begin{aligned}&(20 \text{ ft})(10 \text{ ft})(2 \text{ faces}) \\ &\quad + (10 \text{ ft})\left(\frac{10 \text{ in}}{12 \ \frac{\text{in}}{\text{ft}}}\right)(2 \text{ ends}) = 416.67 \text{ ft}^2\end{aligned}$$

The footage of the forms required per segment for segments 2 through 5 is

$$\begin{aligned}&(20 \text{ ft})(10 \text{ ft})(2 \text{ faces}) \\ &\quad + (10 \text{ ft})\left(\frac{10 \text{ in}}{12 \ \frac{\text{in}}{\text{ft}}}\right)(1 \text{ end}) = 408.33 \text{ ft}^2\end{aligned}$$

The cost of leasing the formwork is

$$(416.67 \text{ ft}^2)\left(2 \ \frac{\$}{\text{ft}^2}\right) + (4)(408.33 \text{ ft}^2)\left(0.50 \ \frac{\$}{\text{ft}^2}\right) = \$1650$$

The labor cost of erecting the formwork is

$$\left(5 \ \frac{\$}{\text{ft}^2}\right)\left(416.67 \text{ ft}^2 + (4)(408.33 \text{ ft}^2)\right) = \$10,250$$

The labor cost of stripping the formwork is

$$\left(2 \ \frac{\$}{\text{ft}^2}\right)\left(416.67 \text{ ft}^2 + (4)(408.33 \text{ ft}^2)\right) = \$4100$$

The total cost of the formwork is

$$\$1650 + \$10,250 + \$4100 = \$16,000$$

The answer is (C).

330. A girder is a beam that is the main horizontal floor framing of a structure, or the large beam that supports smaller beams. In the plan shown, the W14 × 22 sections are the beams and the W24 × 68 sections are the girders. The beams, girders, and horizontal members in the moment frames comprise the horizontal load-carrying framing, which transfers the floor/roof loads to vertical load-carrying members.

The plans show that there are eight bays on the roof floor. Each bay has seven beams. Each beam spans 30 ft. The W14 sections each weigh 22 lbf/ft. The total steel tonnage is

$$\begin{aligned}&(8 \text{ bays})\left(7 \ \frac{\text{beams}}{\text{bay}}\right)(30 \text{ ft})\left(22 \ \frac{\text{lbf}}{\text{ft}}\right)(1.02)\left(\frac{1 \text{ ton}}{2000 \text{ lbf}}\right) \\ &\quad = 18.85 \text{ tons} \quad (19 \text{ tons})\end{aligned}$$

The answer is (C).

Solutions
Exam 4

331. Expansion by minors is used to find the value of a determinant. Using the *NCEES Handbook*, Mathematics section notations, the value of a third-order determinant can be found by expanding the determinant.

$$\begin{vmatrix} a_1 & a_2 & a_3 \\ b_1 & b_2 & b_3 \\ c_1 & c_2 & c_3 \end{vmatrix} = a_1 \begin{vmatrix} b_2 & b_3 \\ c_2 & c_3 \end{vmatrix} - a_2 \begin{vmatrix} b_1 & b_3 \\ c_1 & c_3 \end{vmatrix} + a_3 \begin{vmatrix} b_1 & b_2 \\ c_1 & c_2 \end{vmatrix}$$
$$= a_1b_2c_3 + a_2b_3c_1 + a_3b_1c_2 - a_3b_2c_1 - a_2b_1c_3 - a_1b_3c_2$$

For the third-order determinant, the equation is

$$\begin{vmatrix} 1 & 2 & -1 \\ 2 & 3 & 2 \\ 1 & -2 & -2 \end{vmatrix} = 1\begin{vmatrix} 3 & 2 \\ -2 & -2 \end{vmatrix} - 2\begin{vmatrix} 2 & 2 \\ 1 & -2 \end{vmatrix} - 1\begin{vmatrix} 2 & 3 \\ 1 & -2 \end{vmatrix}$$
$$= 1[3(-2) - (2)(-2)] - 2[2(-2) - 2(1)] - 1[2(-2) - 3(1)]$$
$$= 17$$

The answer is 17.

332. Radioactive elements decay with time, and the formula is based on continuous compounding of decay. In this case,

$$m_0 = 100 \text{ g}$$
$$m = 50 \text{ g}$$
$$t = 10 \text{ yr}$$

Substitute the values into the given formula.

$$m = m_0 e^{kt}$$
$$50 \text{ g} = (100 \text{ g})e^{k(10)}$$
$$0.50 = e^{k(10)}$$

Take the logarithms on both sides of the equation and isolate k.

$$\log 0.5 = \log e^{10k} = 10(k)(\log e)$$
$$k = \frac{\log 0.5}{10 \log e} = \frac{-0.301}{10(0.434)} = -0.069$$

The answer is –0.069.

333. The roots of the quadratic equation $ax^2 + bx + c = 0$ are

$$x_1, x_2 = \frac{-b \pm \sqrt{b^2 - 4ac}}{2a}$$

The two roots are

$$x_1 = \frac{-b + \sqrt{b^2 - 4ac}}{2a}$$
$$x_2 = \frac{-b - \sqrt{b^2 - 4ac}}{2a}$$

Add the two roots.

$$x_1 + x_2 = \frac{-b}{2a} + \frac{-b}{2a} = \frac{-b}{a}$$

The answer is –b/a.

334. The two roots are $x_1 = 12$ and $x_2 = -2$.

By definition,

$$x - x_1 = x - x_2 = 0$$

Therefore,

$$(x - 12)(x + 2) = 0$$
$$x^2 + 2x - 12x - 24 = 0$$
$$x^2 - 10x - 24 = 0$$

The answer is

$$A = 1$$
$$B = -10$$
$$C = -24$$

335. The equation is in quadratic form. To solve it, let $x = \cos\theta$.

The equation becomes $2x^2 + 9x - 5 = 0$.

This is a quadratic equation of the form $ax^2 + bx + c = 0$.

The equation has two roots. Use the factor method with basic algebra.

$$2x^2 + 9x - 5 = (2x - 1)(x + 5) = 0$$

Therefore,

$$2x - 1 = 0 \text{ or } x + 5 = 0$$

The variable x has two values.

$$x = \frac{1}{2} = 0.5 \text{ or } x = -5$$

$$\cos\theta = 0.5 \text{ or } \cos\theta = -5$$

Since the value of $\cos\theta$ varies between -1 and 1, $\cos\theta = 0.5$.

Therefore,

$$\theta = \cos^{-1} 0.5 = 60°$$

The answer is 60°.

336. Two matrices can be added if the matrices have the same size (i.e., if the number of rows and the number of columns in one matrix equals those in the other). Here, both matrices have three rows and three columns. Therefore, addition is possible. The notation c_{23} stands for the term in the second row and third column of the matrix C.

$$c_{23} = a_{23} + b_{23} = 7 + 9 = 16$$

The answer is 16.

337. The radius, R, at a point along the curve is defined as the inverse of the curvature, K, at that point.

$$R = \frac{1}{|K|} = \left| \frac{[1 + (y')^2]^{3/2}}{|y''|} \right|$$

The following is given.

$$y = x - \frac{1}{6}x^2$$

For $x = 3$,

$$y' = 1 - \frac{x}{3}$$

$$y'' = -\frac{1}{3}$$

Substitute the values of y' and y'' in the radius of curvature equation.

$$K = \frac{-\frac{1}{3}}{[1 + (0)^2]^{3/2}} = -\frac{1}{3}$$

$$R = \frac{1}{|K|} = 3$$

The answer is 3.

338. The probability of getting at least four men is the sum of the probabilities of getting four men and getting five men. Use the binomial theorem.

$$\text{probability of success} = p = \frac{8}{17}$$

$$\text{probability of failure} = q = \frac{9}{17}$$

The probability of four successes in five trials is given by

$$P_5(4) = \frac{5!}{4!\,(5-4)!}\left(\frac{8}{17}\right)^4\left(\frac{9}{17}\right)^1 = 0.1298$$

The probability of five successes in five trials is given by

$$P_5(5) = \frac{5!}{5!\,(5-5)!}\left(\frac{8}{17}\right)^5\left(\frac{9}{17}\right)^0 = 0.0231$$

Therefore,

$$P\{\text{at least 4 men}\} = 0.1298 + 0.0231 = 0.1529 \approx 0.15$$

The answer is 0.15.

339. Since there are four points, it is given that $n = 4$.

The mean standard error of estimate is

$$\text{MSE} = S_e^2 = \frac{S_{xx}S_{yy} - S_{xy}^2}{S_{xx}(n-2)}$$

To evaluate the parameters in the regression equations, tabulate the data.

	x	y	xy	x^2	y^2
	2	9	18	4	81
	3	11	33	9	121
	5	15	75	25	225
	9	22	198	81	484
Σ	**19**	**57**	**324**	**119**	**911**

Use the *NCEES Handbook*, Engineering Probability and Statistics section, to determine the three knowns in the MSE equation.

$$S_{\text{xx}} = \sum_{i=1}^{n}(x_i)^2 - \left(\frac{1}{n}\right)\left(\sum x_i\right)^2$$
$$= 119 - \left(\frac{1}{4}\right)(19)^2 = 28.75$$

$$S_{\text{xy}} = \sum_{i=1}^{n} x_i y_i - \left(\frac{1}{n}\right)\left(\sum x_i\right)\left(\sum y_i\right)$$
$$= 324 - \frac{(19)(57)}{4} = 53.25$$

$$S_{\text{yy}} = \sum_{i=1}^{n} y_i^2 - \left(\frac{1}{n}\right)\left(\sum_{i=1}^{n} y_i\right)^2$$
$$= 911 - \frac{(57)^2}{4} = 98.75$$

Substituting the above values into the equation for MSE gives

$$S_e^2 = \frac{S_{\text{xx}}S_{\text{yy}} - S_{\text{xy}}^2}{S_{\text{xx}}(n-2)} = \frac{(28.75)(98.75) - (53.25)^2}{(28.75)(4-2)}$$
$$= 0.061 \quad (0.06)$$

The answer is 0.06.

340. Options A, C, and D are correct. The fitted equation should be used in the range of data that it was based on. No extrapolation is permissible. Since the data used in this regression analysis had values of x between 2 and 21 worker-hours, the fitted equation cannot be used to predict the values beyond the relevant range of x-values.

The answer is (B).

341. Options B, C, and D are correct. The capitalist economic system is based on an open and free market, where people tend to buy the best product at the best price, thereby rewarding the most efficient producer. Bribery corrupts the capitalist economic system, so option A is incorrect.

The answer is (A).

342. *NCEES Model Rules*, Sec. 240.15, Rules of Professional Conduct, § B.8 prohibits a licensee from soliciting or accepting a professional contract from a governmental body in which the licensee serves as a principal or an officer of that organization. This is not the case here. Also, NCEES rule § B.6 requires licensees to make full prior disclosure to their clients of potential conflict of interest or other circumstances that could influence, or appear to influence, their judgment or the quality of their service. The NCEES code does not discuss being a "friend" with government officials on social media, since doing so does not necessarily provide the licensee any influence with those officials. Therefore, the licensee may accept the assignment. Option D is correct.

The answer is (D).

343. See the *NCEES Handbook* section on Ethics and Professional Conduct. It is true that a licensed professional has earned the right to practice his or her profession. However, in practicing the profession, the engineer is obligated to society. Therefore, option A is incorrect. Engineers are obligated to perform the service in accordance with customary standard of care. Knowing and following the state-of-the-art design methods is essential. Therefore, option B is incorrect. Most of the claims arise out of nontechnical issues such as ambiguity, miscommunication, and lack of communication. Therefore, option D is incorrect. Technical ability is needed greatly but is not the only way to succeed in design practice. Nontechnical skill matters a great deal. Option C is correct.

The answer is (C).

344. Option A, option B, and option C violate the code of ethics. Attending a disciplinary hearing with a professional engineering board is ethical, professional conduct. These meetings are open to the public and provide a good experience in learning about how the profession regulates itself. Option D is correct.

The answer is (D).

345. A design-bid delivery system is the appropriate system when the project cost is of primary importance and the schedule is of secondary importance. A design-bid-build contract is appropriate when a complete set of project drawings and specifications are available. When unit quantities are known, a lump-sum contract would be appropriate. The cost-plus method should be limited to emergency situations and for a limited scope of work. For example, when an imminent danger to human life or property damage exists, the cost-plus contract may be the proper type of contract. Option D is correct.

The answer is (D).

346. The *NCEES Handbook*, Engineering Economics section, defines bond value as the present worth of the payments that the purchaser (or holder of the bond) receives during the life of the bond at some interest rate, i. Bond yield equals the computed interest rate of the bond value when compared with the bond cost. The city promises an annual 6% yield on $20,000,000, which equals $1.2M. Other bonds in the market pay 4%, which equals $0.8M annually on a $20M loan—a premium of $400,000 per year.

Bond prices are adjusted by comparing the market yield with a particular bond's face interest rate. Therefore, the investors would be willing to pay more than the face value because they would recoup it through $400,000 annual interest payments by the city. Using the tables in the *NCEES Handbook*, Engineering Economics section, compute the present value of additional payments.

$$\begin{aligned}(P/A, 4.0\%, 10) &= 8.1109 \\ P &= (8.1109)(\$400{,}000) \\ &= \$3{,}244{,}360\end{aligned}$$

$$\begin{aligned}\text{bond value} &= \$20\text{ M} + \$3.24\text{ M} \\ &= \$23.24\text{ M} \quad (\$23\text{ M})\end{aligned}$$

The answer is $23 million.

347. The nominal annual rate is $12(1\%) = 12\%$.

The effective annual rate is

$$\begin{aligned}i_{\text{eff}} &= (1+i)^{12} - 1 \\ &= (1+0.01)^{12} - 1 \\ &= 0.1269 \quad (12.69\%)\end{aligned}$$

The answer is (C).

348. Total cost to manufacture 50,000 units of PPE annually is

$$\begin{aligned}\text{cost} &= \text{initial cost} + (\text{annual units needed}) \\ &\quad \times(\text{unit manufacturing cost})(\text{number of years})\end{aligned}$$

Alternate 1 has a lower initial setup cost but higher manufacturing cost. On the other hand, alternate 2 has higher initial cost but lower manufacturing cost. In the long run, alternate 2 would be more profitable.

Let n be the number of years to break even. Compare total annual cost to manufacture PPE using both alternates.

$$\begin{aligned}C_1 &= \$10{,}000 + \left(50{,}000\ \frac{\text{units}}{\text{yr}}\right)(\$1.5/\text{unit})(n) \\ &= \$10{,}000 + \$75{,}000n\end{aligned}$$

$$\begin{aligned}C_2 &= \$150{,}000 + \left(50{,}000\ \frac{\text{units}}{\text{yr}}\right)(\$0.6/\text{unit})(n) \\ &= \$150{,}000 + \$30{,}000n\end{aligned}$$

For breakeven, $C_1 = C_2$.

Substitute the known values and determine n.

$$\begin{aligned}\$10{,}000 + \$75{,}000n &= \$150{,}000 + \$30{,}000n \\ n &= \frac{\$140{,}000}{\$45{,}000} = 3.11\text{ yr} \quad (3\text{ yr})\end{aligned}$$

The answer is 3 years.

349. Draw the timeline of cash flow as shown.

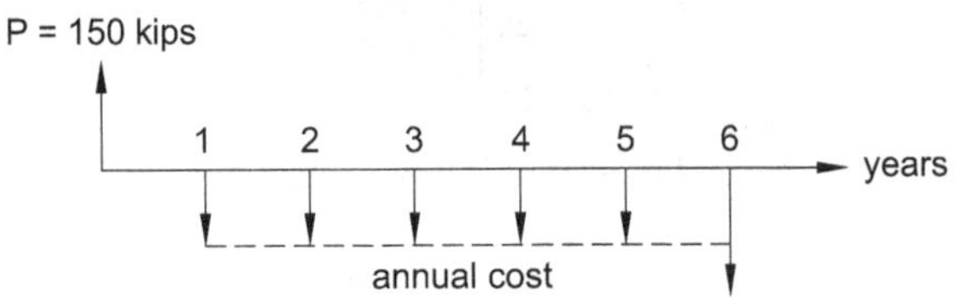

Use the factor tables given in the Engineering Economics section of the *NCEES Handbook*.

$$\begin{aligned}\text{annual cost} &= (\$150{,}000)(A/P, 8\%, 6) \\ &\quad -(\$10{,}000)(A/F, 8\%, 6) \\ &= (\$150{,}000)(0.2163) - (\$10{,}000)(0.1363) \\ &= \$31{,}082 \quad (\$31{,}000)\end{aligned}$$

The answer is $31,000.

350. A loss (or an opportunity loss) is defined as follows.

$$\text{opportunity loss} = \left|\begin{array}{l}\text{given payoff} \\ \quad -\text{the highest possible payoff} \\ \quad\quad \text{under the state of the world}\end{array}\right|$$

The loss is not negative. It is zero when the best decision is made for the situation. For example, if the decision is made to drill and the oil is found,

$$\text{opportunity loss} = \$10{,}000{,}000 - \$10{,}000{,}000 = \$0$$

If the "no drill" decision is made and oil is present,

$$\begin{aligned}\text{opportunity loss} &= \$10{,}000{,}000 - \$2{,}000{,}000 \\ &= \$8{,}000{,}000\end{aligned}$$

If the decision is to drill and no oil is present,

$$\begin{aligned}\text{opportunity loss} &= \$1{,}000{,}000 - (-\$3{,}000{,}000) \\ &= \$4{,}000{,}000\end{aligned}$$

The answer is $4,000,000.

351. An absolute protection from floods is generally uneconomical. An optimal design is one that minimizes the flood risk and the capital expenditure.

$$\begin{aligned}\text{flood risk} &= (\text{flood probability}) \\ &\quad \times(\text{future loss from flood})\end{aligned}$$

$$\begin{aligned}\text{capital recovery factor} &= \left(\frac{A}{P}, 8\%, 40\right) \\ &= 0.0839\end{aligned}$$

The formula for the annual cost is

$$\begin{aligned}\text{annual cost} &= (\text{probability of flood})(\text{flood damage cost}) \\ &\quad +(0.0839)(\text{flood protection cost})\end{aligned}$$

Use the formula to compute the annual cost of all four alternatives.

$$\begin{aligned}\text{cost for alternative 1} &= (0.050)(\$4{,}000{,}000) \\ &\quad +(0.0839)(\$1{,}000{,}000) \\ &= \$283{,}900 \\ \text{cost for alternative 2} &= (0.010)(\$6{,}000{,}000) \\ &\quad +(0.0839)(\$2{,}000{,}000) \\ &= \$227{,}800 \\ \text{cost for alternative 3} &= (0.005)(\$8{,}000{,}000) \\ &\quad +(0.0839)(\$3{,}000{,}000) \\ &= \$291{,}700 \\ \text{cost for alternative 4} &= (0.001)(\$10{,}000{,}000) \\ &\quad +(0.0839)(\$4{,}000{,}000) \\ &= \$345{,}600\end{aligned}$$

The annual cost is the lowest for alternative 2.

The answer is alternative 2.

352. The maximum shear stress is given by

$$\tau_{max} = \frac{Tr}{J}$$

The polar moment of inertia for a solid shaft is given by

$$\begin{aligned}J &= \frac{\pi d^4}{32} \\ &= \frac{\pi(2\ \text{in})^4}{32} \\ &= 1.57\ \text{in}^4\end{aligned}$$

The maximum shear stress is

$$\begin{aligned}\tau_{max} &= \frac{(100\ \text{lbf-ft})\left(12\,\frac{\text{in}}{\text{ft}}\right)(1\ \text{in})}{1.57\ \text{in}^4} \\ &= 764\ \text{psi} \quad (800\ \text{psi})\end{aligned}$$

The answer is (C).

353. The forces are concentric with the eyebolt. Resolve the applied loads into the x- and y-direction components and calculate the angles measured counterclockwise from the x-axis, as shown. For equilibrium, compute the x- and y-components of the force system.

$$\begin{aligned}\sum F_x &= (15\,\text{kN})(\cos 0^\circ) + (20\,\text{kN})(\cos 45^\circ) \\ &\quad +(25\,\text{kN})(\cos 90^\circ) + (30\,\text{kN})(\cos 110^\circ) \\ &\quad +(10\,\text{kN})(\cos 330^\circ) \\ &= 27.55\ \text{kN}\end{aligned}$$

$$\begin{aligned}\sum F_y &= (15\ \text{kN})(\sin 0^\circ) + (20\ \text{kN})(\sin 45^\circ) \\ &\quad +(25\ \text{kN})(\sin 90^\circ) + (30\ \text{kN})(\sin 110^\circ) \\ &\quad +(10\ \text{kN})(\sin 330^\circ) \\ &= 62.33\ \text{kN}\end{aligned}$$

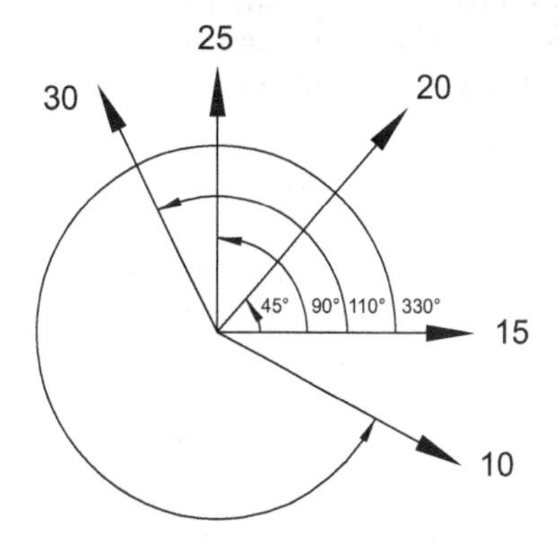

Find the resultant force.

$$\begin{aligned}F &= \sqrt{(F_x)^2 + (F_y)^2} \\ &= \sqrt{(27.55\ \text{kN})^2 + (62.33\,\text{kN})^2} \\ &= 68.15\,\text{kN} \quad (68\,\text{kN})\end{aligned}$$

The direction of force F with respect to the x-axis is

$$\begin{aligned}\theta &= \tan^{-1}\left(\frac{62.33}{27.55}\right) = 1.154\ \text{rad} \\ &= 66.2^\circ \quad (66^\circ\ \text{from}\ x\text{-axis})\end{aligned}$$

The answer is (B) and (D).

354. The forces are concentric with the eyebolt. Resolve the applied loads in the x- and y-directions. For equilibrium, compute the x- and y-components of the force system.

$$\begin{aligned}\sum F_x &= (15\,\text{kN})(\cos 0^\circ) + (20\,\text{kN})(\cos 45^\circ) \\ &\quad + (25\,\text{kN})(\cos 90^\circ) + (30\,\text{kN})(\cos 110^\circ) \\ &\quad + (10\,\text{kN})(\cos 330^\circ) \\ &= 27.55\ \text{kN}\end{aligned}$$

$$\begin{aligned}\sum F_y &= (15\ \text{kN})(\sin 0^\circ) + (20\ \text{kN})(\sin 45^\circ) \\ &\quad + (25\ \text{kN})(\sin 90^\circ) + (30\ \text{kN})(\sin 110^\circ) \\ &\quad + (10\ \text{kN})(\sin 330^\circ) \\ &= 62.33\ \text{kN}\end{aligned}$$

Support line ABC is vertical, and point B lies at the centerline of the support. Compute the moment about point B.

A moment is defined in the *NCEES Handbook*, Statics section, as the cross product of the radius vector and the force. For a system of forces, the moment is

$$\mathbf{M} = \sum (\mathbf{r}_n \times \mathbf{F}_n)$$

The radius vector **r** denotes the perpendicular or shortest distance between the point and the direction of the force. Using the force components,

$$\sum M_B = (F_x)(y) + (F_y)(x)$$

The shortest distances along the axes from support point B to the center of the eyebolt are

$$\begin{aligned}x &= 140\ \text{cm} = 1.4\ \text{m} \\ y &= 100\ \text{cm} + 120\ \text{cm} = 220\ \text{cm} = 2.2\ \text{m}\end{aligned}$$

Assuming counterclockwise moment as positive, determine the moment.

$$\begin{aligned}\sum M_B &= F_x y + F_y x \\ &= -(27.55\ \text{kN})(2.2\ \text{m}) \\ &\quad + (62.33\ \text{kN})(1.4\ \text{m}) \\ &= 26.7\ \text{kN·m}\end{aligned}$$

The positive sign of the resultant indicates that the resultant moment is counterclockwise.

The answer is (B) and (C).

355. The area of cross section of the girder is

$$\begin{aligned}A &= (A_{\text{web}} + A_{\text{top flange}} + A_{\text{bottom flange}}) \\ &= (0.9\,\text{m})(0.004\,\text{m}) \\ &\quad + (0.2\,\text{m})(0.05\,\text{m}) + (0.2\,\text{m})(0.05\,\text{m}) \\ &= 0.0236\ \text{m}^2\end{aligned}$$

The volume, V, of steel per meter length is

$$\begin{aligned}V &= (\text{area})(\text{length}) \\ &= (0.0236\,\text{m}^2)(1\,\text{m}) \\ &= 0.0236\,\text{m}^3\end{aligned}$$

The mass density of steel is 7.8 Mg/m^3.

$$\begin{aligned}\text{mass per meter} &= (0.0236\ \text{m}^3)\left(7.8\ \frac{\text{Mg}}{\text{m}^3}\right) \\ &= 0.1841\ \text{Mg} \quad (184.1\,\text{kg})\end{aligned}$$

The answer is (B).

356. The force mechanism is as shown.

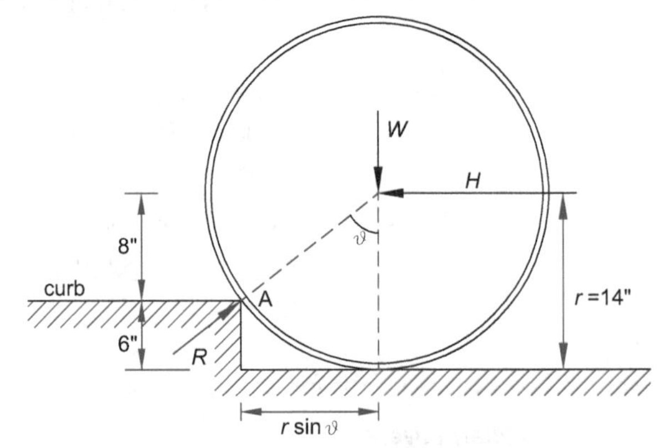

H is the minimum horizontal force applied at the center for the wheel.

For the wheel to roll up onto the curb, the applied moment should exceed the resisting moment about A. At equilibrium,

$$\begin{aligned}\sum M_A &= 0 \\ (H)(8\ \text{in}) - (W)(r\sin\theta) &= 0\end{aligned}$$

For a 28 in diameter wheel, the radius is 14 in.

$$\begin{aligned}\cos\theta &= \frac{8\ \text{in}}{14\ \text{in}} = 0.57 \\ \sin\theta &= \sqrt{1 - \cos^2\theta} \\ &= \sqrt{1 - (0.57)^2} \\ &= 0.82\end{aligned}$$

Find the minimum horizontal force.

$$(H)(8 \text{ in}) - (W)(r\sin\theta) = 0$$

$$H = \frac{Wr\sin\theta}{8 \text{ in}} = \frac{(4000 \text{ lbf})(14 \text{ in})(0.82)}{8 \text{ in}} = 5740 \text{ lbf} \quad (6 \text{ kips})$$

The answer is (C).

357. First, determine the moment about point D.

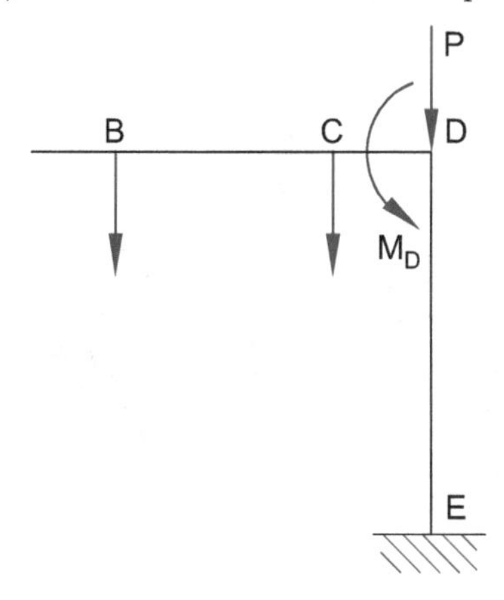

$$M = \sum Fd$$

$$M_D = (1 \text{ kip})(3 \text{ ft}) + (1 \text{ kip})(14 \text{ ft}) = 17 \text{ ft-kips}$$

Next, transfer the force system from D to E. Force P at D is axial in the DE direction and does not induce any moment at E. Thus, the moment M_D will remain unchanged (counterclockwise) when transferred from D to E. The moment about the base, M_E, is 17 ft-kips.

The answer is (B) and (D).

358. The moment at point B due to 2 kips point at point A is 16 ft-kips and it is in a counterclockwise direction. The free-body diagram at point B is shown.

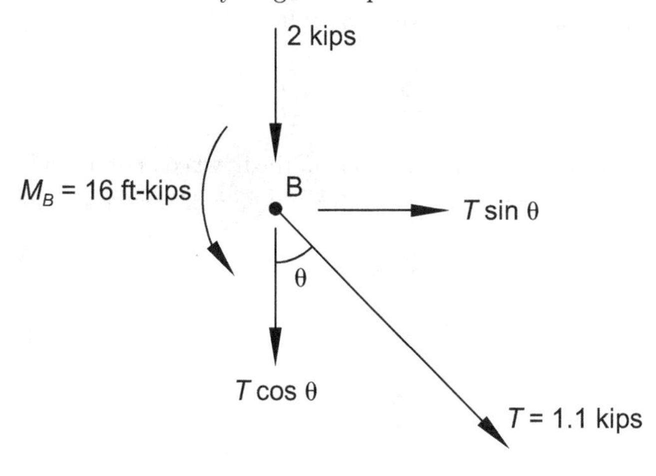

Find the cable angle θ.

$$\tan\theta = \frac{15}{18} = 0.833$$

$$\theta = \tan^{-1}0.833 = 39.8°$$

Assuming counterclockwise moment as positive, the moment at C is

$$M_C = M_B - (T\sin\theta)(18 \text{ ft}) = (2 \text{ kips})(8 \text{ ft}) - (1.1 \text{ kips})(\sin 39.8°)(18 \text{ ft}) = 3.3 \text{ ft-kips} \quad (3 \text{ ft-kips})$$

The answer is (B) and (C).

359. The bars are solid with a rectangular shape. Their stiffness is proportional to the moment of inertia (MOI) of the section. For a rectangular section, the MOI is given by

$$I = \frac{bd^3}{12}$$

The two bars placed on the top of the other act independently and non-compositely. Both carry equal loads and both deflect equally, as shown.

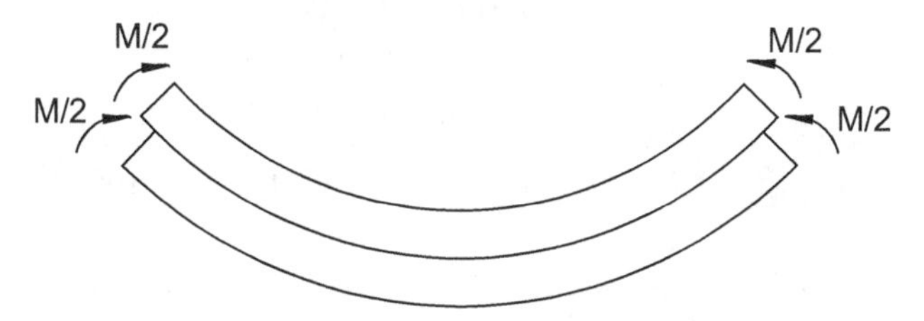

$$I_{\text{noncomp}} = \sum I = \frac{bd^3}{12} + \frac{bd^3}{12} = \frac{2bd^3}{12}$$

After welding, the two sections act as a composite section with a depth of $2d$.

$$I_{\text{comp}} = \frac{b(2d)^3}{12} = \frac{8bd^3}{12} = (4)\left(\frac{2bd^3}{12}\right) = 4I_{\text{noncomp}}$$

The ratio of composite MOI to the non-composite MOI is

$$\frac{I_{\text{comp}}}{I_{\text{noncomp}}} = \frac{4I_{\text{noncomp}}}{I_{\text{noncomp}}} = 4$$

The answer is 4.

360. Consider equilibrium along the x- and y-axes at impending motion between the block and the plane, as shown. Determine the force needed to stop the block from sliding down.

Consider the equilibrium in the y-direction.

$$\sum F_y = 0$$
$$W = 500 \text{ lbf}$$

Find the force perpendicular to the sliding surface.

$$\begin{aligned} N &= W\cos 30^\circ \\ &= (500 \text{ lbf})(0.866) \\ &= 433 \text{ lbf} \end{aligned}$$

Find the frictional force.

$$\begin{aligned} F_{\text{friction}} &= \mu_s N \\ &= (0.5)(433 \text{ lbf}) \\ &= 216.5 \text{ lbf} \end{aligned}$$

Consider the equilibrium in the x-direction.

$$\begin{aligned} \sum F_x &= 0 \\ &= F_{\text{applied}} + F_{\text{friction}} - W_x \\ &= F_{\text{applied}} + F_{\text{friction}} - W\sin\theta \\ &= F_{\text{applied}} + 216.5 \text{ lbf} - (500 \text{ lbf})\sin 30^\circ \\ F_{\text{applied}} &= (500 \text{ lbf})\sin 30^\circ - 216.5 \text{ lbf} = 33.5 \text{ lbf} \end{aligned}$$

For the block not to slide and move in the other direction, the tension T_1 in the cable must be at least 33.5 lbf.

Use the belt friction formula to find the tension T_2 needed to pull the block up the slope. θ is the total angle of contact between the cable and the pulley, in radians.

$$\theta = (90^\circ + 30^\circ)\left(\frac{\pi}{180^\circ}\right) = 2.09 \text{ rad}$$

$$\begin{aligned} T_2 &= T_1 e^{\mu\theta} \\ T_2 &= (33.5 \text{ lbf})\, e^{(0.2)(2.09 \text{ rad})} \\ &= 51 \text{ lbf} \end{aligned}$$

The answer is (A).

361. The area of the sliding surface is not needed in computing the friction force. The impending sliding force is

$$\begin{aligned} F &= \mu N \\ &= (0.7)(145 \text{ lbf}) \\ &= 102 \text{ lbf} \end{aligned}$$

The answer is (C).

362. Assume north direction as the datum. Draw the force vectors AB and BC to scale in a tip-to-toe fashion, as shown.

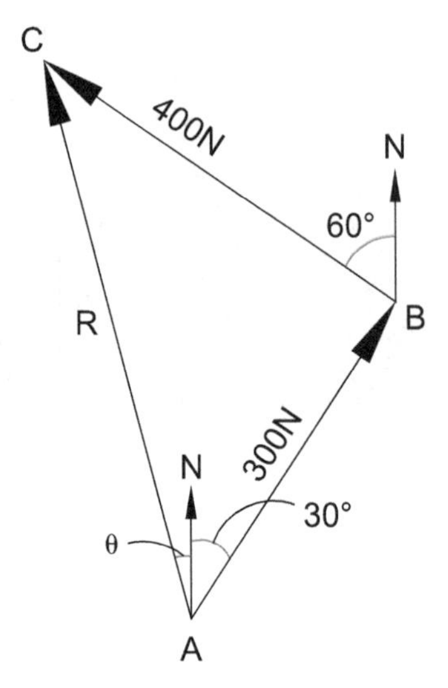

Vector AC is the resultant. Triangle ABC is a 3-4-5 right triangle, so vector AC is 500 N. Find the angle θ.

$$\angle A = \tan^{-1}\left(\frac{400 \text{ N}}{300 \text{ N}}\right) = 53.1^\circ$$
$$\theta = 30^\circ - 53.1^\circ = -23.1^\circ$$

The answer is (A) and (D).

363. When a body is in the x-y plane and is in motion about its z-axis, the torque acting on the body can be determined using the equation for Mass Moment of Inertia in the Dynamics section of the *NCEES Handbook.*

$$\sum M_{ZC} = I_{ZC}\alpha$$

α is the angular acceleration of the flywheel. For a single torque situation, the equation for torque, T, is simplified.

$$T = I_{ZC}\alpha$$

The equation for mass moment of inertia, I_{ZC}, is given in the Mass Moment of Inertia table in the Dynamics section of the *NCEES Handbook*.

$$I_{ZC} = MR^2 = (2000\ \text{kg})(1)^2 = 2000\ \text{kg}\cdot\text{m}^2$$

Rearrange the equation for torque to solve for angular acceleration.

$$T = 2000\ \text{N}\cdot\text{m}$$
$$\alpha = \frac{T}{I_{ZC}} = \frac{2000\ \text{N}\cdot\text{m}}{2000\ \text{kg}\cdot\text{m}^2} = 1\ \text{rad/s}^2$$

The answer is (A).

364. The springs are in series. The total elongation of the system is the sum of elongations of each spring.

$$\begin{aligned}\delta &= \delta_1 + \delta_2\\ &= \frac{P}{k_1} + \frac{P}{k_2}\\ &= \frac{P}{5\ \frac{\text{kN}}{\text{m}}} + \frac{P}{10\ \frac{\text{kN}}{\text{m}}}\\ &= 0.3P\ \text{m/kN}\end{aligned}$$

The stiffness of the system, k, is

$$k = \frac{P}{\delta} = \frac{P}{0.3P\ \text{m/kN}} = 3333\ \text{N/m}$$

The period of vibration, T, is

$$\begin{aligned}T &= 2\pi\sqrt{\frac{m}{k}}\\ &= 2\pi\sqrt{\frac{100\ \text{kg}}{3333\ \frac{\text{N}}{\text{m}}}}\\ &= 1.09\ \text{s}\end{aligned}$$

The answer is (B).

365. Work done by the braking force is equal to the kinetic energy of the roller, which is equal to force times distance. Solve for the force required.

$$\begin{aligned}W &= Fd\\ F &= \frac{W}{d}\\ &= \frac{70{,}000\ \text{N}\cdot\text{m}}{5\ \text{m}}\\ &= 14{,}000\ \text{N}\quad (14\ \text{kN})\end{aligned}$$

The answer is (B).

366. By symmetry, the two-span beam is equivalent to a single-span propped cantilever, as shown, because the middle support in the two-span loading does not rotate and can be considered fixed.

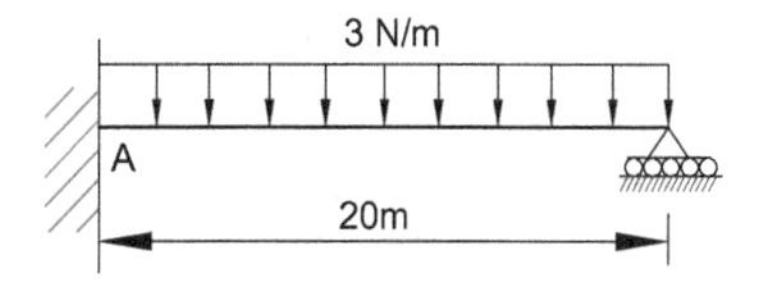

The moment at the fixed end of a propped cantilever is

$$M = \frac{wL^2}{8} = \frac{\left(3\ \frac{\text{N}}{\text{m}}\right)(20\ \text{m})^2}{8} = 150\ \text{N}\cdot\text{m}$$

The answer is (C).

367. The wall thickness must satisfy two stress criteria.

$$\text{tangential (circumferential) stress},\ \sigma_1 = \frac{pd}{2t} = \frac{pr}{t}$$
$$\text{axial (longitudinal) stress},\ \sigma_2 = \frac{pd}{4t} = \frac{pr}{2t}$$

$$d = 1.1\ \text{m}$$

$$p = 3\ \frac{\text{N}}{\text{mm}^2}$$

Let t_1 be the wall thickness required to resist σ_1, and let t_2 be the wall thickness required to resist σ_2.

Use the equation for tangential stress and solve for the required wall thickness.

$$\sigma_1 = \frac{pd}{2t}$$

$$t_1 = \frac{pd}{2\sigma_1} = \frac{\left(3\ \frac{\text{N}}{\text{mm}^2}\right)(1.1\ \text{m})}{(2)\left(60\ \frac{\text{N}}{\text{mm}^2}\right)} = 0.0275\ \text{m}$$

Use the equation for axial stress, and solve for the required wall thickness.

$$\sigma_2 = \frac{pd}{4t}$$

$$t_2 = \frac{pd}{4\sigma_2} = \frac{\left(3\ \frac{\text{N}}{\text{mm}^2}\right)(1.1\ \text{m})}{(4)\left(60\ \frac{\text{N}}{\text{mm}^2}\right)} = 0.0183\ \text{m}$$

For proper design, provide the larger of the two computed thicknesses, 0.0275 m (28 mm).

The answer is (C).

368. For thin-walled tubes, shear stress is constant around the cross section and is

$$\tau = \frac{T}{2A_m t}$$

The mean enclosed area is

$$A = \frac{\pi(d_{\text{mean}})^2}{4} = \frac{\pi(0.3\ \text{m})^2}{4} = 0.071\ \text{m}^2$$

The wall thickness is $t = 30\ \text{mm} = 0.03\ \text{m}$.

$$\begin{aligned} T_{\text{allowable}} &= 2A_m t\,\tau_{\max} \\ &= (2)(0.071\ \text{m}^2)(0.03\ \text{m})\left(5\times 10^6\ \frac{\text{N}}{\text{m}^2}\right) \\ &= 21{,}300\ \text{N·m} \quad (20\ \text{kN·m}) \end{aligned}$$

The answer is (B).

369. See the Simply Supported Beam Slopes and Deflections table in the Mechanics of Materials section of the *NCEES Handbook*. The beam deflections are denoted by letter v in the *Handbook*. In this problem, letter y is used for the vertical deflections.

The maximum deflections of beams A and B are

$$y_{\text{A-max}} = \frac{5wL^4}{384EI}$$

$$y_{\text{B-max}} = \frac{0.006563wL^4}{EI}$$

It is known that both beams have equal L, E, and areas of cross section. Divide beam A's deflection equation by beam B's equation.

$$\begin{aligned} \frac{y_{\text{max-A}}}{y_{\text{max-B}}} &= \left(\frac{5wL^4}{384EI}\right)\left(\frac{EI}{0.006563wL^4}\right) \\ &= \frac{5}{(384)(0.00656)} \\ &= 1.98 \end{aligned}$$

The ratio of deflections at the midspan of the beam is 2, but since the maximum deflection in beam B occurs at $0.4598L$ and not at $0.5L$, the ratio is 1.98.

The answer is (C).

370. The K-factor for a free-fixed column is 2.1. Use the Euler equation to determine the column buckling capacity.

$$\begin{aligned} P_{\text{cr,free-fixed}} &= \frac{\pi^2 EI}{(KL)^2} = \frac{\pi^2 EI}{(2.1L)^2} = \frac{1}{4.41}\left(\frac{\pi^2 EI}{L^2}\right) \\ &= \frac{1}{4.41}\left(P_{\text{cr,pinned-pinned}}\right) \end{aligned}$$

The capacity of a cantilever column is less than 1/4 of the capacity of a nearly identical column with two pinned ends.

The answer is (A).

371. A rectangular column will buckle about its weaker axis, which is its y-axis. The radius of gyration about the y-axis is

$$r_y = \sqrt{\frac{I_y}{A}}$$

The column section is rectangular, so the area is $A = bd$.

The column's moment of inertia about the weaker axis is

$$I_y = \frac{db^3}{12}$$

Substitute the area properties in the equation for r_y.

$$\begin{aligned} r_y &= \sqrt{\frac{I_y}{A}} = \sqrt{\frac{db^3}{12bd}} = \frac{b}{\sqrt{12}} \\ &= \frac{3.5\ \text{in}}{\sqrt{12}} \\ &= 1.01\ \text{in} \quad (1\ \text{in}) \end{aligned}$$

The answer is (A).

372. The MOI about the z-axis is the polar moment of inertia, J or I_p. The theorem of perpendicular axes states that the polar MOI of a section is the sum of the moments of inertia about the x- and y-axes.

$$I_p = I_x + I_y$$

This can be verified for the section properties listed in the *NCEES Handbook*, Statics section.

The answer is (A).

373. The water/cement (w/c) ratio affects the mix strength the most significantly. The concrete strength, f_c', decreases with an increase in the w/c ratio. The relationship is nonlinear, as shown.

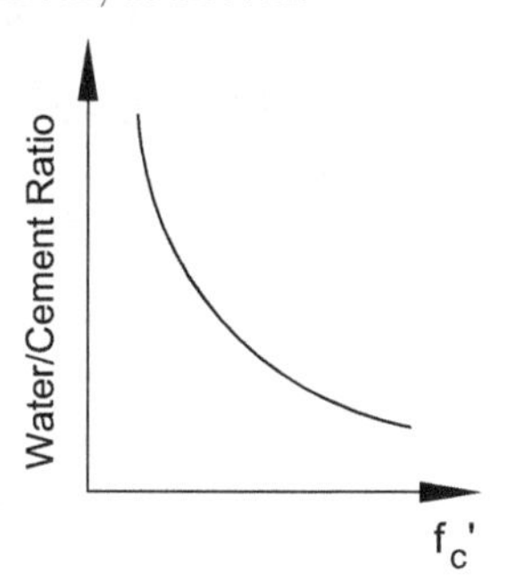

The answer is (C).

374. Moist-cured cylinders gain more strength than the air-cured cylinders from the same batch of concrete mix. Concrete strength increases with time. Therefore, the moist-cured cylinders at the age of 180 days exhibit more strength than the cylinder at 28 days.

The answer is (C).

375. It is known that 611.7 lbf of cement is needed to produce one cubic yard of concrete. The w/c ratio is 0.4. Therefore,

$$\text{weight of water} = (0.4)(611.7 \text{ lbf}) = 245 \text{ lbf}$$

The weight of 1 gal of water is 8.345 lbf, so the number of gallons of water needed to produce 1 yd^3 of concrete is

$$n = \frac{245 \text{ lbf}}{8.345 \text{ lbf/gal}} = 29.4 \text{ gal} \quad (30 \text{ gal})$$

The answer is (A).

376. The tensile strength of concrete is much lower than its compressive strength. The illustration shows the relationship between the compressive and tensile strengths, with compressive stress designated as positive.

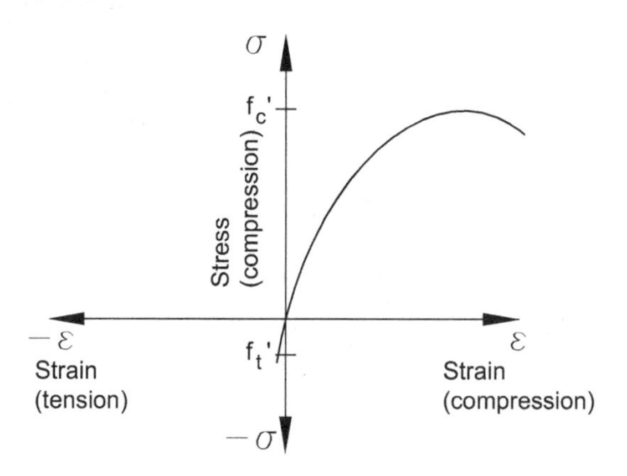

For a compressive strength of 3000 psi,

$$f_t' = 7.5\sqrt{3000 \text{ psi}} = 411 \text{ psi}$$

When the compressive strength is doubled,

$$f_t' = 7.5\sqrt{6000 \text{ psi}} = 581 \text{ psi}$$

The strength increase is

$$\frac{581 - 411}{411} = 0.41 = 41\% \quad (40\%)$$

The answer is (B).

377. The modulus of resilience is the area under the stress-strain diagram up to the yield point. It is the area of a right triangle.

$$U_R = \frac{f_y \epsilon_y}{2} = \frac{(60 \text{ ksi})(0.002)}{2} = 0.06 \text{ ksi}$$

The answer is (B).

378. In this problem, discharge and area of the pipe are given. Use *NCEES Handbook*, Fluid Mechanics section, to determine fluid velocity, $v = Q/A$.

The problem uses several units. Using the *NCEES Handbook*, Units and Conversion Factors section, convert all units to feet and seconds. To convert MGD (million gallons per day) to cubic feet per second, use the relation 1 cfs = 0.646317 MGD.

The discharge rate is

$$Q = \frac{3 \text{ MGD}}{0.646317 \text{ MGD/cfs}} = 4.64 \text{ cfs}$$

The area of the pipe cross section is

$$A = \pi r^2 = \pi\left(\frac{6 \text{ in}}{12 \text{ in/ft}}\right)^2 = 0.785 \text{ ft}^2$$

The average velocity of the oil is

$$\begin{aligned} v &= \frac{Q}{A} \\ &= \frac{4.64 \text{ cfs}}{0.785 \text{ ft}^2} \\ &= 5.91 \text{ ft/sec} \quad (6 \text{ ft/sec}) \end{aligned}$$

The answer is (B).

379. Equate the energy level between points A and C, using point C as the datum.

$$\frac{p_A}{\gamma} + \frac{v_A^2}{2g} + z_A = \frac{p_C}{\gamma} + \frac{v_C^2}{2g} + z_C$$

The velocity at point A, v_A, is zero. Both points A and C are subject to atmospheric pressure, so p_A and p_C are zero. The elevation at point C, z_C, is zero. The initial head, z_A, is 10 m. The energy equation simplifies to

$$10 \text{ m} = \frac{v_C^2}{2g}$$

$$v_C = \sqrt{2\left(9.81 \ \frac{\text{m}}{\text{s}^2}\right)(10 \text{ m})} = 14 \text{ m/s}$$

The answer is (C).

380. The velocity is zero for plate A and 1.2 m/s for plate B. For the gap between the plates, assume a linear variation in velocity, as shown.

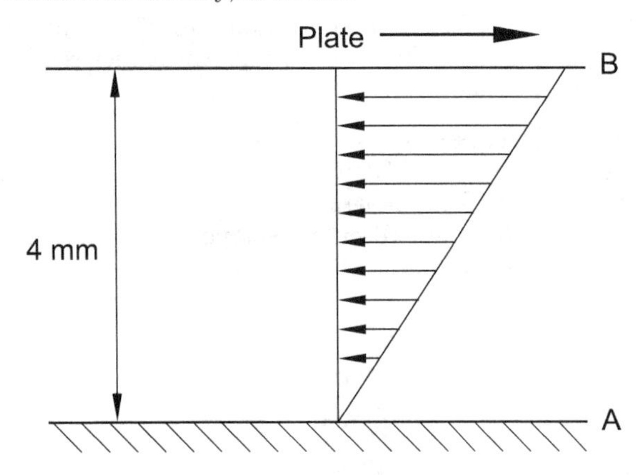

The velocity gradient is

$$\frac{dv}{dy} = \frac{1.2 \ \frac{\text{m}}{\text{s}}}{0.004 \text{ m}} = 300 \text{ s}^{-1}$$

Find the shear stress on the bottom plate. From the conversion table in the *NCEES Handbook*, 1 poise = 0.1 Pa·s.

$$\tau = \mu \frac{dv}{dy} = (0.2 \text{ poise})\left(\frac{0.1 \text{ Pa·s}}{1 \text{ poise}}\right)(300 \text{ s}^{-1}) = 6 \text{ Pa}$$

The answer is (A).

381. Venturi meters, manometers, pitot tubes, nozzles, and orifices are used to measure flow in pipes. Weirs and flumes are used to measure flow in open channels.

The answer is (A) and (B).

382. Use Newton's second law, given in the Dynamics section of the *NCEES Handbook.*

$$F = \frac{d(mv)}{dt} = \frac{\dot{W}}{g}(v_2 - v_1)$$

$\dot{W}$ is the weight of the fluid flowing per second. Using the *NCEES Handbook,* Fluid Mechanics section, the resultant force in a given direction on water is the rate of change of the momentum of the water.

$$\sum F = Q_2\rho_2 v_2 - Q_1\rho_1 v_1$$

Since water density and flow rate remain unchanged after the impact, $\rho_1 = \rho_2 = \rho$ and $Q_1 = Q_2 = Q$. Therefore,

$$F = Q\rho(v_2 - v_1)$$

After impact, the water flow in the 10 in diameter pipe splits into two directions so that the velocity of water in the 10 in diameter pipe is completely destroyed. Therefore, the impact force equation reduces to

$$F = \left(\frac{w}{g}\right)(Q_{10})(v_{10})$$

The water density is $w = 62.4 \text{ lbm/ft}^3$. The flow, Q, is 10 cfs. The cross-sectional area of the pipe is

$$A = \frac{\pi}{4}\left(\frac{10 \text{ in}}{12}\right)^2 = 0.55 \text{ ft}^2$$

The water velocity in the 10 in diameter pipe is

$$v_{10} = \frac{Q_{10}}{A_{10}} = \frac{10 \ \frac{\text{ft}^3}{\text{sec}}}{0.55 \text{ ft}^2} = 18.18 \text{ ft/sec}$$

The resultant force is

$$F = \left(\frac{62.4 \ \frac{\text{lbm}}{\text{ft}^3}}{32.2 \ \frac{\text{lbm-ft}}{\text{lbf-sec}^2}}\right)\left(10 \ \frac{\text{ft}^3}{\text{sec}}\right)\left(18.18 \ \frac{\text{ft}}{\text{sec}}\right) = 352 \text{ lbf} \quad (350 \text{ lbf})$$

The answer is (B).

383. The total weight of the loaded ship, W, must equal the volume of water, V, displaced by the ship. It is expressed as $W = \gamma V$.

Although the total depth of the vessel is 38 ft, only 28 ft of the depth should be used to fully load the ship. Thus, the volume of displaced water is defined as the base area of the vessel multiplied by the allowable depth of 28 ft.

$$\begin{aligned} W &= \gamma V \\ &= (64\ \text{pcf})\big((400\ \text{ft})(60\ \text{ft})(28\ \text{ft})\big)\left(\frac{1\ \text{ton}}{2000\ \text{lbm}}\right) \\ &= 21{,}504\ \text{tons} \end{aligned}$$

The answer is 21,504 tons.

384. The degree of a curve, D, is the intersection angle that subtends an arc of 100 ft, as shown. This is called the *arc definition* or the highway definition.

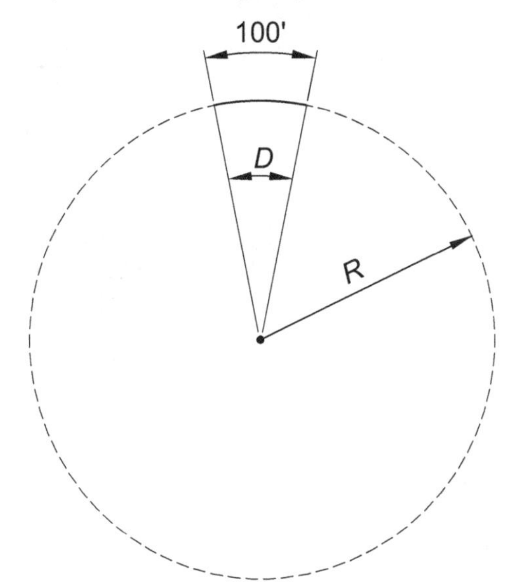

The formula to compute the degree of curve given in the *NCEES Handbook*, Civil Engineering/Horizontal Curves section is

$$R = \frac{5729.58}{D}$$

For $R = 600$ ft,

$$D = \frac{5729.58}{R} = \frac{5729.58}{600\ \text{ft}} = 9.55°$$

The answer is (C).

385. Refer to the Horizontal Curves section of the *NCEES Handbook*. PT = PI + curve length subtended by a 20° intersection angle. The degree of curve is $D =$ 10°. By definition, the curve length subtended by a 10° angle is 100 ft.

$$\begin{aligned} \text{curve length subtended by } 20° \text{ angle} &= \frac{100\ \text{ft}\ (20°)}{10°} \\ &= 200\ \text{ft} \end{aligned}$$

$$\begin{aligned} \text{PT} &= \text{sta } 12{+}34.00 + 200\ \text{ft} \\ &= \text{sta } (12{+}34.00) + (2{+}00) \\ &= \text{sta } 14{+}34.00 \end{aligned}$$

The answer is (C).

386. The reservoir with a slope on all four sides is shown in the illustration. There are two methods given in the *NCEES Handbook*, Civil Engineering: Earthwork Formulas section. The average end area method is simpler and requires two end areas to compute the volume. The prismoidal formula requires three sections: the middle section and two end sections. The prismoidal method is more accurate and is used here.

$$V = \frac{L}{6}(A_1 + 4A_m + A_2)$$

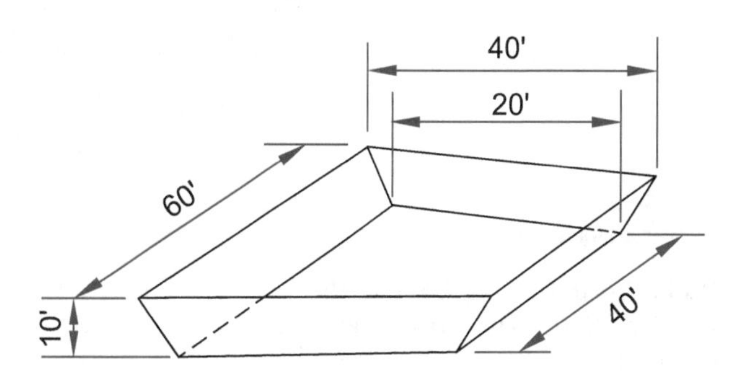

(not to scale)

The depth is given as

$$L = 10\ \text{ft}$$

Use the reservoir base section as A_1 and the ground level section as A_2. The middle section, A_m, is at the mid-height of 5 ft from the base.

Find the area for section A_1,

$$\begin{aligned} B_1 &= 20\ \text{ft} \\ D_1 &= 40\ \text{ft} \\ A_1 &= (20\ \text{ft})(40\ \text{ft}) \\ &= 800\ \text{ft}^2 \end{aligned}$$

Find the area for section A_2.

$$\begin{aligned} B_2 &= 20 \text{ ft} + (2)(10 \text{ ft}) \\ &= 40 \text{ ft} \\ D_2 &= 40 \text{ ft} + (2)(10 \text{ ft}) \\ &= 60 \text{ ft} \\ A_2 &= (40 \text{ ft})(60 \text{ ft}) \\ &= 2400 \text{ ft}^2 \end{aligned}$$

Find the area of the middle section, A_m.

$$\begin{aligned} B_m &= 20 \text{ ft} + (2)(5 \text{ ft}) \\ &= 30 \text{ ft} \\ D_m &= 40 \text{ ft} + (2)(5 \text{ ft}) \\ &= 50 \text{ ft} \\ A_m &= (30 \text{ ft})(50 \text{ ft}) \\ &= 1500 \text{ ft}^2 \end{aligned}$$

Determine the volume in cubic yards.

$$\begin{aligned} V &= \left(\frac{10 \text{ ft}}{6}\right)\left(800 \text{ ft}^2 + (4)(1500 \text{ ft}^2) + 2400 \text{ ft}^2\right) \\ &= 15{,}333 \text{ ft}^3 \\ &= \frac{15{,}333 \text{ ft}^3}{\left(27 \frac{\text{ft}^3}{\text{yd}^3}\right)} \\ &= 567.89 \text{ yd}^3 \quad (568 \text{ yd}^3) \end{aligned}$$

The answer is 568 yd³.

387. Both the radius and the angle are given. Use *NCEES Handbook*, Mathematics section to determine the arc length along the surface of the earth.

$$\begin{aligned} L &= r\theta \\ &= (3960 \text{ mi})(42.0383°)\left(\frac{\pi \text{ rad}}{180°}\right) \\ &= 2905 \text{ mi} \end{aligned}$$

The answer is 2905 mi.

388. The back tangent slope is negative, and the forward slope is positive. Therefore, the vertical curve is a sag curve. Use *NCEES Handbook*, Civil Engineering: Vertical Curves section. The illustration given in the section applies. The point of vertical tangent is denoted as PVT. Since the forward tangent is a sloping line, any point on the tangent can be found by using the equation

$$Y_{\text{PVT}} = Y_{\text{PVI}} + g_2\left(x - \frac{L}{2}\right)$$

The slope is

$$g_2 = 2\% = 0.02$$

The distance, x, for PVT is equal to the length, L.

$$\begin{aligned} Y_{\text{PVT}} &= Y_{\text{PVI}} + g_2\left(L - \frac{L}{2}\right) \\ &= 2540 \text{ m} + (0.02)\left(1000 \text{ m} - \frac{1000 \text{ m}}{2}\right) \\ &= 2550 \text{ m} \end{aligned}$$

The answer is (D).

389. The curve is a crest curve because the back slope is positive and the forward slope is negative. The rate of change of the grade of the curve is given by the equation

$$r = \frac{g_2 - g_1}{L}$$

For simplicity, the grades are expressed in percentages, and the curve length, L, is expressed in stations. In route surveying, 1 station = 100 ft. The back tangent slope, g_1, is 2%, and the forward tangent slope, g_2, is −4%.

The curve length is $L = 1000 \text{ ft} = 10 \text{ sta}$

$$r = \frac{(-4) - 2}{10 \text{ sta}} = -0.6$$

The answer is (A).

390. The curve number (CN) is an abbreviation of the runoff curve number. CN is an empirical number used in hydrology for predicting direct runoff or infiltration from rainfall excess. The runoff of an area or region depends on factors such as the land cover, the soil's hydrologic soil group (HSG), and the drainage area of each type. The land cover, such as bare soil, vegetation, and paved surfaces, establishes runoff production potential. In this method, the soil is divided into four HSGs, depending on the soil texture and its infiltration rate.

CN ranges from 30 to 100; lower numbers indicate low runoff potential while larger numbers are for increasing runoff potential. The more permeable the soil is, the lower its CN. The less pervious the soil is, the higher its CN. For example, a paved parking lot, roof, or driveway would be assigned a CN of 98. On the other hand, an agricultural wooded land would be assigned a CN of 30. CN does not depend on the amount or intensity of the rainfall. Therefore, options B and C are incorrect. The initial abstraction losses to the runoff are due to the depression storage, evaporation, and interception. The initial abstraction affects the storage capacity of the drainage area, but does not affect CN. Therefore, option D is incorrect.

The answer is (A).

391. The hydraulic radius is defined as

$$R_H = \frac{\text{cross-sectional area}}{\text{wetted perimeter}} = \frac{A}{p}$$

In this case, the pipe is half full. Therefore, the wetted perimeter is the length of the 180° arc of 12 in radius. Similarly, the cross-sectional area is the area of one-half of the circular pipe. The pipe radius is

$$r = \frac{\text{diameter}}{2} = \frac{24 \text{ in}}{2} = 12 \text{ in}$$

Find the area and perimeter.

$$p = \pi r = \pi(12 \text{ in}) = 37.7 \text{ in}$$
$$A = \left(\frac{1}{2}\right)\pi r^2 = \left(\frac{1}{2}\right)\pi(12 \text{ in})^2 = 226.2 \text{ in}^2$$

Find the pipe hydraulic radius.

$$R_H = \frac{A}{p} = \frac{226.2 \text{ in}^2}{37.7 \text{ in}} = 6.0 \text{ in}$$

The answer is (B).

392. The energy equation for incompressible flow is

$$\frac{p_1}{\gamma} + \frac{v_1^2}{2g} + z_1 = \frac{p_2}{\gamma} + \frac{v_2^2}{2g} + z_2 + h_f + h_{f,\text{fitting}}$$

The difference in elevation of two reservoirs connected by a pipe equals the losses in the pipeline and in the entry and exit fittings. The head loss and fitting losses are given.

$$h_f = 30 \text{ ft}$$
$$h_{f,\text{ fitting}} = 2 \text{ ft} + 3 \text{ ft} + 5 \text{ ft} = 10 \text{ ft}$$

For reservoirs A and B, the parameters in the energy equation are

$$p_A = v_A = 0$$
$$z_A = 100 \text{ ft}$$
$$p_B = v_B = 0$$

At both points A and B, the water velocity is zero, and the pressure heads are equal (i.e., open to air). Substitute the values into the energy equation.

$$0 \text{ ft} + 0 \text{ ft} + 100 \text{ ft} = 0 \text{ ft} + 0 \text{ ft} + z_B + 30 \text{ ft} + 10 \text{ ft}$$
$$z_B = 60 \text{ ft}$$

The answer is 60 ft.

393. The applicable chart is the *NCEES Handbook* Moody diagram for Flow in Closed Conduits. To use the Moody diagram, first determine the head loss due to pipe friction, using the Darcy-Weisbach equation.

$$h_f = f\frac{L}{D}\frac{v^2}{2g}$$

The pipe diameter is

$$D = 1 \text{ ft}$$

Find the cross-sectional area of the pipe.

$$A = \frac{\pi D^2}{4} = \frac{\pi (1 \text{ ft})^2}{4} = 0.79 \text{ ft}^2$$

The discharge rate is

$$Q = 2 \text{ ft}^3/\text{sec}$$

Find the velocity.

$$v = \frac{Q}{A} = \frac{2 \frac{\text{ft}^3}{\text{sec}}}{0.79 \text{ ft}^2} = 2.53 \text{ ft/sec}$$

For galvanized steel pipe, use a roughness factor, ϵ, of 0.0005 ft.

Find the relative roughness of a pipe.

$$\frac{\epsilon}{D} = \frac{0.0005 \text{ ft}}{1 \text{ ft}} = 0.0005$$

For kinematic viscosity, see the *NCEES Handbook*, Fluid Mechanics section, Properties of Water table. The kinematic viscosity of water at 57°F is 1.22×10^{-5} ft²/sec. The Reynolds number is

$$\text{Re} = \frac{D\text{v}}{\upsilon} = \frac{(1\ \text{ft})\left(2.53\ \frac{\text{ft}}{\text{sec}}\right)}{\left(1.22 \times 10^{-5}\ \frac{\text{ft}^2}{\text{sec}}\right)} = 2 \times 10^5$$

Use the Moody diagram. (1) For the Reynolds number, determine the point that is between $\epsilon/D = 0.0004$ and 0.0006 curves and represents $\epsilon/D = 0.0005$. (2) Read the corresponding point on the y-axis on the left side of the chart. It indicates that the friction factor value, f, is 0.0185. See the illustration shown.

The answer is

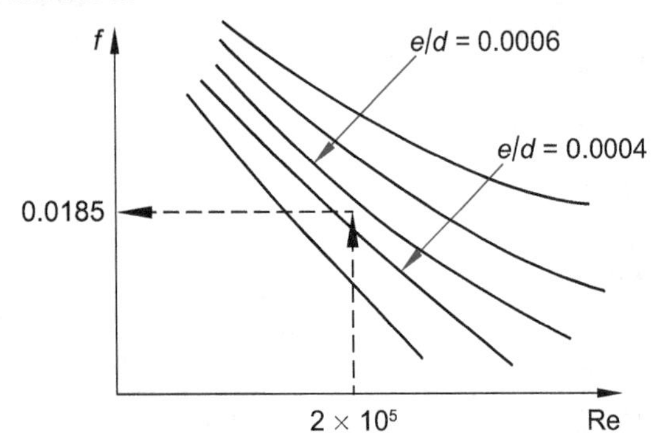

394. Use the *NCEES Handbook*, Hydrology/Water Resources section. The Hazen-Williams equation is

$$V = k_1 C R_H^{\ 0.63} S^{0.54}$$

In this case, $k_1 = 1.318$ for USCS units.

For a new cast-iron pipe, the roughness coefficient is $C = 130$.

From the Fluid Mechanics section of the *NCEES Handbook*, the hydraulic radius is defined as

$$R_H = \frac{\text{cross-sectional area}}{\text{wetted perimeter}}$$

For a pipe running full, the hydraulic radius is

$$\begin{aligned} R_H &= \frac{\pi r^2}{2\pi r} = \frac{r}{2} = \frac{D}{4} \\ &= \frac{12\ \text{in}}{(4)\left(12\ \frac{\text{in}}{\text{ft}}\right)} = 0.25\ \text{ft} \end{aligned}$$

Alternately, use the *NCEES Handbook* chart "Hydraulic-Elements Graph for Circular Sewers" to determine the hydraulic radius.

The water velocity using the Hazen-Williams equation is

$$\begin{aligned} V &= 11\ \frac{\text{ft}}{\text{sec}} \\ &= (1.318)(130)(0.25\ \text{ft})^{0.63} S^{0.54} \\ &= 71.54 S^{0.54} \end{aligned}$$

Solve for S, the slope of the pipe.

$$\begin{aligned} S^{0.54} &= \frac{11\ \frac{\text{ft}}{\text{sec}}}{71.54} = 0.154 \\ S &= 0.154^{\left(\frac{1}{0.54}\right)} \\ &= 0.031\ \text{ft/ft} \end{aligned}$$

Since *slope* is defined as the change in length per unit length, the total change in elevation is given by

$$\begin{aligned} \text{total head loss} &= (\text{slope})(\text{pipe length}) \\ &= \left(0.031\ \frac{\text{ft}}{\text{ft}}\right)(1000\ \text{ft}) \\ &= 31\ \text{ft} \end{aligned}$$

The answer is (B).

395. Use the *NCEES Handbook*, Hydrology/Water Resources section. Darcy's equation applies. The discharge rate is

$$Q = -KA\left(\frac{dh}{dx}\right)$$

The illustration shows the groundwater flow described in the problem.

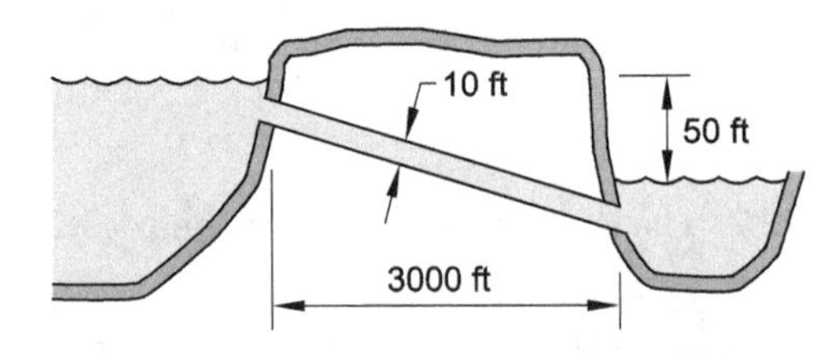

The hydraulic conductivity, K, is 0.2 ft/hr.

The aquifer cross-sectional area is

$$A = (10\ \text{ft})(100\ \text{ft}) = 1000\ \ \text{ft}^2$$

The aquifer length, L, is 3000 ft. The elevations h_1 and h_2 are 500 ft and 450 ft respectively.

Therefore, the change in elevation in water surfaces at two reservoirs is

$$\Delta h = h_2 - h_1 = 450\ \text{ft} - 500\ \text{ft} = -50\ \text{ft}$$

The elevation is negative. The difference in elevation per unit length is

$$\frac{dh}{dx} = \frac{\Delta h}{L} = \frac{-50 \text{ ft}}{3000 \text{ ft}}$$

Substitute the values into the Darcy equation and determine the discharge rate.

$$\begin{aligned} Q &= -KA\left(\frac{dh}{dx}\right) = -\frac{\left(0.2 \ \frac{\text{ft}}{\text{hr}}\right)(1000 \text{ ft}^2)(-50 \text{ ft})}{3000 \text{ ft}} \\ &= 3.33 \text{ ft}^3/\text{hr} \quad (3.3 \text{ ft}^3/\text{hr}) \end{aligned}$$

The answer is (C).

396. The molecular weight of CO_2 is

$$\text{MW} = 16 \text{ g} + 16 \text{ g} + 12 \text{ g} = 44 \text{ g}$$

Use the Units and Conversion Factors section of the *NCEES Handbook.*

$$1 \text{ m}^3 = 1000 \text{ L}$$

The molar volume of an ideal gas at 1 atm pressure is 22.4 L.

Therefore, 44 g of CO_2 occupies 22.4 L at STP.

The volume needed for 100 μg of CO_2 is

$$V = \left(\frac{100 \ \frac{\mu\text{g}}{\text{m}^3}}{44 \text{ g}}\right)(22.4 \text{ L}) = 51 \ \mu\text{L/m}^3$$

Convert to ppm.

$$V = \frac{51 \ \frac{\mu\text{L}}{\text{m}^3}}{\left(1000 \ \frac{\text{L}}{\text{m}^3}\right)} = 0.051 \ \mu\text{L/L} \quad (0.051 \text{ ppm})$$

The answer is (A).

397. Total solids (TS) include both the dissolved and suspended matter in the water.

$$\text{TS} = \frac{\text{residue mass after drying at } 103^\circ\text{C}}{\text{sample volume}}$$

The residue mass is

$$\text{residue mass} = 50.029 \text{ g} - 50.002 \text{ g} = 0.027 \text{ g}$$

The sample volume is 20 mL. The total solids concentration of the sample is

$$\text{TS} = \left(\frac{0.027 \text{ g}}{20 \text{ mL}}\right)\left(1000 \ \frac{\text{mg}}{\text{g}}\right)\left(1000 \ \frac{\text{mL}}{\text{L}}\right) = 1350 \text{ mg/L}$$

The answer is (D).

398. The overflow or hydraulic loading rate is the flow rate per unit of surface area in the *NCEES Handbook,* Environmental Engineering section. It is used in sedimentation theory. The depth of the clarifier is not used in determining the hydraulic loading rate.

$$\text{overflow rate} = \frac{\text{flow rate, } Q}{\text{cross-sectional area}}$$

The flow rate, Q, is 2 MGD.

Convert the flow rate from gal/day to ft^3/hr using the conversion rate 1 gal = 0.134 ft^3. The flow rate is

$$\begin{aligned} Q &= \left(2{,}000{,}000 \ \frac{\text{gal}}{\text{day}}\right)\left(0.134 \ \frac{\text{ft}^3}{\text{gal}}\right) \\ &= 268{,}000 \text{ ft}^3/\text{day} \\ &= \left(\frac{268{,}000 \ \frac{\text{ft}^3}{\text{day}}}{24 \ \frac{\text{hr}}{\text{day}}}\right) = 11{,}167 \ \ \text{ft}^3/\text{hr} \end{aligned}$$

Find the cross-sectional (surface) area.

$$A_{\text{surface}} = \frac{\pi(50 \text{ ft})^2}{4} = 1963.4 \text{ ft}^2$$

Find the overflow rate.

$$\frac{Q}{A} = \frac{11{,}167 \ \frac{\text{ft}^3}{\text{hr}}}{1963.4 \text{ ft}^2} = 5.69 \text{ ft/hr} \quad (6 \text{ ft/hr})$$

The answer is (B).

399. The list contains monovalent and divalent cations. Monovalent cations do not contribute to water hardness, so both sodium and potassium cations are eliminated. Therefore, the three cations causing water hardness are calcium, magnesium, and ferrous iron (2, 3, 4).

The answer is (B).

400. The BOD in the stream at the point of discharge is the weighted sum of the BODs of the three effluents and the stream. Since the stream flow is twice the sum of the three lines, the total stream flow is

$$\begin{aligned} Q_{\text{effluent}} &= \left(5\times10^6\ \frac{\text{L}}{\text{day}} + 3\times10^6\ \frac{\text{L}}{\text{day}} + 2\times10^6\ \frac{\text{L}}{\text{day}}\right) \\ &= 10\ \text{ML/day} \end{aligned}$$

$$Q_{\text{stream}} = (2)\left(10\ \frac{\text{ML}}{\text{day}}\right) = 20\ \text{ML/day}$$

The stream flow increases after the effluent is discharged into the stream.

$$\begin{aligned} Q_{\text{total}} &= Q_{\text{effluent}} + Q_{\text{stream}} = \left(10\ \frac{\text{ML}}{\text{day}} + 20\ \frac{\text{ML}}{\text{day}}\right) \\ &= 30\ \text{ML/day} \end{aligned}$$

The total BOD is the weighted ratio considering the three wastewater sewer lines discharges and the BOD concentrations and the stream's original BOD.

$$\begin{aligned} \text{BOD}_{\text{total}} &= \frac{\Sigma(Q_i(\text{BOD}_i))}{Q_{\text{total}}} \\ &= \frac{\begin{array}{c}\left(5\ \frac{\text{ML}}{\text{day}}\right)\left(300\ \frac{\text{mg}}{\text{L}}\right) + \left(3\ \frac{\text{ML}}{\text{day}}\right)\left(350\ \frac{\text{mg}}{\text{L}}\right) \\ + \left(2\ \frac{\text{ML}}{\text{day}}\right)\left(500\ \frac{\text{mg}}{\text{L}}\right) + \left(20\ \frac{\text{ML}}{\text{day}}\right)\left(10\ \frac{\text{mg}}{\text{L}}\right)\end{array}}{30\ \frac{\text{ML}}{\text{day}}} \\ &= 125\ \text{mg/L} \end{aligned}$$

The answer is (B).

401. Both supports A and B are hinges and can develop both horizontal and vertical reactions. The truss has four unknown reactions, three joints, and two members. Each member has an unknown axial force. For a statically determinate truss, the relationship between bars, reactions, and joints must satisfy the following criteria.

$$r + m = 2j$$

r = the number of reactions = 4

m = the number of members = 2

j = total number of joints = 3

Substitute the values in the criteria equation.

$$4 + 2 = (2)(3)$$

Since the criterion is satisfied, the truss is statically determinate.

The answer is (B).

402. To determine the reaction at support B, take the moment about point A.

$$\begin{aligned} \sum M_{\text{A}} &= 0 \\ &= (-30\ \text{kips})(34.64\ \text{ft}) + R_{\text{B}}(23.09\ \text{ft}) \\ R_{\text{B}} &= 45\ \text{kips} \end{aligned}$$

The positive sign indicates that R_{B} is an upward force. Using equilibrium in the vertical direction, the force is

$$\begin{aligned} \sum F_y &= R_{\text{A}} + R_{\text{B}} = 30\ \text{kips} \\ R_{\text{A}} &= 30\ \text{kips} - R_{\text{B}} \\ &= 30\ \text{kips} - 45\ \text{kips} \\ &= -15\ \text{kips} \end{aligned}$$

A reaction of −15 kips indicates an uplift at support A.

The answer is −15 kips and 45 kips.

403. The column has pinned-pinned end conditions so that its effective length factor, K, is 1.0.

The column's weak (y-) axis is braced at 10 ft intervals. Therefore, the effective length in the y-direction is

$$KL_y = (1.0)(10\ \text{ft})\left(12\ \frac{\text{in}}{\text{ft}}\right) = 120\ \text{in}$$

The column is unbraced along its major (x-) axis. Therefore, the effective length in the x-direction is

$$KL_x = (1.0)(20\ \text{ft})\left(12\ \frac{\text{in}}{\text{ft}}\right) = 240\ \text{in}$$

The radii of gyration of W10 × 45 given in the W Shapes Dimensions and Properties table in the Civil Engineering section of the *NCEES Handbook* are

$$\begin{aligned} r_x &= 4.32\ \text{in} \\ r_y &= 2.01\ \text{in} \end{aligned}$$

Find the slenderness ratios for the x- and y-axes.

$$\begin{aligned} \frac{KL_x}{r_x} &= \frac{240\ \text{in}}{4.32\ \text{in}} = 55.6 \\ \frac{KL_y}{r_y} &= \frac{120\ \text{in}}{2.01\ \text{in}} = 59.7\quad(60) \end{aligned}$$

Out of the two, the larger slenderness ratio controls the column buckling capacity.

The answer is (B).

404. Use the unit force method given in the *NCEES Handbook*, Civil Engineering: Truss Deflection by Unit Load Method section. The vertical displacement of the joint C caused by the loading shown can be determined by applying a unit vertical load at joint C and using the expression

$$\Delta_C = \sum_{i=1}^{\text{members}} f_i(\Delta L)_i$$

ΔL is the change in a member length caused by member force F due to the applied loading.

$$\Delta L = \frac{FL}{AE}$$

The steel's modulus of elasticity is 29.0 Mpsi, as given in the *NCEES Handbook*, Mechanics of Materials section.

$$\begin{aligned} A &= 5 \text{ in}^2 \\ E &= 29{,}000 \text{ ksi} \\ AE &= (5 \text{ in}^2)(29{,}000 \text{ ksi}) \\ &= 145{,}000 \text{ kips} \end{aligned}$$

The member force, F, and member length, L, are tabulated as shown. For consistency, use inch units for member lengths.

In this case, the applied load and the unit load are congruent. The applied load is 90 kips and the unit load is 1 kip, both acting at the same point and in the same direction. There is no other applied load. Therefore, the force, f, caused by unit load is

$$f = \frac{1}{90} \text{ of the force } F \text{ in the respective member}$$

member	F (kips)	L (in)	$\Delta L = \frac{FL}{AE}$	f (kips)	$f(\Delta L)$
AC	90	480	0.298	90 kips /90 = 1	0.298
BC	−155.7	277	−0.297	−155.7 kips/90 = −1.73	0.514
				total	0.812

Find the vertical deflection.

$$\Delta_C = 0.812 \text{ in} \quad (0.8 \text{ in})$$

The positive sign denotes displacement in the direction of the applied load.

The answer is (B).

405. The deflection caused by the UDL is upward, as shown.

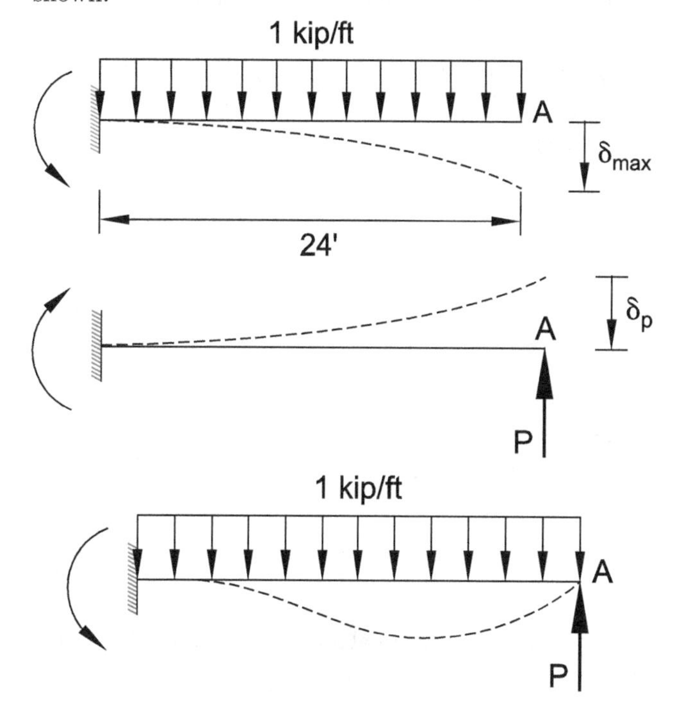

The tip deflection is

$$\delta_{\text{max}} = \frac{wL^4}{8EI}$$

The propping action requires an upward force, P, to negate the beam deflection so that the net deflection is zero. For a cantilever under a point load at a distance, a, from the fixed end, the tip deflection is

$$\delta_P = \frac{Pa^2}{6EI}(3L - a)$$

For the load, P, located at the tip of the cantilever, the distance a is equal to L.

$$\delta_P = \frac{PL^2}{6EI}(3L - L) = \frac{PL^3}{3EI}$$

Since the UDL, w, and point load, P, are in opposite directions, equate the two deflections numerically.

$$\begin{aligned} \delta_P &= \delta_{\text{max}} \\ \frac{PL^3}{3EI} &= \frac{wL^4}{8EI} \\ P &= \frac{3wL}{8} = \frac{(3)\left(1 \frac{\text{kip}}{\text{ft}}\right)(24 \text{ ft})}{8} \\ &= 9 \text{ kips} \end{aligned}$$

The answer is 9 kips.

406. To determine forces T_1 and T_2 in bars 1 and 2, respectively, take the moment about support A.

$$\begin{aligned}\sum M_A &= (90\text{ kN})(18\text{ m}) - T_1(3\text{ m}) - T_2(13\text{ m}) \\ &= 0 \\ T_1 &= 540 - 4.33\,T_2\end{aligned}$$

For deflection compatibility, use similar triangles ABB′ and ACC′.

$$\frac{\Delta L_1}{\Delta L_2} = \frac{\text{AB}}{\text{AC}} = \frac{3}{13} = 0.23$$

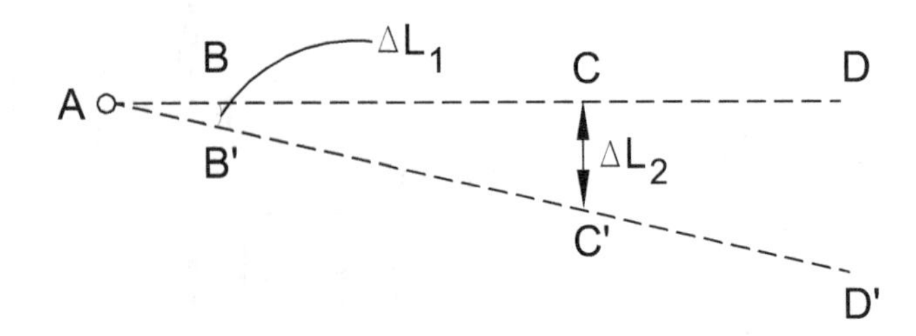

Since both rods have equal area, length, and E, the tensile force ratio is

$$\frac{T_1}{T_2} = 0.23$$

Find the axial tensile force in rod C.

$$\begin{aligned}T_1 = 0.23\,T_2 &= 540 - 4.33\,T_2 \\ (0.23 + 4.33)\,T_2 &= 540 \\ T_2 &= 118.4\text{ kN} \quad (120\text{ kN})\end{aligned}$$

The answer is (D).

407. The maximum reaction occurs when the rolling load passes over the support A or B.

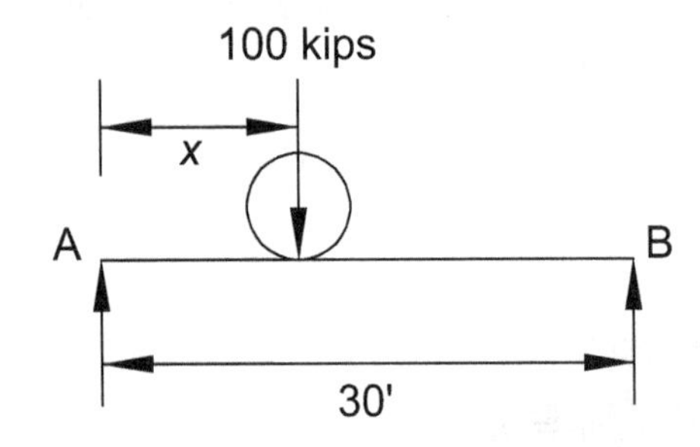

As shown in the illustration, the maximum reaction corresponds to two positions, $x = 0$ or $x = 30$ ft.

The corresponding magnitude of reactions is the entire load.

$$\begin{aligned}&\text{for } x = 0\text{ ft}, \\ &\quad R_A = 100\text{ kips} \\ &\quad R_B = 0\text{ kips} \\ &\text{for } x = 30\text{ ft}, \\ &\quad R_A = 0\text{ kips} \\ &\quad R_B = 100\text{ kips}\end{aligned}$$

The answer is (D).

408. Determine the maximum bending moment in the beam. As shown in the illustration, the maximum moment is

$$M_{\text{max}} = \frac{PL}{4} = \frac{(50\text{ kips})(20\text{ ft})}{4} = 250\text{ ft-kips}$$

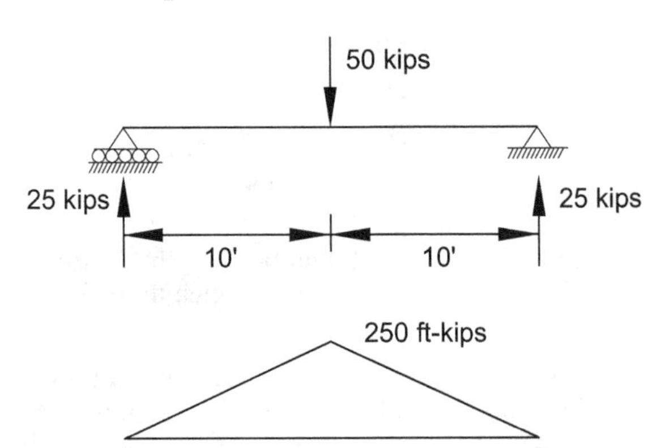

Use the W Shapes Dimensions and Properties table given in the *NCEES Handbook*, Civil Engineering section, and select a beam section that meets two requirements given below.

(1) The beam's allowable moment capacity, $\phi_b M_{rx}$, should be slightly greater than the applied moment.

(2) The section meets the depth limitation.

The section W18 × 55 has a depth of 18.1 in and a flexural strength of

$$\phi_b M_{rx} = 258\text{ ft-kips}$$

Select the W18 × 55 section.

The answer is (B).

409. The maximum bending moment in the beam is

$$M_{\text{max}} = (30\text{ kips})(12\text{ ft}) = 360\text{ ft-kips}$$

Since the loads are factored, the bending strength needed is

$$\phi M_n = 360\text{ ft-kips}$$

Since the unbraced length is 6 ft, the selected section should be able to carry the moment with a larger unbraced length. Use the Available Moment vs. Unbraced Length diagram in the Civil Engineering section of the *NCEES Handbook*. Several sections qualify. The lightest section that qualifies is a W21 × 48. Use the table W Shapes—Selection by Z_x to verify the selection. The selected section is fully compact.

$$L_p = 6.09 \text{ ft} > 6 \text{ ft}$$
$$\phi_b M_{px} = 398 \text{ ft-kips} > 360 \text{ ft-kips}$$

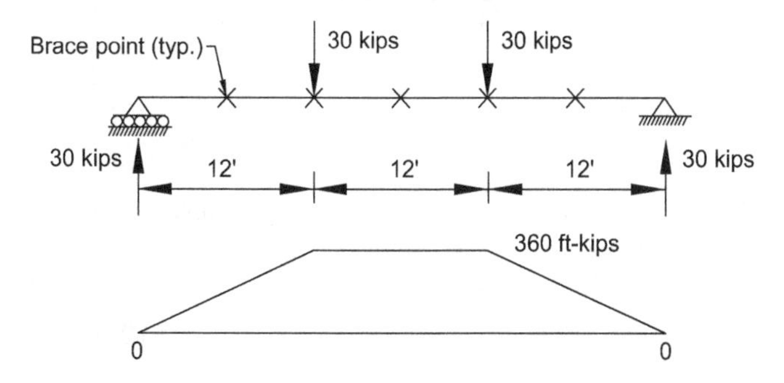

The answer is (D).

410. The moment capacity modification factor, C_b, can be found by using the AISC formulas or the Values of C_b for Simply Supported Beams table. For the beam loading and bracing configuration, the C_b factor is 1.14.

The answer is (C).

411. Use the *NCEES Handbook*, Civil Engineering section. The design of reinforced concrete components is based on ACI 318-14 code. The factored moment can be determined by the equation

$$M_n = A_s f_y\left(d - \frac{a}{2}\right)$$

For steel bar area, see the *NCEES Handbook*, ASTM Standard Reinforcement Bars table. The values for areas and force are given.

$$A_s = 3 \text{ \#9 bars} = 3 \text{ in}^2$$
$$f_y = 60 \text{ ksi}$$

The effective depth is

$$d = 22 \text{ in} - 2 \text{ in} = 20 \text{ in}$$

The lever arm is

$$a = \frac{A_s f_y}{0.85 b f_c'} = \frac{(3 \text{ in}^2)(60 \text{ ksi})}{(0.85)(12 \text{ in})(5 \text{ ksi})} = 3.53 \text{ in}$$

Find the moment capacity.

$$\begin{aligned} M_n &= (3 \text{ in}^2)(60 \text{ ksi})\left(20 \text{ in} - \frac{3.53 \text{ in}}{2}\right) \\ &= 3280 \text{ in-kips} \end{aligned}$$

For flexure, $\phi = 0.9$, and the design moment strength is

$$\begin{aligned} \phi M_n &= (0.9)(3280 \text{ in-kips}) = 2952 \text{ in-kips} \\ &= 246 \text{ ft-kips} \quad (240 \text{ ft-kips}) \end{aligned}$$

Alternate method:

For a rectangular section, assume a lever arm, a, of $0.9d$. The moment capacity is

$$\begin{aligned} \phi M_n &= \phi(0.9d)(A_s)(f_y) = (0.9)(0.9)(20 \text{ in})(3 \text{ in}^2)(60 \text{ ksi}) \\ &= 2916 \text{ in-kips} = 243 \text{ ft-kips} \\ &\quad [1\% \text{ off on conservative side}] \quad (240 \text{ ft-kips}) \end{aligned}$$

The answer is (A).

412. The factored shear force, V_u, can be determined using the governing load combinations.

$$\begin{aligned} V_u &= 1.2D + 1.6L = (1.2)(10 \text{ kips}) + (1.6)(15 \text{ kips}) \\ &= 36 \text{ kips} \end{aligned}$$

The beam effective depth, d, is

$$\begin{aligned} d &= \text{total depth} - \text{concrete cover} \\ &= 23 \text{ in} - 3 \text{ in} \\ &= 20 \text{ in} \end{aligned}$$

The nominal strength provided by concrete, V_c, can be determined as follows.

$$\begin{aligned} V_c &= 2\sqrt{f_c'}\, bd = 2\sqrt{5000 \text{ psi}}\ (12 \text{ in})(20 \text{ in}) \\ &= 33{,}941 \text{ lbf} \\ &= 33.94 \text{ kips} \end{aligned}$$

The value of ϕ is 0.75.

$$\phi V_c = (0.75)(33.94 \text{ kips}) = 25.46 \text{ kips} < V_u$$

Since the applied shear is more than the available concrete capacity, shear steel is needed to resist V_s.

$$V_s = \frac{V_u}{\phi} - V_c = \frac{36 \text{ kips}}{0.75} - 33.94 \text{ kips} = 14.06 \text{ kips}$$

Use maximum allowed stirrup spacing.

$$\frac{d}{2} = 10 \text{ in}$$

For stirrup design, the spacing is

$$s = \frac{A_v f_y d}{V_s}$$

Rearrange the equation and find the area, A_v.

$$A_v = \frac{(10\text{ in})(14.06\text{ kips})}{(60\text{ ksi})(20\text{ in})} = 0.12\text{ in}^2$$

Since the stirrup has two legs, the required area per leg is 0.06 in^2.

Use #3 stirrups. Each stirrup leg has an area of 0.11 in^2, which is adequate.

The answer is (A).

413. The particle size of 75 microns (0.075 mm) equals the mesh size of U.S. sieve no. 200. Determine the particle percentage that passes through sieve no. 200 by finding out the percentage retained on all sieves coarser than and including sieve no. 200.

sieve no.	retained (%)	total retained (%)	passed (%)
4	0	0	100
16	20	20	100 − 20 = 80
50	25	20 + 25 = 45	100 − 45 = 55
80	20	45 + 20 = 65	100 − 65 = 35
200	25	65 + 25 = 90	100 − 90 = 10

The answer is (A).

414. See the *NCEES Handbook*, Civil Engineering section. Draw the phase diagram, as shown in the illustration. The degree of saturation is the ratio of the volume of water to the total voids in a soil mass, expressed as a percentage. It measures the extent to which the voids have been filled with water.

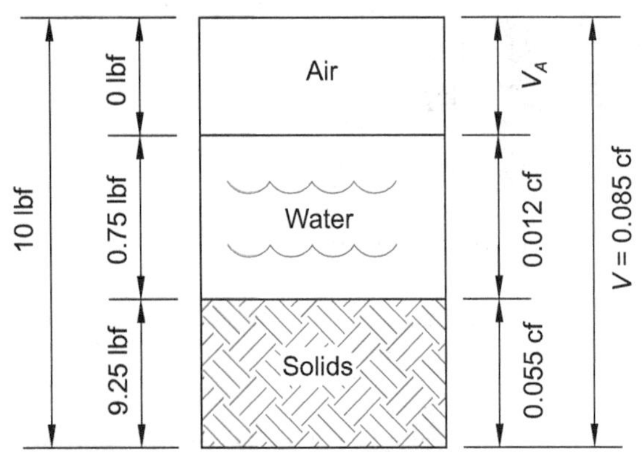

The weight of the sample, solids, and water in the sample are given.

The moist sample weight is

$$W = 10\text{ lbf}$$

The weight of solids is equal to the oven-dried sample weight.

$$W_S = 9.25\text{ lbf}$$

The weight of water in the sample is

$$W_W = 10\text{ lbf} - 9.25\text{ lbf} = 0.75\text{ lbf}$$

The degree of saturation, S, is the ratio of the volume of water to the volume of all voids present in the soil sample. If all voids are filled with water, then the degree of saturation is 100%.

$$S = \left(\frac{V_W}{V_V}\right)(100\%)$$

Convert the given weight of water and solids to the volume they occupy.

$$V_W = \frac{\text{weight of water}}{\text{density of water}} = \frac{0.75\text{ lbf}}{62.4\ \frac{\text{lbf}}{\text{ft}^3}} = 0.012\text{ ft}^3$$

$$V_S = \frac{\text{weight of solid}}{\text{density of solid}} = \frac{W_S}{\gamma_S} = \frac{W_S}{G_S\gamma_W}$$
$$= \frac{9.25\text{ lbf}}{(2.70)\left(62.4\ \frac{\text{lbf}}{\text{ft}^3}\right)}$$
$$= 0.055\text{ ft}^3$$

The total volume is

$$V = V_S + V_W + V_A$$

where V_A = volume of air in the sample, V_V = volume of voids in the sample, and $V_V = V_A + V_W$.

In dry soil, since there is no water, the volume V_A equals V_V. The volume is

$$V = V_S + V_V$$

Rearrange the equation and find the volume of the voids.

$$V_V = V - V_S = 0.085\text{ ft}^3 - 0.055\text{ ft}^3 = 0.03\text{ ft}^3$$

Find the degree of saturation.

$$S = \frac{V_W}{V_V} = \left(\frac{0.012\text{ ft}^3}{0.03\text{ ft}^3}\right)(100\%) = 40\%$$

The answer is (B).

415. Using the constant head test, the coefficient of permeability is

$$k = \frac{Q}{iAt_e}$$

Using the falling head test, the coefficient of permeability is

$$k = 2.303\left(\frac{aL}{At_e}\right)\log_{10}\frac{h_1}{h_2}$$

The coefficient of permeability, k, is also called the hydraulic conductivity. Therefore, options A, B, and C are correct. The coefficient of permeability, k, can be determined using Darcy's law.

$$Q = -kA\frac{dh}{dx}$$

From Darcy's law, it is clear that coefficient k is not dimensionless. The unit is length/time.

The answer is (D).

416. The effective overburden pressure is defined as the intensity of intergranular pressure on any horizontal plane at or below the foundation level before the construction starts. Mathematically,

$$\begin{aligned}\text{effective overburden pressure} &= \text{total overburden pressure} \\ &\quad -\text{pore water pressure} \\ &= \text{total overburden pressure} \\ &\quad -\gamma_{\text{water}} \times h_{\text{water}}\end{aligned}$$

Find the unsaturated soil height.

$$z - h = 6 \text{ ft} - 3.5 \text{ ft} = 2.5 \text{ ft}$$

The total overburden at 2.5 ft depth is

$$p_1 = \gamma(z-h) = \left(120 \ \frac{\text{lbf}}{\text{ft}^3}\right)(2.5 \text{ ft}) = 300 \text{ psf}$$

The total overburden at footing base is

$$\begin{aligned}p &= p_1 + \gamma_{\text{sat}}h \\ &= 300 \text{ psf} + (128 \text{ pcf})(3.5 \text{ ft}) \\ &= 748 \text{ psf}\end{aligned}$$

To determine the effective overburden pressure, deduct pore water pressure from the total overburden pressure.

$$\begin{aligned}\begin{array}{c}\text{effective overburden} \\ \text{at footing base}\end{array} &= 748 \ \frac{\text{lbf}}{\text{ft}^2} - \left(62.4 \ \frac{\text{lbf}}{\text{ft}^3}\right)(3.5 \text{ ft}) \\ &= 530 \text{ psf}\end{aligned}$$

The answer is (A).

417. The active pressure causes overturning and the passive pressure provides resistance. The active pressure coefficient is

$$\begin{aligned}k_a &= \tan^2\left(45° - \frac{\phi}{2}\right) \\ &= \tan^2\left(45° - \frac{30°}{2}\right) \\ &= 0.33\end{aligned}$$

The passive pressure coefficient is

$$\begin{aligned}k_p &= \tan^2\left(45° + \frac{\phi}{2}\right) \\ &= \tan^2\left(45° + \frac{30°}{2}\right) \\ &= 3\end{aligned}$$

The active pressure behind the sheet piling wall segment BC is

$$P_a = 0.5(0.33)(\gamma)(H^2) = 0.167(\gamma)(H^2)$$

The passive pressure at sheet piling wall segment AB is

$$P_p = 0.5(3)(\gamma)(h^2) = 1.5(\gamma)(h^2)$$

Use point A as the rotation point and determine resisting and overturning moments.

$$M_{\text{OT}} = \text{active pressure} \times \text{lever arm} = P_a\left(\frac{H}{3}\right)$$

$$M_R = \text{passive pressure} \times \text{lever arm} = P_p\left(\frac{h}{3}\right)$$

Find the factor of safety, FS, against rotation.

$$\begin{aligned} \text{FS} &= \frac{M_R}{M_{\text{OT}}} = \frac{P_p\left(\frac{h}{3}\right)}{P_a\left(\frac{H}{3}\right)} = \left(\frac{P_p}{P_a}\right)\left(\frac{h}{H}\right) \\ &= \left(\frac{1.5\gamma h^2}{1.67\gamma H^2}\right)\left(\frac{h}{H}\right) = 9\left(\frac{h}{H}\right)^3 = (9)\left(\frac{15 \text{ ft}}{30 \text{ ft}}\right)^3 \\ &= 1.125 \quad (1.1) \end{aligned}$$

The answer is (A).

418. In a drained test, the sample ends are capped with porous end plates and water is allowed to drain out as the load is applied. Use the Mohr-Coulomb Failure diagram given in the *NCEES Handbook*, Geotechnical section and the equation. The shear stress at failure is

$$\tau_F = c + \sigma_N \tan\phi$$

ϕ is the angle of internal friction.

Find the slope of the failure line. Given two points on a straight line, the line slope is

$$m = \frac{y_2 - y_1}{x_2 - x_1}$$

$$\begin{aligned} \tan\phi &= \frac{\tau_2 - \tau_1}{\sigma_2 - \sigma_1} \\ &= \frac{2500 \frac{\text{lbf}}{\text{ft}^2} - 700 \frac{\text{lbf}}{\text{ft}^2}}{4000 \frac{\text{lbf}}{\text{ft}^2} - 1000 \frac{\text{lbf}}{\text{ft}^2}} \\ &= \frac{1800}{3000} \\ &= 0.6 \end{aligned}$$

Find the angle of internal friction.

$$\begin{aligned} \phi &= \arctan 0.6 \\ &= 31° \end{aligned}$$

The answer is (C).

419. See the *NCEES Handbook*, Geotechnical section, for the ultimate bearing capacity equation. Using units of ft-lbf, the strip footing ultimate capacity is given by

$$q_{\text{ult}} = cN_c + \gamma' D_f N_q + \frac{1}{2}\gamma' B N_\gamma$$

Since the water table is well below the footing, the value of γ' in the above equation is replaced by γ, which is 110 pcf. Thus, the bearing capacity equation becomes

$$\begin{aligned} q_{\text{ult}} &= cN_c + \gamma D_f N_q + \frac{1}{2}\gamma B N_\gamma \\ &= (200 \text{ psf})(25.13) + (110 \text{ pcf})(4 \text{ ft})(12.72) \\ &\quad + \left(\frac{1}{2}\right)(110 \text{ pcf})(4 \text{ ft})(8.34) \\ &= 12{,}458 \text{ psf} \end{aligned}$$

Using a safety factor of 3.0, determine the allowable footing capacity.

$$\begin{aligned} q_{\text{allow}} &= \frac{q_{\text{ult}}}{\text{FS}} = \frac{12{,}458 \text{ psf}}{3.0} = 4152 \text{ psf} \\ &= \frac{(4152 \text{ psf})(1 \text{ ksf})}{1000 \text{ psf}} = 4.15 \text{ ksf} \quad (4 \text{ ksf}) \end{aligned}$$

The answer is 4 ksf.

420. Consider the unit length of the strip footing.

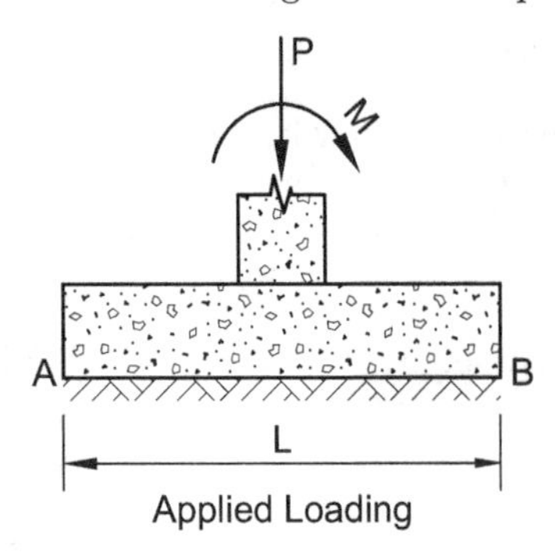

Applied Loading

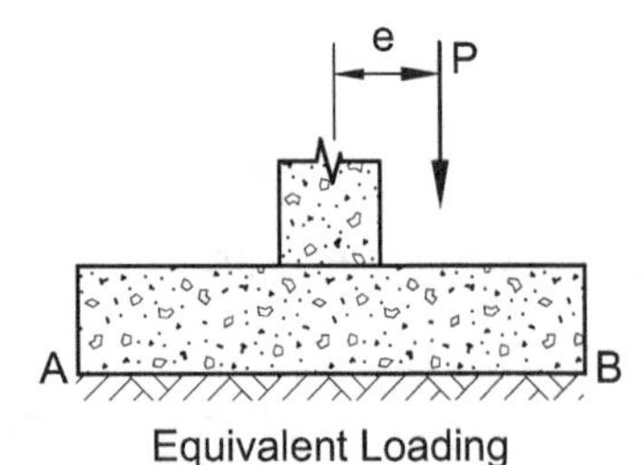

Equivalent Loading

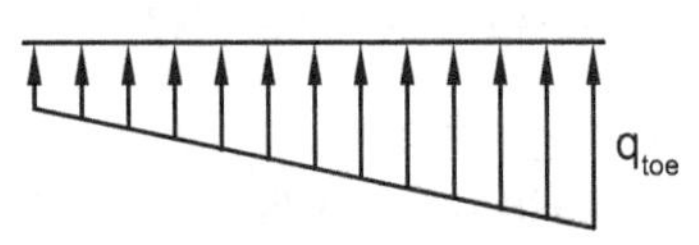

Soil Stress Distribution

The eccentricity is

$$e = \frac{M}{P} = \frac{10 \text{ ft-kips}}{10 \text{ kips}} = 1 \text{ ft} = 12 \text{ in}$$

As a result, the soil pressure on one edge would be greater than the pressure on the other edge. The general equation for determining the pressure along the footing is

$$\sigma = \frac{P}{A} \pm \frac{Mc}{I} = \frac{P}{A} \pm \frac{Pec}{I}$$

Point B on the soil stress distribution diagram denotes the toe of the footing and would have the maximum compressive pressure. The pressure at the toe is given by the expression

$$q_{\text{toe}} = \frac{\sum V}{BL}\left(1 + \frac{6e}{B}\right)$$

It is given that the footing width B is 5 ft. The unit length, L, of the wall is 1 ft. Find the soil pressure at the toe.

$$q_{\text{toe}} = \frac{10 \text{ kips}}{(5 \text{ ft})(1 \text{ ft})}\left(1 + \frac{(6)(1 \text{ ft})}{5 \text{ ft}}\right) = 4.4 \text{ ksf}$$

The answer is (D).

421. The soil is overconsolidated. The wall is laterally supported on both ends and the back of the wall is smooth. The backfill is level and has no cohesion. Therefore, the wall is subjected to at-rest pressure. Since the soil is overconsolidated, the at-rest pressure coefficient is

$$K_0 = (1 - \sin\phi)(\text{OCR})^{\sin\phi}$$

The overconsolidation ratio (OCR) is the overburden ratio.

$$\begin{aligned} \text{OCR} &= \frac{\begin{array}{c}\text{pressure under which deposit was} \\ \text{fully consolidated in the past}\end{array}}{\text{present overburden pressure}} \\ &= \frac{125 \ \frac{\text{lbf}}{\text{ft}^2}}{75 \ \frac{\text{lbf}}{\text{ft}^2}} \\ &= 1.667 \end{aligned}$$

Find the coefficient of at-rest pressure.

$$\begin{aligned} K_0 &= (1 - \sin 25°)(\text{OCR})^{\sin 25°} \\ &= (1 - 0.42)(1.667)^{0.42} \\ &= (0.58)(1.242) \\ &= 0.72 \end{aligned}$$

The answer is (C).

422. The table in the *NCEES Handbook*, Civil Engineering: Geotechnical section, titled "Variation of time factor with degree of consolidation," shows the degree of consolidation, U, as it varies with elapsed time since application of the consolidation load. For the degree of consolidation between 0% and 60%, the table is based on the time factor, T_v.

$$T_v = \frac{\pi}{4} U^2 \quad \text{[Eq.1]}$$

The time factor is also defined as

$$T_v = \frac{C_v}{H_{dr}^2} t \quad \text{[Eq. 2]}$$

The sand layers are on both the top and bottom ends of the clay stratum.

$$\begin{aligned} 2H_{dr} &= 10 \text{ m} \\ H_{dr} &= 5 \text{ m} \\ &= 500 \text{ cm} \end{aligned}$$

For 50% consolidation, solve the two equations.

$$T_v = \frac{C_v}{H_{dr}^2} t = \left(\frac{\pi}{4}\right) U^2 = \left(\frac{\pi}{4}\right)(0.5)^2 = 0.197$$

Rearrange the equation and solve for time.

$$\begin{aligned} t &= \frac{0.197 H_{dr}^2}{C_v} = \frac{0.197(500 \text{ cm})^2}{0.002 \ \frac{\text{cm}^2}{\text{s}}} \\ &= 24{,}625{,}000 \text{ s} \\ &= 285 \text{ days} \end{aligned}$$

The answer is (D).

423. Soil erosion loss can be due to precipitation, wind, and ice. Coastal areas are affected by waves impacting the shorelines. Universal equations are available for soil loss caused by both wind and water. This problem concerns erosion due to runoff water. Water erosion of soil starts as splash erosion when raindrops hit bare soil. The splashed particle rises and moves up from the point of impact, blocking the spaces between soil aggregates. It reduces infiltration and increases the runoff. Sheet erosion is the removal of soil in thin layers. Surface water flow that causes sheet erosion gives rise to rills after flowing for a few feet. Rills are shallow drainage lines less than a foot deep. The rills, if unremedied, become gullies, which may grow, resulting in tunnel erosion.

Using Darcy's formula, runoff increases with area. The longer the unsheltered terrain, the greater the runoff is. Also, the steeper the terrain, the faster the runoff. Therefore, option A is incorrect. Not all rainfalls cause the same amount of erosion. For example, the rainfalls in the western part of the United States cause more erosion than those in the Midwest. Rainfall is categorized based on the erosion it can cause. The higher the rainfall erosivity factor, the more soil loss there is. Therefore, option B is incorrect.

Option D is incorrect because, in coastal areas, waves may cause coastal erosion.

Soil composition plays a significant part in soil loss. Soils are typically graded based on their resistance to erosion, called an *erodibility factor*. The universal equation has a table of soil erodibility factors depending on the soil composition. The organic contents in the soil have an adhesive effect in keeping the soil together. The more organic content in the soil, the less soil is eroded.

The answer is (C).

424. Use the stopping sight distance formula given in the Civil Engineering: Transportation section of the *NCEES Handbook*.

$$\text{SSD} = 1.47Vt + \frac{V^2}{30\left(\left(\frac{a}{32.2}\right) \pm G\right)}$$

V = design speed = 55 mph

t = driver reaction time = 1.5 sec

a = deceleration rate = 10 ft/sec^2

G = percentage grade divided by 100 = 0.02

Find the SSD.

$$\begin{aligned}\text{SSD} &= (1.47)(55 \text{ mph})(1.5 \text{ sec}) + \frac{(55 \text{ mph})^2}{30\left(\left(\frac{10 \frac{\text{ft}}{\text{sec}^2}}{32.2}\right) + 0.02\right)} \\ &= 121.3 + \frac{3025}{9.92} \\ &= 426 \text{ ft} \quad (425 \text{ ft})\end{aligned}$$

The answer is (C).

425. First, determine if the curve is a sag curve or a crest curve. In this case, since the back tangent has a 5% grade followed by a −5% grade, the curve is a crest curve.

Since the height of the driver's eyes and the object's height meet standard headlight criteria for a crest curve analysis, apply the formula.

$$L_{\text{min}} = \frac{AS^2}{2158}$$

The difference in the grades and the length of the curve are given.

$$\begin{aligned}A &= |g_1 - g_2| = |-5\% - (+5\%)| = 10 \\ L_{\text{min}} &= 300 \text{ ft}\end{aligned}$$

Rearrange the equation and find the sight distance.

$$\begin{aligned}L_{\text{min}} &= \frac{AS^2}{2158} = \frac{10S^2}{2158 \text{ ft}} = 300 \text{ ft} \\ S^2 &= \frac{(2158 \text{ ft})(300 \text{ ft})}{10} = 64{,}740 \text{ ft}^2 \\ S &= 254 \text{ ft}\end{aligned}$$

Use the stopping sight distance formula.

$$S = 1.47Vt + \frac{V^2}{30\left(\left(\frac{a}{32.2}\right) \pm G\right)}$$

$t = 1.5$ sec

$a = 10 \text{ ft/sec}^2$

$G = -0.05$ (worst-case slope)

Rearrange the stopping sight distance equation to find the speed.

$$\begin{aligned}S &= 254 \text{ ft} \\ &= (1.47)V(1.5 \text{ sec}) \\ &\quad + \frac{V^2}{30\left(\left(\frac{10 \frac{\text{ft}}{\text{sec}^2}}{32.2}\right) - 0.05\right)}\end{aligned}$$

$$\begin{aligned}2.21V + 0.13V^2 &= 254 \\ 0.13V^2 + 2.21V - 254 &= 0\end{aligned}$$

Using the solution for a quadratic equation, find the speed.

$$x = \frac{-b \pm \sqrt{b^2 - 4ac}}{2a}$$

$$\begin{aligned} V &= \frac{-2.21 \pm \sqrt{(2.21)^2 + 4(0.13)(254)}}{2(0.13)} \\ &= \frac{-2.21 + 11.7}{0.26} \\ &= 36.5 \text{ mph} \quad (35 \text{ mph}) \end{aligned}$$

The answer is (A).

426. The stopping sight distance formula is

$$\text{SSD} = 1.47Vt + \frac{V^2}{30\left(\left(\frac{a}{32.2}\right) \pm G\right)}$$

Since the grade changes constantly on a vertical curve, use the worst-case value for grade.

$$\begin{aligned} G &= -0.05 \\ V &= 45 \text{ mph} \\ t &= 2.5 \text{ sec} \\ a &= 10 \text{ ft/sec}^2 \end{aligned}$$

Find the SSD.

$$\begin{aligned} \text{SSD} &= (1.47)(45 \text{ mph})(2.5 \text{ sec}) + \frac{(45 \text{ mph})^2}{30\left(\left(\frac{10 \frac{\text{ft}}{\text{sec}^2}}{32.2}\right) - 0.05\right)} \\ &= 165 \text{ ft} + 259 \text{ ft} \\ &= 424 \text{ ft} \quad (420 \text{ ft}) \end{aligned}$$

The answer is (D).

427. The AASHTO structural number equation is

$$\text{SN} = a_1D_1 + a_2D_2m_2 + \cdots + a_nD_nm_n$$

a_i = layer coefficient

D_i = thickness of layer (inches)

m_i = drainage coefficient

The drainage coefficient is given as 1.0, so the AASHTO structural number equation can be simplified as

$$\text{SN} = a_1D_1 + a_2D_2 + \cdots + a_nD_n$$

Use the values from the table to find the structural number.

$$\begin{aligned} \text{SN} &= (0.4)(4 \text{ in}) + (0.14)(5 \text{ in}) + (0.11)(8.5 \text{ in}) \\ &= 3.24 \quad (3) \end{aligned}$$

The answer is (C).

428. See the *NCEES Handbook*, Civil Engineering: Transportation section. The peak hourly factor, PHF, is defined as

$$\text{PHF} = \frac{\text{volume during peak hour}}{(4)(\text{volume during peak 15 min interval})}$$

In this method, the hourly traffic volume and the peak 15 min segment traffic volume in the hour are compared.

$$\begin{aligned} V_{15 \text{ min}} &= \text{peak 15 in volume, vehicles per 15 min} \\ &= 400 + 425 + 465 + 450 \\ &= 1740 \text{ vehicles} \end{aligned}$$

Find the PHF.

$$\text{PHF} = \frac{1740 \text{ vehicles}}{(4)(465 \text{ vehicles})} = 0.935$$

The answer is (A).

429. A vehicle is considered to be located within a dilemma zone when it can neither stop nor continue and clear the intersection safely, legally, and comfortably. The dilemma zone can be eliminated by either changing the speed limit or selecting an appropriate minimum duration for the yellow signal phase.

The answer is (D).

430. Traffic flow is defined as the number of vehicles crossing a point per hour per lane. See "Traffic Flow Relationships" diagrams given in the *NCEES Handbook*, Civil Engineering: Transportation section. According to the Greenshields model, as the traffic speed increases, the flow initially increases and then decreases after reaching an optimal maximum speed. Option C is correct.

The answer is (C).

431. See the *NCEES Handbook*, Civil Engineering: Transportation section. To compare traffic crashes at different locations, the number of crashes at different locations are counted on an annual basis. The crash rate per million entering vehicles in a given year at a location can be determined using the equation

$$\text{RMEV} = \frac{A(1{,}000{,}000)}{V}$$

The number of crashes occurring in a single year at the intersection is

$$A = 18$$

The average daily traffic entering the intersection is

$$\text{ADT} = 10{,}000$$

Compute the annual traffic volume.

$$\begin{aligned} V &= (\text{ADT})(365 \text{ days}) \\ &= (10{,}000)(365 \text{ days}) \\ &= 3{,}650{,}000 \text{ vehicles} \end{aligned}$$

Find the intersection's crash rate.

$$\text{RMEV} = \frac{(18)(1{,}000{,}000)}{3{,}650{,}000} = 4.9 \quad (5.0)$$

The answer is (C).

432. The expected number of crashes if countermeasures are not implemented and if the traffic volume remains the same is $N = 25$.

The crash reduction factor is $\text{CR} = 30\% = 0.30$.

Find the number of crashes prevented, CP, by the countermeasures.

$$\begin{aligned} \text{CP} &= N(\text{CR})\left(\frac{\text{ADT after countermeasures}}{\text{ADT before countermeasures}}\right) \\ &= (25)(0.30)\left(\frac{10{,}000}{8000}\right) \\ &= 9 \end{aligned}$$

The answer is (C).

433. An LDP is a person authorized to practice engineering and/or architecture, as defined by the professional licensing laws of a state or jurisdiction. Construction documents are written and graphic documents prepared for communicating the project design, for construction, and for administering the construction contract. These documents typically include a set of drawings of the proposed construction project, specifications for the construction, procurement documents, which inform bidders of the procedures for submitting bids, and contracting documents, which describe the proposed construction project and the terms and conditions of the project.

Tasks such as preparing shop drawings, designing formwork, shoring, and re-shoring are means and methods of construction. A contractor performs these tasks upon award of a construction contract, and the tasks are not part of the construction documents an LDP is required to prepare.

The answer is (C).

434. Construction sites are considered hazardous. The workers are most often struck by vehicles, falling or flying objects, or concrete and masonry walls that collapse while under construction. The three types of hazards are the leading causes of construction-related deaths. While the hazards exist, there are control measures that can be taken to ensure safety. Options A, B, and C are correct. A criminal act by a third party on a construction site is not considered a struck-by hazard.

The answer is (D).

435. The cost variance of work performed, CV, is defined in the Civil Engineering: Construction section of the *NCEES Handbook* as

$$\text{CV} = \text{BCWP} - \text{ACWP}$$

BCWP and ACWP denote the budgeted and actual costs of the work performed, respectively. If the CV is positive, it implies that the contractor has been profitable. On the other hand, if the ACWP exceeds the budgeted cost, the contractor is over budget and experiencing a loss. In this case, the CV is

$$\begin{aligned} \text{CV} &= \text{BCWP} - \text{ACWP} \\ &= \$150{,}000 - \$210{,}000 \\ &= -\$60{,}000 \end{aligned}$$

The contractor's cost variance for the work performed is −$60,000.

The answer is (A).

436. The balanced region is defined as the distance between two stations where the cut and fill materials are equal in mass. The volumes of cut and fill may differ due to variations in their respective densities, but their masses should be the same. In a balanced region, no hauling of the material into or out of the region is needed since all excavated fill is placed throughout the site.

The answer is (A).

437. Formwork development requires serious and detailed engineering consideration. Generally, the cost of concrete formwork varies from 20% to 70% of the cost of a concrete structure.

The answer is (D).

438. A CPM arrow diagram is also called an activity-on-arrow network. In this diagram, the nodes are events, and the arrows represent activities. To complete an event, all activities culminating at the node must be completed. The dotted lines represent dummy activities, which consume no time or resource. The critical path is the longest. The time required to finish the project is 22 days. The activities on the critical path are 1-2-4-6-7-8-9. The minimum number of days needed to complete the project is

$$\text{1-2-4-6-7-8-9} = 3+2+3+4+6+4 = 22 \text{ days}$$

Second, the effect of dummy activities should be evaluated. The dummy activity 3-4 imposes a restriction that event 4 cannot occur until event 3 can occur. Per the CPM arrow diagram, it takes 7 days for event 3, and 5 days for event 4, to complete. This means that it would take 2 additional days to complete the project, increasing the number of days to 24 to complete the project.

Now consider the effect of dummy activity 5-7. It imposes a restriction that event 7 cannot occur until event 5 can occur. In this case, both events 5 and 7 take the same number of days to complete. Therefore, there is no further impact on completing the project.

The answer is (D).

439. The wall is designed to be laterally supported at its top end and at the base. The wall can resist the soil lateral pressure if both the top and bottom slabs are in place to act as the supports. The wall needs both slabs to act as a simply supported beam. If backfill is placed before the top slab, the wall would try to act as a cantilever and collapse. If the slab on grade is absent, the wall would slide or bulge at the base.

The answer is 1: A, 2: B, 3: D, 4: C.

440. When all four activities are happening simultaneously, $6+4+3+7=20$ workers would be needed. This would happen in week 15, as shown in the line diagram.

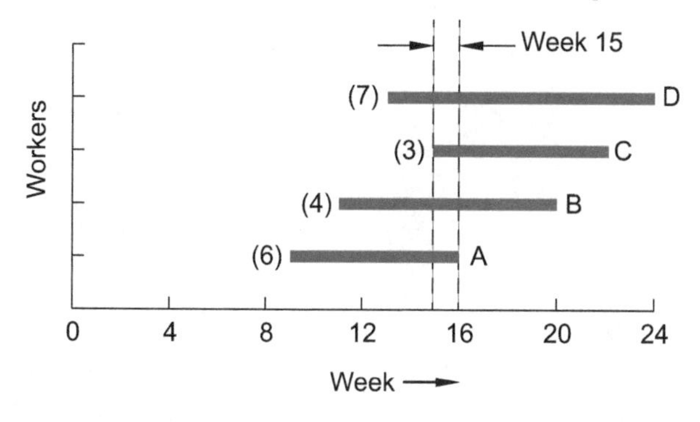

The answer is week 15.

Solutions
Exam 5

441. The quadratic equation $ax^2 + bx + c = 0$ has two roots.

$$x_1 = \frac{-b + \sqrt{b^2 - 4ac}}{2a}$$
$$x_2 = \frac{-b - \sqrt{b^2 - 4ac}}{2a}$$

Multiply the roots.

$$x_1 x_2 = \left(\frac{-b + \sqrt{b^2 - 4ac}}{2a}\right)\left(\frac{-b - \sqrt{b^2 - 4ac}}{2a}\right) = \frac{1}{4a^2}\left(-b + \sqrt{b^2 - 4ac}\right)\left(-b - \sqrt{b^2 - 4ac}\right)$$

Use the identity

$$(u + v)(u - v) = u^2 - v^2$$

Simplify to find the roots.

$$x_1 x_2 = \frac{1}{4a^2}\left((-b)^2 - (b^2 - 4ac)\right) = \frac{1}{4a^2}(b^2 - b^2 + 4ac) = \frac{4ac}{4a^2} = \frac{c}{a} \quad (c/a)$$

The answer is (D).

442. To determine if the given sequence of plant production forms an arithmetic progression series, subtract each number from the following number. If the differences are equal, the series is arithmetic. In this case, the plant casts five pieces during the first week, 10 during the second week, and 15 during the third week. The common difference is five each week. Since the differences are equal, express the numbers as an arithmetic progression series where the first term, a, is 5.

The common difference, d, is computed as

$$d = (\text{pieces cast during } i^{\text{th}} \text{ week}) - \left(\text{pieces cast during}(i-1)^{\text{th}} \text{ week}\right) = 5$$

The number of terms, n, is 10.

The sum of n terms, s, can be determined.

$$s = \frac{n\left(2a + (n-1)d\right)}{2} = \frac{10\left(2(5) + (10-1)(5)\right)}{2} = 275 \text{ pieces}$$

The answer is 275 pieces.

443. To multiply two matrices, the number of columns in the first matrix should equal the number of rows in the second matrix. In this case, the condition is satisfied because matrix A has two columns and matrix B has two rows. Therefore, the multiplication of matrices A and B is possible. Option D is incorrect. The product matrix will have as many rows as the first matrix and as many columns as the second matrix. Therefore, the product matrix C will have three rows and three columns.

The answer is (C).

444. The relationships between the polar and xy-coordinates are

$$x = r\cos\theta$$
$$y = r\sin\theta$$
$$x^2 + y^2 = r^2$$

r was given in the problem statement. Multiply both sides of the equation by r.

$$r = 4\cos\theta$$
$$r[r = r(4\cos\theta)]$$
$$r^2 = 4r\cos\theta$$

Therefore,

$$r^2 = x^2 + y^2 = 4(r\cos\theta) = 4x$$
$$x^2 - 4x + y^2 = 0$$

Add 4 to both sides of the equation to complete the square in x, and then rearrange.

$$(x^2 - 4x + 4) + y^2 = 4$$
$$(x-2)^2 + y^2 = (2)^2$$

The equation pertains to a circle whose center is located at (2, 0) and whose radius is 2.

The answer is (B).

445. A donut or torus has a tire shape. It is generated by rotating a disk (circle) along the centerline of a ring. Two radii are given in the illustrations.

The radius of the disk circle, r, is 2.

The radius of the ring (or revolution), R, is 4.

A disk cross section is a circle with the equation

$$(x-4)^2 + y^2 = 4$$

The disk circle's center is located at (4, 0). The equation is

$$\begin{aligned} y^2 &= 4 - (x-4)^2 \\ y &= \sqrt{4-(x-4)^2} \end{aligned}$$

The circle is rotated 360° (or 2 π rad). It is given that the radius of revolution, R, which is the distance from the origin to the center of the circle, is 4. The volume of a torus is expressed as

$$\begin{aligned} V &= \int_2^6 (2\pi R)y\,dx = \int_2^6 (2\pi)(R)\sqrt{4-(x-4)^2}\,dx \\ &= (2\pi)(R)\int_2^6 \sqrt{4-(x-4)^2}\,dx \end{aligned}$$

The expression within the integral is the area of the circular disk.

Therefore,

$$\begin{aligned} V &= (2\pi R)(A_{\text{circle}}) \\ A_{\text{circle}} &= \pi r^2 = \pi(2)^2 = 4\pi \\ V &= (2\pi R)(A_{\text{circle}}) = (2\pi(4))(4\pi) \\ &= 32\pi^2 \end{aligned}$$

The answer is (C).

446. Solve the integral.

$$\begin{aligned} &\int_{-2}^{1} (3x^2 + 2x - 9)\,dx \\ &= x^3 + x^2 - 9x\Big|_{-2}^{1} \\ &= \left(1^3 + 1^2 - 9(1)\right) - \left((-2)^3 + (-2)^2 - 9(-2)\right) \\ &= -21 \end{aligned}$$

The answer is –21.

447. For addition of vectors, see the *NCEES Handbook*, Mathematics: Vectors section. The resultant of two vectors is given by

$$\mathbf{A} + \mathbf{B} = (a_x + b_x)\mathbf{i} + (a_y + b_y)\mathbf{j} + (a_z + b_z)\mathbf{k}$$

Similarly,

$$\begin{aligned} \mathbf{R} &= \mathbf{F_1} + \mathbf{F_2} + \mathbf{F_3} \\ \mathbf{R} &= (\mathbf{4i} + \mathbf{7j} + \mathbf{2k}) + (\mathbf{5i} + \mathbf{8j} + \mathbf{3k}) + (\mathbf{6i} + \mathbf{9j} - \mathbf{4k}) \\ &= (\mathbf{4} + \mathbf{5} + \mathbf{6})\mathbf{i} + (\mathbf{7} + \mathbf{8} + \mathbf{9})\mathbf{j} + (\mathbf{2} + \mathbf{3} - \mathbf{4})\mathbf{k} \\ &= \mathbf{15i} + \mathbf{24j} + \mathbf{k} \end{aligned}$$

To determine the magnitude of the resultant force, see the *NCEES Handbook*, Statics section.

The magnitude of the resultant vector is

$$R = \sqrt{x^2 + y^2 + z^2} = \sqrt{15^2 + 24^2 + 1^2} = 28.3$$

The answer is 28.3.

448. The equation for probability is

$$P = \frac{\#\text{ of ways it can succeed}}{\#\text{ of ways it can succeed} + \#\text{ of ways it can fail}}$$

For two events A and B in which neither $P(\text{A})$ nor $P(\text{B})$ is zero, compound probability of both events is given by the equation from the Engineering Probability and Statistics section of the *NCEES Handbook*.

$$P(\text{A, B}) = P(\text{A})P(\text{B}|\text{A}) = P(\text{B})P(\text{A}|\text{B})$$

There are two events: rolling a 5 and rolling a 6. Each event is independent of the other. The probability of each event is 1/6.

$$\text{joint probability} = \left(\frac{1}{6}\right)\left(\frac{1}{6}\right) = \frac{1}{36}$$

The answer is (A).

449. For each coin, there are two possible outcomes: heads (H) or tails (T). Each toss is an independent event. Assume that getting heads is a success and getting tails is a failure. Use the binomial distribution equation from the Engineering Probability and Statistics section of the *NCEES Handbook*.

$$P_n(x) = C(n, x)p^x q^{n-x} = \frac{n!}{x!\,(n-x)!}p^x q^{n-x}$$

Let

$$p(H) = \frac{1}{2}$$
$$p(T) = \frac{1}{2}$$

n is the total number of trials, or 5 coin flips. x is the number of successes, 3 heads. The probability is

$$\begin{aligned} p_5(3) = C(5,3)p^3q^{(5-3)} &= \frac{5!}{3!\,(5-3)!}\left(\frac{1}{2}\right)^3\left(\frac{1}{2}\right)^2 \\ &= \frac{5(4)}{1(2)}\left(\frac{1}{8}\right)\left(\frac{1}{4}\right) \\ &= \frac{5}{16} \end{aligned}$$

The answer is (D).

450. Arrange the data in either descending or ascending order of wage rate. For 100 workers, n is 100. The data in ascending order of wages is shown.

no.	no. of workers	wage ($/hr)	cumulative frequency	notes
1	10	$10.00	10	
2	20	$12.00	$10 + 20 = 30$	
3	20	$15.00	$30 + 20 = 50$	50th ordered value
4	25	$20.00	$50 + 25 = 75$	51st ordered value
5	15	$25.00	$75 + 15 = 90$	
6	10	$30.00	$90 + 10 = 100$	total workers, $n = 100$

Since n is an even number, consider the two middle values: the 50th value is $15 and the 51st value is $20.

The median is the average.

$$\begin{aligned} \text{median} &= \frac{\frac{\$15.00}{\text{hr}} + \frac{\$20.00}{\text{hr}}}{2} \\ &= \$17.50/\text{hr} \end{aligned}$$

The answer is $17.50.

451. Drawing a line between a gift and a bribe may be difficult in certain occasions. One rule is: if a gift is perceived as a bribe, then it is a bribe. The gifts mentioned in options B, C, and D may be important for cultivating business relationships; however, a licensee should report them to his/her employer. A raffle prize is not perceived as a bribe, but rather an incentive to attract conference attendees to see certain products exhibited.

The answer is (A).

452. Statements A, B, and D are correct. Competitive bidding has two categories: Open competitive bidding and closed competitive bidding. In an open competitive bidding, the sealed bids are opened in full view of all who may wish to witness the bid opening. In a closed competitive bidding, the sealed bids are opened in the presence of the authorized personnel only. The terms and conditions of the bidding process generally describe how the bids are opened.

The answer is (C).

453. The grounds for taking disciplinary action against licensees are given in Section 150.10 of the *NCEES Rules of Professional Conduct.*

Specifically, Section 150.10.A(2) states

> *The board shall have the power to suspend, revoke, place on probation, fine, recover costs, and/or reprimand, or to refuse to issue, restore, or renew a license or intern certification to any licensee or intern that is found guilty of:*
>
> *... 2. Any negligence, incompetence, or misconduct in the practice of engineering or surveying ...*

Therefore, the licensee can be disciplined under the terms of Section 150.10.A(2).

The answer is Sec. 150.10.A(2).

454. According to the contract, the owner's budget amount is irrelevant.

The contractor is entitled to collect

$$\begin{aligned} \text{payment} &= (\$2{,}556{,}900)\ (1.05) \\ &= \$2{,}684{,}745 \quad (\$2{,}685{,}000) \end{aligned}$$

The answer is (D).

455. The contract price is based on unit cost and actual crack lineage. The unit cost includes the contractor's profit. Because of the terms of the contract, the owner's estimate and budget are not relevant.

The contract price is the actual cracks lineage (ft) multiplied by the unit repair cost ($/ft).

$$\text{contract price} = (30{,}000\ \text{ft})\ (\$5/\text{ft}) = \$150{,}000$$

The answer is (C).

456. The interest is 12%. Use the net present worth formulas from the *NCEES Handbook* tables given in the Engineering Economics section.

$$\begin{aligned} P &= -\left(\frac{P}{A}, 12\%, 2\right)(\$120{,}000) \\ &\quad +\left(\frac{P}{F}, 12\%, 2\right)\left(\frac{P}{A}, 12\%, 8\right)(\$240{,}000) \\ &= -(1.6901)(\$120{,}000) \\ &\quad +(0.7972)(4.9676)(\$240{,}000) \\ &= \$747{,}629 \quad (\$747{,}000) \end{aligned}$$

The answer is (A).

457. The loan rate is 0.5% monthly. Use the net present worth formulas from the *NCEES Handbook*, Engineering Economics section tables. The loan receipt and payout timeline are shown. In the first 24 months, the engineer draws the loan on a monthly basis. From 25th month on, the engineer starts paying off the loan in monthly payments. Thus, the problem is divided into the two periods.

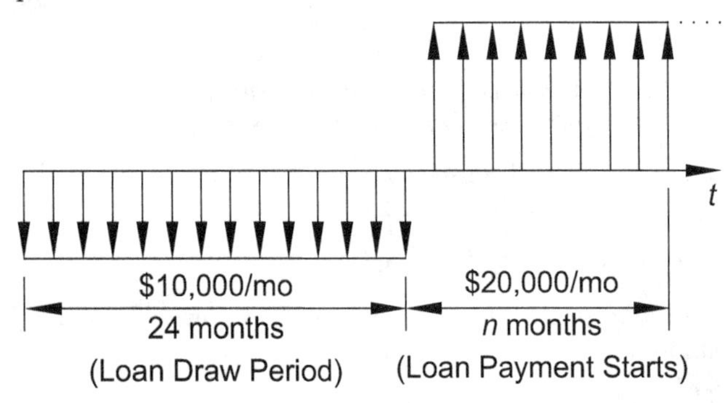

First compute the future value of 24 monthly loan payments received.

$$\begin{aligned} F/A_{24} &= (\$10{,}000)\left(\frac{F}{A}, 0.5{,}24\right) = (\$10{,}000)(25.4320) \\ &= \$254{,}320 \end{aligned}$$

The annual earning of the firm will be $240,000, or $20,000 monthly from 25th month on. To pay the loan as soon as possible, the engineer would pay $20,000 every month until fully paid-off. Use the first payment month as the present period P and compute the number of months.

$$\begin{aligned} (\$20{,}000)\left(\frac{P}{A}, 0.5, n\right) &= \$254{,}320 \\ \left(\frac{P}{A}, 0.5, n\right) &= \frac{\$254{,}320}{20{,}000} = 12.7160 \end{aligned}$$

From the table, the interest rate per period, i, is 0.5.

$$\begin{aligned} \frac{P}{A} &= 12.5562 \quad [\text{for } n = 13] \\ \frac{P}{A} &= 13.4887 \quad [\text{for } n = 14] \quad (\text{governs}) \end{aligned}$$

Therefore, it would take the engineer 14 monthly payments of $20,000 to pay off the debt in full, his 4th year after opening the firm.

The answer is (C).

458. By observation, the break-even exceeds manufacturing 100,000 units. A detailed analysis is given below. The manufacturing cost has three components. Initial set-up cost is fixed regardless of the number of units produced. The labor and material costs are variable, they increase linearly with the number of units produced. The marketing cost is linear up to first 100,000 units. From marketing analysis, the market would saturate after 100,000 units are sold, To sell additional units, the marketing cost would double. The total cost, C, is the sum of the fixed, variable and incremental costs. This can be expressed mathematically as

$$\begin{aligned} C_{\text{total}} &= C_{\text{fixed}} + C_{\text{variable}} + C_{\text{incremental}} \\ &= \$100{,}000 + \left(\frac{\$1.5}{\text{unit}} + \frac{\$0.5}{\text{unit}}\right)(n \text{ units}) \\ &\quad +(\$0.3)(100{,}000 \text{ units}) \\ &\quad +(\$0.6)(n - 100{,}000 \text{ units}) \\ &= \$100{,}000 + \$2n + \$30{,}000 \\ &\quad +\$0.6n - \$60{,}000 \\ &= \$70{,}000 + \$2.6n \end{aligned}$$

$$\text{total income} = \left(\frac{\$2.95}{\text{unit}}\right)(n \text{ units}) = \$2.95n$$

For break-even point, cost equals income.

$$\begin{aligned} \$2.95n &= \$70{,}000 + \$2.6n \\ \$0.35n &= \$70{,}000 \\ n &= \frac{70{,}000}{0.35} \\ &= 200{,}000 \text{ units} \end{aligned}$$

The answer is 200,000 units.

459. For the project to be feasible, its present worth must be positive (greater than zero). Let the annual income from the project be X.

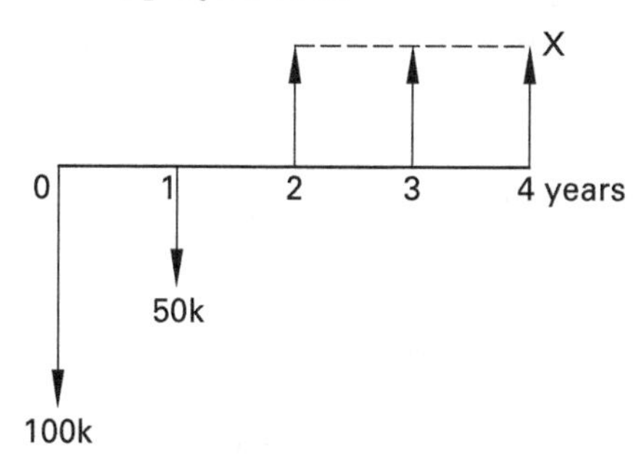

For a 10% annual return, use factor tables. The present worth is

$$\begin{aligned}P &= -\$100{,}000 - \$50{,}000\left(\frac{P}{F},\ 10\%,\ 1\right)\\ &\quad +X\left(\frac{P}{F},\ 10\%,\ 2\right)\\ &\quad +X\left(\frac{P}{F},\ 10\%,\ 3\right) + X\left(\frac{P}{F},\ 10\%,\ 4\right)\\ &= -\$100{,}000 - \$50{,}000(0.9091)\\ &\quad +X(0.8264) + X(0.7513) + X(0.6830)\\ &= -100{,}000 - 45{,}455 + 2.2607X > 0\end{aligned}$$

Therefore, for the present value, P, to be positive,

$$X > \$64{,}341 \quad (\$65{,}000)$$

The answer is (B).

460. The land value and the construction cost are present values. Convert other figures to present values. Both the project life and the annual rate of interest are given. Annual cost/benefits are known. Covert both to present values. Write two equations: one for cost and other for benefits. The present worth of costs is

$$\begin{aligned}C &= \$1{,}000{,}000 + \$11{,}000{,}000 + 500{,}000\left(\frac{P}{A}, 6\%, 25\right)\\ &= \$12{,}000{,}000 + \$500{,}000(12.7834)\\ &= \$18{,}390{,}000\end{aligned}$$

The present worth of benefits is

$$\begin{aligned}B &= (\$1{,}300{,}000)\left(\frac{P}{A}, 6\%, 25\right) = (\$1{,}300{,}000)(12.7834)\\ &= \$16{,}620{,}000\end{aligned}$$

Therefore, the benefit-to-cost ratio is

$$\frac{B}{C} = \frac{\$16{,}620{,}000}{\$18{,}390{,}000} = 0.9$$

The answer is 0.9.

461. See the *NCEES Handbook*, Engineering Economics section for expected value.

$$EV = (C_1)(p_1) + (C_2)(p_2) + \ldots$$

In this case, 99% of concrete conforms and 1% does not.

$$\begin{aligned}EV &= \left(\frac{99}{100}\right)(\$20) + \left(\frac{1}{100}\right)(-\$500)\\ &= \$19.80 - \$5.00\\ &= \$14.80\end{aligned}$$

The answer is $14.80.

462. The equivalent applied loading is the resultant force and accompanying moment of the four applied loads. The reactions induced by the applied loads are not part of the applied loads. The equivalent applied force, P_{total} is the algebraic sum of all applied forces at point A, as shown in the illustration below. Let upward force be positive.

$$\begin{aligned}P_{\text{total}} &= \sum P = 100\text{ kN} - 200\text{ kN} + 400\text{ kN} - 50\text{ kN}\\ &= 250\text{ kN}\end{aligned}$$

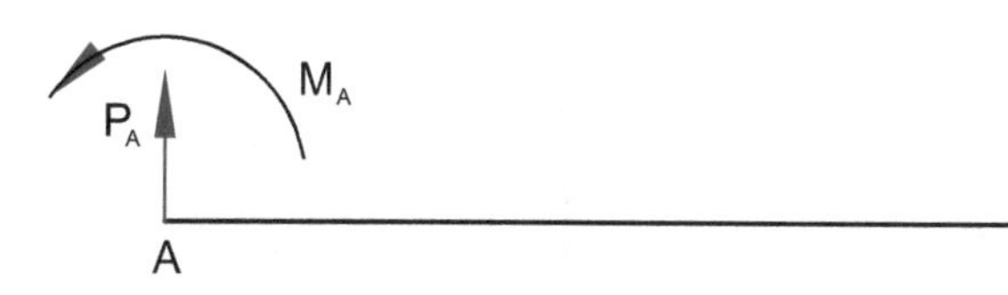

Next, take the moments of all applied loads about A. As the result is positive, the moment is counterclockwise.

$$\begin{aligned}M_A &= \sum Px = (100\text{ kN})(0) - (200\text{ kN})(5\text{ m})\\ &\quad +(400\text{ kN})(8\text{ m}) - (50\text{ kN})(13\text{ m})\\ &= 1550\text{ kN}\cdot\text{m}\end{aligned}$$

The answer is (A).

463. The equivalent applied loading is the resultant force and accompanying moment of the four applied loads. The reactions induced at points B and D by the applied loads are not part of the applied loads. The resultant force is

$$\begin{aligned}\overline{P} &= 100\text{ kN} - 200\text{ kN} + 400\text{ kN} - 50\text{ kN}\\ &= 250\text{ kN}\end{aligned}$$

Its location, x, is determined by taking the moments about A and equating it with an equivalent moment.

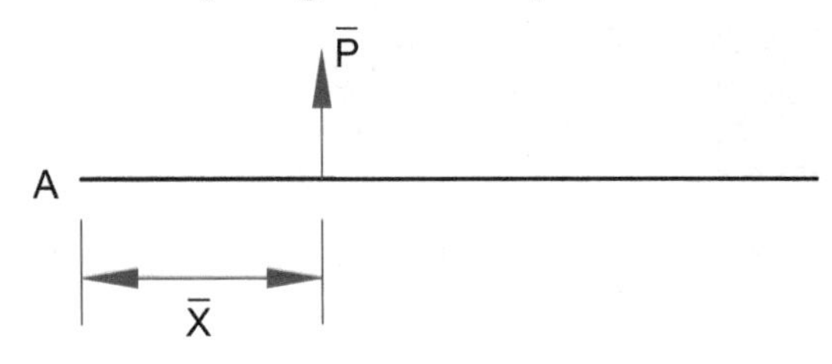

$$\begin{aligned} M_A = \overline{P}\overline{x} &= \sum P_i x_i \\ &= (100\ \text{kN})(0\ \text{m}) - (200\ \text{kN})(5\ \text{m}) \\ &\quad +(400\ \text{kN})(8\ \text{m}) - (50\ \text{kN})(13\ \text{m}) \\ &= 1550\ \text{kN}\cdot\text{m} \end{aligned}$$

$$\begin{aligned} x &= \frac{M_A}{\overline{P}} = \frac{1550\ \text{kN}\cdot\text{m}}{250\ \text{kN}} \\ &= 6.2\ \text{m} \end{aligned}$$

The answer is (A) and (E).

464. See the *NCEES Handbook*, Statics section. Statements A and B are correct. Statement D is the definition of the polar moment of inertia. Statement C is incorrect.

The answer is (C).

465. Construct the free-body diagram.

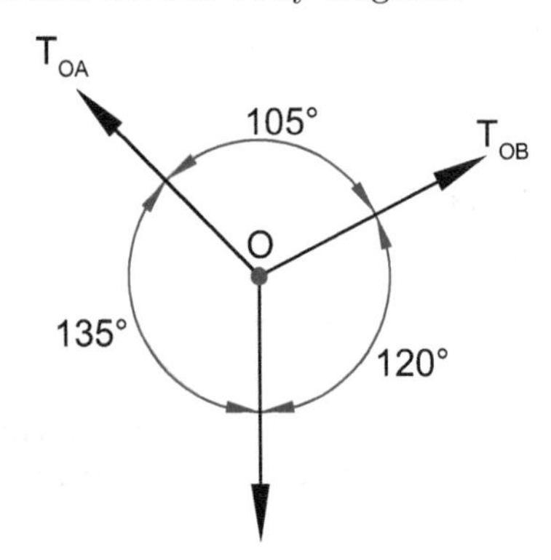

Use Lami's theorem. When a body is in equilibrium under the three coplanar vectors, then each force is proportional to the sine of the angle between the other two. The equation is

$$\frac{F_{\text{OA}}}{\sin 120^\circ} = \frac{F_{\text{OB}}}{\sin 135^\circ} = \frac{500\ \text{N}}{\sin 105^\circ} = \frac{500\ \text{N}}{0.966} = 517.6\ \text{N}$$

Determine F_{OB}.

$$\frac{F_{\text{OB}}}{\sin 135^\circ} = \frac{F_{\text{OB}}}{0.707} = 517.6\ \text{N}$$

Therefore,

$$\begin{aligned} F_{\text{OB}} &= (517.6\ \text{N})(0.707) \\ &= 366\ \text{N} \end{aligned}$$

Determine F_{OA} similarly.

$$\begin{aligned} F_{\text{OA}} &= (517.6\ \text{N})(0.866) \\ &= 448\ \text{N} \end{aligned}$$

The answer is (D) and (E).

466. The degree of indeterminacy for a truss structure is determined by

$$\text{DI} = m + r - 2j$$

The tower is a planar (2-D) truss with two supports. Each support can develop horizontal and vertical reactions.

The number of reactions is

$r = 2(2) = 4$

The number of bars, m, is 17 and the number of joints, j, is 9.

The degree of indeterminacy is

$$\text{DI} = 17 + 4 - 2(9) = 3$$

The answer is 3.

467. The average force equals the change in momentum per unit time. It is expressed as

$$F = \frac{d(mv)}{dt} = ma$$

In this case, the ball decelerated to a complete stop (that is, v_2 is 0). By definition

$$a = \frac{v_2 - v_1}{t} = \frac{30\ \frac{\text{m}}{\text{s}} - 0\ \frac{\text{m}}{\text{s}}}{0.1\ \text{s}} = 300\ \text{m/s}^2$$

Therefore,

$$\begin{aligned} F = ma &= (0.149\ \text{kg})\left(300\ \frac{\text{m}}{\text{s}^2}\right) \\ &= 44.7\ \text{N} \quad (45\ \text{N}) \end{aligned}$$

The answer is (B).

468. The truss tower is a rigid body, and its reactions can be determined using the equilibrium conditions.

$$\sum F_x = 0$$
$$\sum F_y = 0$$
$$\sum M = 0$$

The applied vertical load is 500 kN. Therefore, the sum of the vertical reactions at A and B should be 500 kN. Due to symmetry, the reactions at supports A and B are equal. Therefore, both reactions are 250 kN.

The answer is 250 kN.

469. First, determine the moment about point D.

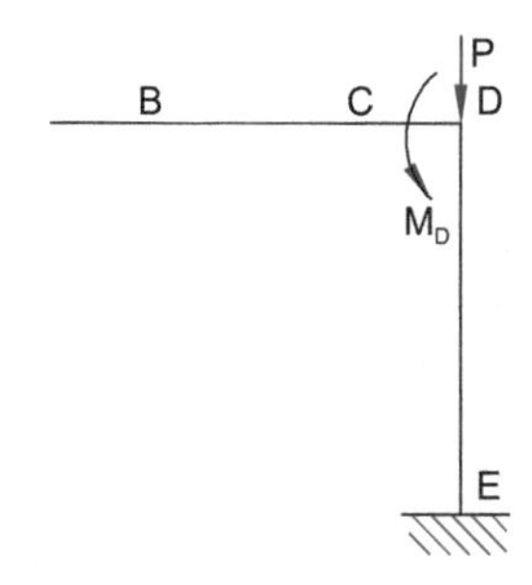

$$M = FD$$
$$M_D = (1 \text{ kip})(3 \text{ ft}) + (1 \text{ kip})(14 \text{ ft}) = 17 \text{ ft-kips}$$

Next, transfer the force system from D to E. Force P at D is axial in the DE direction and does not induce any moment at E. Thus, the moment M_D will remain unchanged when transferred from D to E. It is acting counterclockwise, therefore, the moment is positive.

$$M_D = 17 \text{ ft-kips}$$

The answer is 17 ft-kips.

470. Draw the equilibrium diagram for point B.

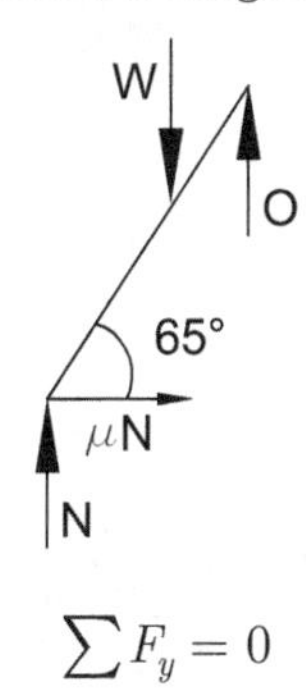

$$\sum F_y = 0$$

Therefore,

$$N = W = 200 \text{ lbf}$$

By definition, the force of friction is the coefficient of friction multiplied by the normal force.

$$F_{\text{Friction}} = \mu N = \mu(200 \text{ lbf})$$

Given the geometry,

$$\frac{N}{\mu N} = \tan 65°$$
$$\mu = \frac{1}{\tan 65°} = 0.47 \quad (0.5)$$

The answer is (D).

471. The general angular velocity-acceleration equation is

$$\omega^2 = \omega_0{}^2 + 2\alpha_0(\theta - \theta_0)$$

By definition,

$$\omega_0 = 2\pi N = 2\pi \left(\frac{900 \frac{\text{rev}}{\text{min}}}{60 \frac{\text{sec}}{\text{min}}} \right) = 30\pi \text{ rad/sec}$$

Given that the initial acceleration is -10 rad/sec^2 and the initial angle is 0, the general equation becomes

$$0 = \left(30\pi \ \frac{\text{rad}}{\text{sec}^2} \right)^2 + 2\left(-10 \ \frac{\text{rad}}{\text{sec}^2} \right)(\theta - 0)$$
$$\theta = 45\pi^2$$

The number of revolutions is

$$\text{rev} = \frac{1}{2\pi}(45\pi^2) = 70.65 \text{ revolutions} \quad (70 \text{ revolutions})$$

The answer is (A).

472. The torsional frequency is given by the expression

$$\omega_n = \sqrt{\frac{GJ}{IL}}$$

For steel,

$$G = 80 \text{ GPa} = 80 \times 10^9 \text{ Pa}$$

Given that

$$L = 1 \text{ m}$$
$$J = \frac{\pi d^4}{32} = \frac{\pi(0.2 \text{ m})^4}{32} = 157 \times 10^{-6} \text{ m}^4$$

Since

$$\begin{aligned} R &= 1 \text{ m} \\ I &= mR^2 \\ &= (1000 \text{ kg})(1 \text{ m})^2 \\ &= 1000 \text{ kg} \cdot \text{m}^2 \end{aligned}$$

Substituting the above values,

$$\begin{aligned} \omega_n &= \sqrt{\frac{(80 \times 10^9 \text{ Pa})(157 \times 10^{-6} \text{m}^4)}{(1000 \text{ kg} \cdot \text{m}^2)(1 \text{ m})}} \\ &= 112.1 \frac{\text{rad}}{\text{sec}} = \frac{112.1}{2\pi} \text{ Hz} \\ &= 17.85 \text{ Hz} \quad (18 \text{ Hz}) \end{aligned}$$

The answer is (B).

473. The total kinetic energy, KE, of a roller includes its KE of translation and KE of rotation.

$$E_{\text{total}} = \frac{1}{2} m\text{v}^2 + \frac{1}{2} I\omega^2$$

The kinetic energy of translation (motion) depends on the mass of the entire body in motion and its velocity. The mass is 10,000 kg. The velocity is

$$\text{v} = 10 \frac{\text{km}}{\text{hr}} = \frac{10{,}000 \frac{\text{m}}{\text{hr}}}{3600 \frac{\text{s}}{\text{hr}}} = 2.78 \text{ m/s}$$

Therefore, the kinetic energy of translation is

$$E_1 = \frac{1}{2} m\text{v}^2 = \frac{1}{2}(10{,}000 \text{ kg})\left(2.78 \frac{\text{m}}{\text{s}}\right)^2 = 38{,}580 \text{ N·m}$$

The kinetic energy of rotation depends on the mass moments of inertia of the rotating parts and their angular velocity. In this problem, the masses of the front-axle wheel and rear-axle wheel and their angular velocities are equal.

$$\begin{aligned} I &= \sum Mr^2 \\ &= (5000 \text{ kg} + 5000 \text{ kg})(0.5 \text{ m})^2 \\ &= 2500 \text{ kg·m}^2 \end{aligned}$$

The wheel radius is

$$r_{\text{wheel}} = \frac{1.1 \text{ m}}{2} = 0.55 \text{ m}$$

The angular velocity is

$$\omega = \frac{\text{v}}{r} = \frac{2.78 \frac{\text{m}}{\text{s}}}{0.55 \text{ m}} = 5.05 \text{ rad/s}$$

Determine the angular energy.

$$E_2 = \frac{1}{2} I\omega^2 = \frac{1}{2}(2500 \text{ kg·m}^2)\left(5.05 \frac{\text{rad}}{\text{s}}\right)^2 = 31{,}885 \text{ N·m}$$

The total energy is

$$\begin{aligned} E_{\text{total}} &= E_1 + E_2 = 38{,}580 \text{ N·m} + 31{,}885 \text{ N·m} \\ &= 70{,}465 \text{ N·m} \quad (70{,}000 \text{ N·m}) \end{aligned}$$

The answer is (D).

474. The wheel's kinetic energy will be transformed into potential energy as it rolls up the slope. This can be expressed as

$$mgh = \frac{1}{2} m\text{v}^2 + \frac{1}{2} I\omega^2$$

Since the tire mass is known to be well-distributed, its mass moment of inertia is

$$I = \frac{1}{2} mR^2$$

By definition, angular velocity is

$$\omega = \frac{\text{v}}{R}$$

The translational velocity is

$$\text{v} = \left(100 \frac{\text{km}}{\text{hr}}\right)\left(\frac{1 \text{ hr}}{3600 \frac{\text{s}}{\text{hr}}}\right)\left(\frac{1000 \text{ m}}{1 \text{ km}}\right) = 27.78 \text{ m/s}$$

Substitute the value into the energy equation.

$$mgh = \frac{1}{2} m\text{v}^2 + \frac{1}{2}\left(\frac{mR^2}{2}\right)\left(\frac{\text{v}}{R}\right)^2 = \frac{3m\text{v}^2}{4}$$

The change in elevation is

$$h = \frac{3v^2}{4g} = \frac{3\left(27.78\ \frac{\text{m}}{\text{s}}\right)^2}{4\left(9.81\ \frac{\text{m}}{\text{s}^2}\right)} = 59\ \text{m}$$

The horizontal distance, L, is

$$L = \frac{h}{\text{slope}} = \frac{59\ \text{m}}{0.05} = 1180\ \text{m}$$

The answer is (C).

475. The shaft is carrying a uniformly distributed load (UDL) of 250 N/m (0.25 kN/m). The bending moment under a UDL occurs at midspan. Recall that it is given by

$$M_{\text{max}} = \frac{wL^2}{8} = \frac{\left(250\ \frac{\text{N}}{\text{m}}\right)(6\ \text{m})^2}{8} = 1125\ \text{N·m}$$

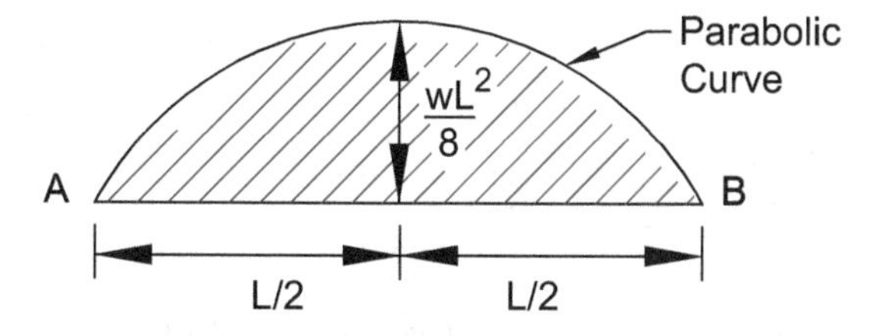

See the *NCEES Handbook*, Mechanics of Materials section, for the bending stress formula. The maximum bending stress of a shaft section with moment M is

$$\sigma_{max} = \frac{Mc}{I}$$

The equation for the moment of inertia is

$$I = \frac{\pi r^4}{4}$$

The distance from the neutral axis to the outermost fiber is

$$c = r = \frac{200\ \text{mm}}{2} = 100\ \text{mm} = 0.1\ \text{m}$$

The moment of inertia for the steel shaft is

$$I = \frac{\pi(0.1\ \text{m})^4}{4} = 0.785 \times 10^{-4}\,\text{m}^4$$

The beam's maximum bending stress is

$$\begin{aligned}\sigma_{\text{max}} &= \frac{(1125\ \text{N·m})(0.1\ \text{m})}{0.785\ \times 10^{-4}\ \text{m}^4} \\ &= 1.43 \times 10^6\ \frac{\text{N}}{\text{m}^2} = 1.43 \times 10^6\ \text{Pa} \\ &= 1.43\ \text{MPa} \quad (1.4\ \text{MPa})\end{aligned}$$

The answer is (B).

476. To determine whether the bar is in equilibrium, establish an axis parallel to the centerline on the bar, as shown.

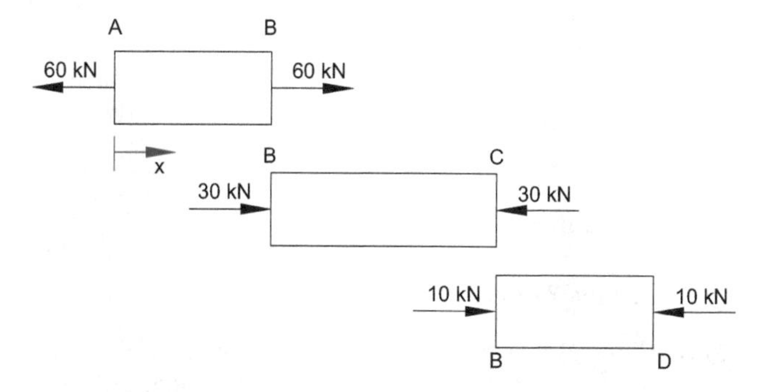

Determine if the sum of the forces in one direction equals the sum of the forces in the other direction. The forces are 60 kN, 20 kN, and 10 kN in one direction, totaling 90 kN, which equals the force acting in the other direction at B. Therefore, the bar is in equilibrium. To determine the force in each segment, start from each end of the bar and determine the segments needed to balance the end force. Then, determine the equilibrium in the middle zone.

The force in segment BC is

$$\begin{aligned}F_{\text{BC}} &= 20\ \text{kN} + 10\ \text{kN} \\ &= 30\ \text{kN}\end{aligned}$$

The answer is (C).

477. The axial force distribution along the bar length is shown.

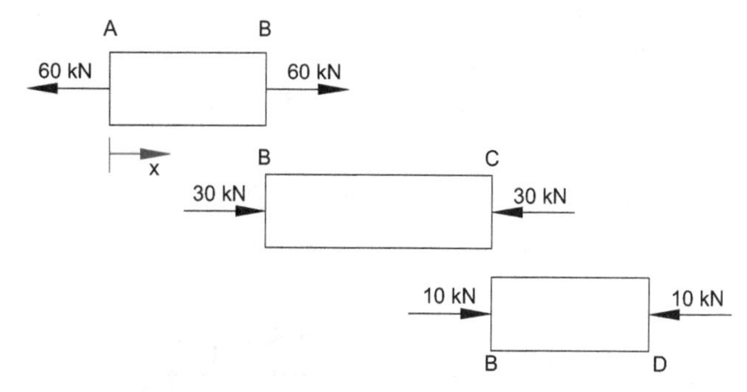

Segment AB is in tension, and segments BC and BD are in compression. Let tension forces be positive and compression forces be negative. The change in length of the bar is

$$(\Delta L)_{AD} = \Sigma\frac{PL}{AE} = \left(\frac{PL}{AE}\right)_{AB} + \left(\frac{PL}{AE}\right)_{BC} + \left(\frac{PL}{AE}\right)_{BD}$$
$$= \frac{1}{AE}\left((PL)_{AB} + (PL)_{BC} + (PL)_{BD}\right)$$

Substitute the values into the equation. AE is 1 for all segments.

$$(\Delta L)_{AD} = \frac{1}{AE}\left((PL)_{AB} + (PL)_{BC} + (PL)_{BD}\right)$$
$$= \left(\frac{1}{1000 \text{ kN}}\right)\left((60 \text{ kN})(1 \text{ m}) + (-30 \text{ kN})(1.2 \text{ m}) + (-10 \text{ kN})(2.2 \text{ m})\right)$$
$$= 2 \text{ mm}$$

The answer is 2 mm.

478. Plane B is at the interface of two segments. Plane B is subjected to two loads, 15,000 lbf each, totaling 30,000 lbf. The area at the interface is 150 in^2. Use the expression given in the Mechanics of Materials section of the *NCEES Handbook* to compute the axial stress on a section.

$$\sigma = \frac{P}{A} = \frac{30{,}000 \text{ lbf}}{150 \text{ in}^2}$$
$$= 200 \text{ lbf/in}^2 \quad (200 \text{ psi})$$

The answer is 200 psi.

479. The truss is symmetrical. Therefore, both members resist the load equally. From statics, the vertical and horizontal forces at joint B are in equilibrium. The horizontal components are equal and opposite. The vertical components are given by

$$\sum F_y = 2F_{AB}\left(\frac{4}{5}\right) - 50 \text{ kips} = 0$$
$$F_{AB} = \frac{(50 \text{ kips})5}{8} = 31.25 \text{ kips}$$

For steel, the modulus of elasticity is 29,000 ksi.

The elongation along member length is

$$\Delta L = \frac{PL}{AE} = \frac{(31.25 \text{ kips})(30 \text{ ft})\left(12 \frac{\text{in}}{\text{ft}}\right)}{(14.7 \text{ in}^2)\left(29{,}000 \frac{\text{kips}}{\text{in}^2}\right)}$$
$$= 0.0264 \text{ in}$$

Using the geometrical relation, the vertical deflection is

$$(\Delta L)_y = \frac{\Delta L}{\cos\theta} = \Delta L\left(\frac{5}{4}\right) = 0.0264 \text{ in}\left(\frac{5}{4}\right)$$
$$= 0.033 \text{ in}$$

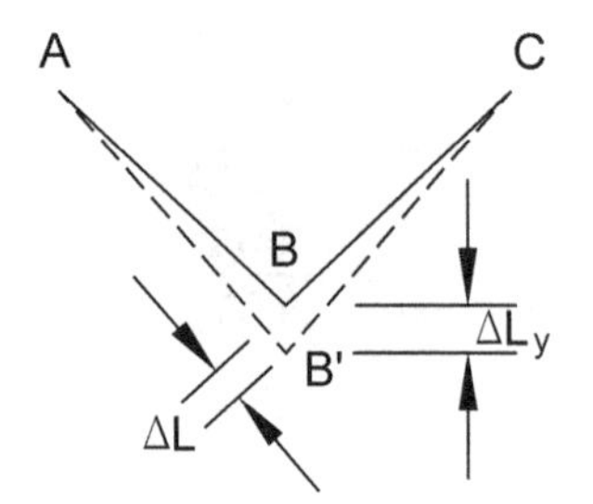

The answer is (D).

480. See the table in the *NCEES Handbook*, Statics section for polar moment of inertia. The angle of rotation (radians) of the tube is given by

$$\phi = \frac{TL}{GJ}$$

The shear modulus, G, for steel is 11 Mpsi. The polar moment of inertia, J, for a hollow tube is given by

$$J = \frac{\pi(d_2{}^4 - d_1{}^4)}{32}$$

The outer diameter of the tube, d_2, is 12 in. The inner diameter of the tube, d_1, can be found by the difference between the outer diameter and twice the wall thickness.

$$d_1 = d_2 - 2(T_{\text{wall}}) = 12 \text{ in} - 2(1 \text{ in}) = 10 \text{ in}$$

The polar moment of inertia is

$$J = \frac{\pi\left((12 \text{ in})^4 - (10 \text{ in})^4\right)}{32} = 1053 \text{ in}^4$$

The angle of twist is

$$\phi = \frac{(100 \text{ ft-kips})\left(12 \frac{\text{in}}{\text{ft}}\right)(10 \text{ ft})\left(12 \frac{\text{in}}{\text{ft}}\right)}{\left(11 \times 10^3 \frac{\text{kips}}{\text{in}^2}\right)(1053 \text{ in}^4)} = 0.0124 \text{ rad}$$
$$= \frac{0.0124 \text{ rad}(180^\circ)}{\pi}$$
$$= 0.71^\circ \quad (0.7^\circ)$$

The answer is (C).

481. See the *NCEES Handbook*, Mechanics of Materials section. It is given that the load is being applied gradually and therefore is not an impact load. The stress level is low, and therefore the load is in the elastic range. The strain energy of the bar is the area under its stress-strain diagram. The strain energy for a bar of volume V is given by

$$U = \frac{\sigma^2}{2E}V$$

By definition, volume is

$$V = A \times L$$

The area of the steel bar is

$$A = \frac{\pi d^2}{4} = \frac{\pi(50\ \text{mm})^2}{4} = 625\pi\ \text{mm}^2$$

The volume of the bar is

$$V = (625\pi\ \text{mm}^2)(4000\ \text{mm}) = 2.5\pi \times 10^6\,\text{mm}^3$$

Young's modulus of elasticity for steel is

$$E = 200\ \text{GPa} = 200 \times 10^9\ \frac{\text{N}}{\text{m}^2} = 2 \times 10^5\ \frac{\text{N}}{\text{mm}^2}$$

The stress on the steel bar is

$$\sigma = \frac{P}{A} = \frac{50{,}000\ \text{N}}{625\pi\ \text{mm}^2} = 25.5\ \text{N/mm}^2$$

The energy absorbed into the bar is

$$U = \frac{\sigma^2}{2E}V = \frac{\left(25.5\ \frac{\text{N}}{\text{mm}^2}\right)^2}{2\left(2 \times 10^5\ \frac{\text{N}}{\text{mm}^2}\right)}(2.5\pi \times 10^6\,\text{mm}^3)$$
$$= 12{,}767\ \text{N·mm} = 12.77\ \text{N·m} = 12.77\ \text{J} \quad (12\ \text{J})$$

The answer is (C).

482. The fixed-end moments at supports A and B are given in the *NCEES Handbook*, Civil Engineering/Structural Analysis section. At the plastic stage, the collapse mechanism of the beam is shown.

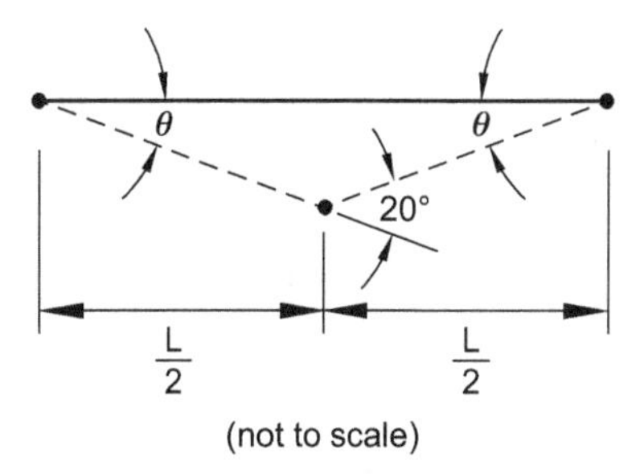

Owing to symmetry, the end plastic hinges are equal and are represented by rotation θ. Thus, the rotation at the midspan would be 2θ.

The work performed by the total external load is the load multiplied by the average deflection.

At nominal capacity of the beam, let

$$w = w_n$$
$$M = M_n$$

The moment around the midspan is

$$M_n(\theta + 2\theta + \theta) = w_n L\left(\frac{1}{2} \times \theta \times \frac{L}{2}\right)$$

Therefore,

$$M_n = \frac{w_n L^2}{16}$$

The answer is 16.

483. The concrete mix strength increases as the water/cement ratio is decreased.

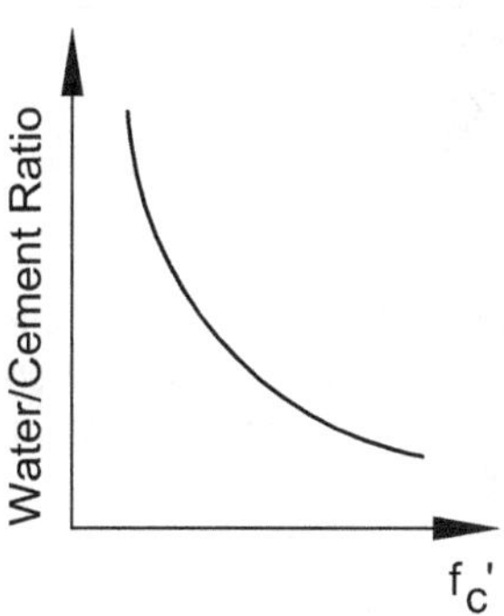

The concrete mix strength increases as the entrained air is decreased. Both relationships are plotted in the *NCEES Handbook*, Materials Science/Structure of Matter section.

The answer is (A).

484. The slump test is performed using a 12 in tall slump cone mold. The slump is the difference between the mold height and the resulting concrete height upon removal of the mold.

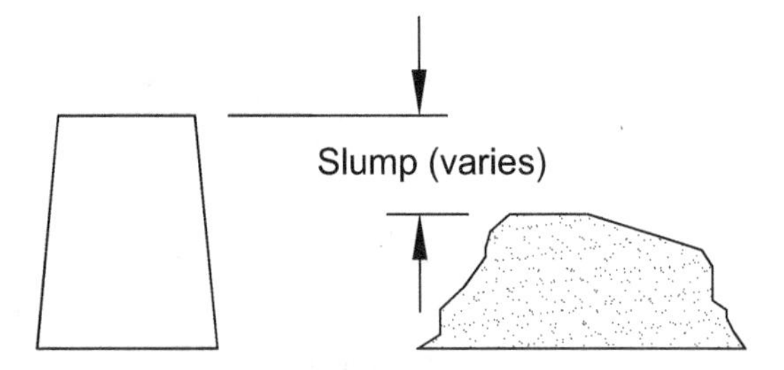

A concrete mix that does not show an appreciable slump is called *a stiff mix.* On the other hand, a mix that has a large drop is termed a *watery* or *soupy mix.* The maximum slump for a concrete mix is nearly 11 inches.

The answer is (C).

485. The voids in the mix are of two types. Some are filled with the asphalt binder, while others are filled with air. The question concerns the voids that have been filled with the binder versus those not filled with the binder.

The percentage of voids filled by the asphalt binder (VFA) for a mixture is defined as

$$\begin{aligned}\text{VFA} &= \frac{\text{volume of binder filling the voids}}{\text{total volume of voids}}\\ &= \frac{V_b}{V_b + V_v}(100\%)\end{aligned}$$

Given that the void volume filled with the binder, V_b, is 0.12 ft^3 and the void volume unfilled, V_v, is 0.04 ft^3, the percentage of voids filled by the asphalt binder is

$$\begin{aligned}\text{VFA} &= \frac{V_b}{V_b + V_v} = \frac{0.12 \text{ ft}^3}{0.12 \text{ ft}^3 + 0.04 \text{ ft}^3}(100\%)\\ &= 75\%\end{aligned}$$

The answer is (C).

486. The garage is located in a severe exposure zone. Entrained air is required for durable concrete that can resist the freezing and thawing cycles. Concrete should have at least 6% entrained air in it. To be effective, the air voids should be well distributed and not clustered.

The answer is (C) and (E).

487. When there is no well-defined yielding point, the yield strength is defined at the 0.2% strain offset. To determine the yield stress, draw a line at 0.2% strain, running parallel to the initial stress-strain curve. The intersection of the line and the stress-strain curve is defined as the yield stress of the material, as shown.

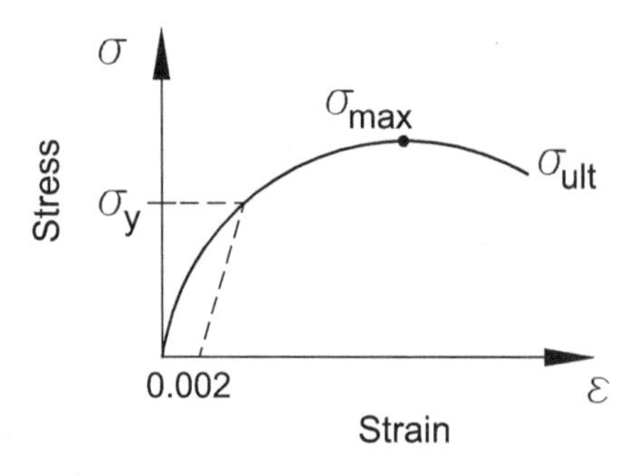

Options A, B, and C are true and are given in the uniaxial stress-strain section of the *NCEES Handbook,* Materials Science/Structure of Matter section. The true stress of a metal is

$$\text{true stress} = \frac{\text{load}}{\text{actual cross-sectional area}}$$

True stress differs from the engineering stress, as the cross-sectional area reduces and "necks" after yielding.

The answer is (D).

488. The Reynolds number is dimensionless and is expressed as

$$\text{Re} = \frac{\text{v}D}{\nu}$$

According to the *NCEES Handbook*, the flow is considered laminar if its Reynolds number is less than 2100. The velocity at which the flow changes from viscous (or laminar) to turbulent is called the critical velocity. In calculating the Reynolds number, it is important that units are consistent, pertaining to only one system.

The pipe diameter, D, is 1 in or 2.54 cm (0.0254 m). The kinematic viscosity, v, is given as 0.0003 m^2/s, and the flow velocity is 10 m/s. Substituting into the Reynolds number equation, the Reynolds number of the pipe is

$$\begin{aligned}\text{Re} &= \frac{\left(10 \ \frac{\text{m}}{\text{s}}\right)(0.0254 \text{ m})}{0.0003 \ \frac{\text{m}^2}{\text{s}}}\\ &= 846 \quad (850)\end{aligned}$$

The answer is (B).

489. See the *NCEES Handbook*, Fluid Mechanics section for the energy equation. It is given that the head is constant. Since energy is consumed by the turbine, the energy equation becomes

$$\frac{p_A}{\gamma} + \frac{v_A^2}{2g} + z_A = \frac{p_C}{\gamma} + \frac{v_C^2}{2g} + z_C + H_{\text{turbine}}$$

The velocity at point A (v_A) is zero, and both points A and C are subject to atmospheric pressure.

$$p_A = p_C = 0$$

Assume point C as the datum.

$$z_C = 0$$

Therefore,

$$z_A = 10 \text{ m}$$

Simplifying the energy equation, the water head energy consumed by the turbine is

$$0 + 0 + 10 \text{ m} = \frac{\left(5 \ \frac{\text{m}}{\text{s}}\right)^2}{2g} + H_{\text{turbine}}$$

$$H_{\text{turbine}} = 10 \text{ m} - \frac{\left(5 \ \frac{\text{m}}{\text{s}}\right)^2}{2\left(9.81 \ \frac{\text{m}}{\text{s}^2}\right)} = 8.73 \text{ m}$$

The area of the pipe is

$$A = \pi r^2 = \pi\left(\frac{0.15 \text{ m}}{2}\right)^2 = 0.0177 \text{ m}^2$$

The volumetric flow rate of water through the turbine is

$$Q = Av = (0.0177 \text{ m}^2)\left(5 \ \frac{\text{m}}{\text{s}}\right) = 0.088 \text{ m}^3/\text{s}$$

Use the fluid power formula.

$$\dot{W} = \rho g H Q = Q\gamma H$$

The specific weight of water at ordinary temperature and pressure is 9.81 kN/m^3.

See the *NCEES Handbook*, Units and Conversion Factors section. 1 kN·m/s is equal to 1.34 hp.

The power consumed by the turbine is

$$\dot{W} = \left(0.088 \ \frac{\text{m}^3}{\text{s}}\right)\left(9.81 \ \frac{\text{kN}}{\text{m}^3}\right)(8.73 \text{ m})\left(1.34 \ \frac{\text{hp}}{\frac{\text{kN}\cdot\text{m}}{\text{s}}}\right) = 10.1 \text{ hp} \quad (10 \text{ hp})$$

The answer is (B).

490. Use the *NCEES Handbook*, Fluid Mechanics section. The force exerted by the jet in the x-direction is

$$-F_x = -Q\rho(v_1 - v)(1 - \cos\alpha)$$
$$F_x = Q\rho(v_1 - v)(1 - \cos\alpha)$$

The above equation is based on the principle

$$\begin{aligned}\text{force} &= (\text{mass})(\text{acceleration}) \\ &= (\text{mass flow per second}) \\ &\quad (\text{change of velocity in the direction of force})\end{aligned}$$

With the vane moving, calculate the relative velocity.

$$v_{\text{rel}} = v_1 - v = 300 \ \frac{\text{ft}}{\text{sec}} - 100 \ \frac{\text{ft}}{\text{sec}} = 200 \text{ ft/sec}$$

The discharge rate from the pipe is

$$Q = Av = \left(\frac{\pi d^2}{4}\right)\left(300 \ \frac{\text{ft}}{\text{sec}}\right) = \left(\frac{\pi}{4}(1 \text{ in})\left(\frac{1 \text{ ft}}{12 \text{ in}}\right)^2\right)\left(300 \ \frac{\text{ft}}{\text{sec}}\right) = 1.64 \text{ ft}^3/\text{sec}$$

If the force is to be measured in pounds, the mass must be in slugs.

Mass flow rate per second is

$$Q(\rho) = \left(1.64 \ \frac{\text{ft}^3}{\text{sec}}\right)\left(\frac{62.4 \ \frac{\text{lbm}}{\text{ft}^3}}{32.2 \ \frac{\text{lbm}}{\text{slug}}}\right) = 3.17 \text{ slugs/sec}$$

The force in the x-direction is

$$F_x = \left(3.17 \ \frac{\text{slugs}}{\text{sec}}\right)\left(200 \ \frac{\text{ft}}{\text{sec}}\right)(1 - \cos 60°) = 317 \text{ lbf}$$

To compute the power of the turbine in horsepower units, use the *NCEES Handbook*, Units and Conversion Factors section.

$$\begin{aligned} W = vF_x &= (317\ \text{lbf})\left(100\ \frac{\text{ft}}{\text{sec}}\right) \\ &= \left(31{,}700\ \frac{\text{ft-lbf}}{\text{sec}}\right)\left(\frac{1\ \text{hp}}{550\ \frac{\text{ft-lbf}}{\text{sec}}}\right) \\ &= 57.6\ \text{hp} \quad (58\ \text{hp}) \end{aligned}$$

The answer is (D).

491. The words *torque* and *torsion* are used interchangeably. From the *NCEES Handbook*, Mechanics of Materials section, by definition, the torsion, or torque, T, is the shear stress multiplied by the area and lever arm.

The area is

$$\begin{aligned} A &= \text{circumferential area of cylinder} \\ &= 2\pi(\text{radius})(\text{length}) \\ &= 2\pi(0.04\ \text{m})(0.06\ \text{m}) \\ &= 0.01507\ \text{m}^2 \end{aligned}$$

The lever arm is

$$LA = \text{shaft radius} = 0.04\ \text{m}$$

From the torque equation, determine the shear stress, τ.

$$\begin{aligned} T = 2\ \text{N·m} &= (\tau)(0.01507\ \text{m}^2)(0.04\ \text{m}) \\ &= (\tau)(6.032\times 10^{-4}\ \text{m}^3) \\ \tau &= 3317.85\ \text{N/m}^2 \end{aligned}$$

Since the shear stress at the interface is caused by the viscosity of the fluid, use the *NCEES Handbook*, Fluid Mechanics section.

The shear stress can be found by

$$\tau = \mu\frac{dv}{dy}$$

For Newtonian fluid film, simplify the above equation.

$$\begin{aligned} \tau &= \mu\frac{dv}{dy} = \mu\frac{v}{\delta} \\ &= \mu\left(\frac{\text{velocity of shaft on film}}{\text{thickness of fluid film}}\right) \\ &= \mu\left(\frac{r\times\omega}{0.001\ \text{m}}\right) \end{aligned}$$

The shaft is rotating. Therefore, it has an angular velocity, ω. Convert the angular velocity to linear velocity.

$$v = r\times\omega = (\text{shaft radius})(\text{angular velocity})$$

$$\tau = \mu\frac{(0.04\ \text{m})\left(1000\ \frac{\text{rad}}{\text{s}}\right)}{0.001\ \text{m}} = 40{,}000\mu = 3317.85\ \text{N/m}^2$$

The viscosity is

$$\mu = \frac{3317.85\ \frac{\text{N}}{\text{m}^2}}{40{,}000\ \frac{\text{rad}}{\text{s}}} = 0.083\ \text{N·s/m}^2$$

The answer is 0.083 N·s/m².

492. The water pressure on the dam has two components: vertical and horizontal. The vertical pressure is due to the weight of water on the curved area of the dam. The horizontal pressure is the lateral fluid thrust.

To determine the volume of the water, use the area table in the *NCEES Handbook*, Mathematics section.

The vertical component of water pressure is equal to the weight of volume ABC with 1 ft run along the dam.

The volume is

$$\begin{aligned} V_{ABC} &= \left(\frac{2}{3}\right)(\text{circumscribing rectangle})(1\ \text{ft}) \\ &= \left(\frac{2}{3}\right)(30\ \text{ft})(60\ \text{ft})(1\ \text{ft}) \\ &= 1200\ \text{ft}^3 \end{aligned}$$

The unit weight of water is 62.4 lbf/ft^3.

The vertical component of the water weight acting as downward pressure per foot of the dam is

$$P_V = W_{\text{water}} = (1200\ \text{ft}^3)\left(62.4\ \frac{\text{lbf}}{\text{ft}^3}\right) = 74{,}880\ \text{lbf}$$

See the *NCEES Handbook*, Horizontal Stress Profiles and Forces section, to determine the hydraulic lateral pressure on a retaining wall.

The horizontal component of the pressure is the lateral hydrostatic pressure per 1 ft length of the dam.

$$\begin{aligned} P_H &= \left(\frac{\gamma_w H^2}{2}\right)(1\ \text{ft}) = (1\ \text{ft})\left(62.4\ \frac{\text{lbf}}{\text{ft}^3}\right)\frac{(60\ \text{ft}^2)}{2} \\ &= 112{,}320\ \text{lbf} \end{aligned}$$

The resultant pressure is

$$P = \sqrt{(P_V)^2 + (P_H)^2} = \sqrt{(74{,}880 \text{ lbf})^2 + (112{,}320 \text{ lbf})^2}$$
$$= 134{,}991 \text{ lbf} \quad (135 \text{ kips})$$

The inclination of the resultant force, P, to the horizontal axis is

$$\theta = \tan^{-1}\frac{P_V}{P_H} = \frac{74{,}880}{112{,}320} = \tan^{-1}0.667$$
$$= 33.69° \quad (34°)$$

The answer is 135 kips and 34°.

493. Calculate the Reynolds number for the pole.

The kinematic viscosity is

$$\nu = 1.6 \times 10^{-5} \text{ m}^2/\text{s}$$

The wind velocity is

$$\text{v} = \left(65 \frac{\text{km}}{\text{h}}\right)\frac{(1000 \text{ m})}{1 \text{ km}}\left(\frac{1 \text{ hr}}{3600 \text{ s}}\right) = 18.06 \text{ m/s}$$

The Reynolds number is

$$\text{Re} = \frac{\text{v}D}{\nu} = \frac{\left(18.06 \frac{\text{m}}{\text{s}}\right)(0.2 \text{ m})}{1.47 \times 10^{-5} \frac{\text{m}^2}{\text{s}}} = 2.46 \times 10^5$$

Use the illustration given in the Fluid Mechanics section of the *NCEES Handbook* to determine the drag coefficient for spheres, disks, and cylinders. The corresponding factor is 1.0.

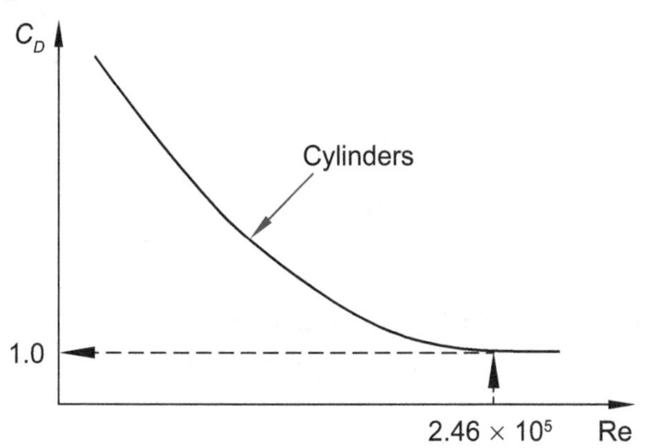

The answer is 1.0.

494. Use the law of sines because one side and two angles of a triangle are given. The convention (lowercase for sides and uppercase for angles) for the law of sines is shown.

The law of sines can be represented as

$$\frac{a}{\sin \text{A}} = \frac{b}{\sin \text{B}} = \frac{c}{\sin \text{C}}$$

Angle C can be found using

$$180° = \angle\text{A} + \angle\text{B} + \angle\text{C}$$
$$\angle\text{C} = 180° - \angle\text{A} - \angle\text{B} = 180° - 55° - 45°$$
$$= 80°$$

The length of AB is

$$\overline{\text{AB}} = c = 120 \text{ m}$$

Use the formula

$$\frac{b}{\sin \text{B}} = \frac{c}{\sin \text{C}}$$
$$b = \sin \text{B}\left(\frac{c}{\sin \text{C}}\right) = \frac{(\sin 45°)(120 \text{ m})}{\sin 80°}$$
$$= 86.15 \text{ m}$$

The answer is (A).

495. A township is described as 6 miles by 6 miles less convergence. Townships are divided into 36 sections. Therefore, each section is 1 square mile in area.

$$\text{area} = \left(\frac{1}{2}\right)\left(\frac{1}{4}\right)\left(\frac{1}{4}\right)\left(\frac{1}{4}\right) = \frac{1}{128} \text{ mi}^2$$
$$= \frac{1}{128} \text{ mi}^2\left(\frac{640 \text{ acres}}{1 \text{ mi}^2}\right)$$
$$= 5 \text{ acres}$$

The answer is 5 acres.

496. Latitude and longitude are imaginary lines that help us survey every place on the surface of the earth. The most important line of latitude is the equator, which runs east-west around the fattest part of the earth. The most important longitude line is the Prime Meridian, which runs north-south and goes through Greenwich, England. The primary unit in which longitude and latitude are given is degrees (°). There are 360° of longitude (180° E and 180° W) and 180° of latitude (90° N and 90° S). Because the earth is a sphere, locations can be precisely measured in degrees, using an angle centered in the earth's core.

The road runs north-south along a meridian. Second, its coordinates change from north to south, implying that the road crosses the equator. The road length is an arc on the surface of the earth. Use the *NCEES Handbook*, Civil Engineering: Horizontal Curves section. The length of the curve is

$$L = RI\frac{\pi}{180}$$

The intersection angle, I, is

$$\begin{aligned} I &= (1°10'20'') + (5°20'23'') \\ &= 6°30'43'' \\ &= 6° + \left(\frac{30'}{\left(\frac{60'}{1°}\right)}\right) + \left(\frac{43''}{\left(\frac{3600''}{1°}\right)}\right) \\ &= 6.5119° \end{aligned}$$

The road length is

$$\begin{aligned} L &= (20{,}906{,}000 \text{ ft})(6.5119°)\left(\frac{\pi}{180}\right)\left(\frac{1 \text{ mile}}{5280 \text{ ft}}\right) \\ &= 450 \text{ miles} \end{aligned}$$

The answer is 450 miles.

497. See the *NCEES Handbook*, Civil Engineering section. The stationing of the route generally follows the arc of the curve in laying out the roads. Determine the arc length subtended by an intersection angle of 66° and a radius of 350 ft.

$$\begin{aligned} L &= RI\left(\frac{\pi}{180}\right) \\ &= \frac{(350 \text{ ft})(66°)\,\pi}{180} = 403.17 \text{ ft} \end{aligned}$$

Find the PT station of the curve.

$$\begin{aligned} \text{PT location} &= \text{sta } 10{+}67.50 + \text{sta } 4{+}03.17 \\ &= \text{sta } 14{+}70.67 \end{aligned}$$

The answer is (A).

498. See the *NCEES Handbook*, Civil Engineering section. The given data pertains to a sag curve because its back slope is negative and its forward slope is positive. A typical sag curve is shown.

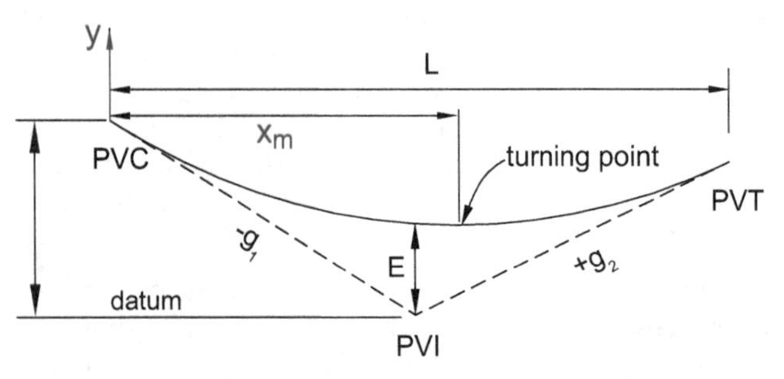

The curve is a parabola and begins at the PVC at

$$\begin{aligned} \text{PVC} &= \text{PVI} - 0.5L = (12 \text{ sta} + 50) - (5 \text{ sta} + 00) \\ &= 7 \text{ sta} + 50 \end{aligned}$$

The elevation of the PVC is

$$\begin{aligned} Y_{\text{PVC}} &= Y_{\text{PVI}} - \frac{g_1 L}{2} = 2540 \text{ ft} - \frac{(-0.02)(1000 \text{ ft})}{2} \\ &= 2550 \text{ ft} \end{aligned}$$

To find the elevation at x, follow these steps.

step 1: Find the number of stations from the PVI to point x.

$$x = 12.50 \text{ sta} - 10 \text{ sta} = 2.5 \text{ sta}$$

step 2: Determine the parabolic curve constant.

$$a = \frac{g_2 - g_1}{2L} = \frac{4\% - (-2\%)}{2(10 \text{ sta})} = 0.3$$

step 3: Find the curve elevation at point $x = 2.5$ sta on the curve from the PVC.

$$\begin{aligned} Y_{\text{PVC}|\,2.5} &= Y_{\text{PVC}} + g_1 x + ax^2 \\ &= 2550 \text{ ft} + (-2)(2.5 \text{ sta}) + (0.3)(2.5 \text{ sta})^2 \\ &= 2546.875 \quad (2547 \text{ ft}) \end{aligned}$$

The answer is (D).

499. The problem involves a sag curve, as shown.

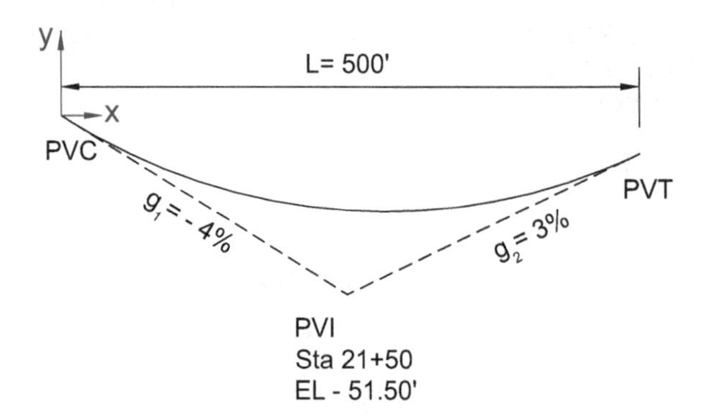

The formulas for both crest and sag type vertical curves are the same. The location of the PVI station is determined using the following relationships.

$$\begin{aligned} \text{PVC station} &= \text{PVI} - 0.5L \\ &= (\text{sta } 21{+}50) - (0.5)(500 \text{ ft}) \\ &= (\text{sta } 21{+}50) - (\text{sta } 2{+}50) \\ &= \text{sta } 19{+}00 \end{aligned}$$

$$\begin{aligned}\text{PVT station} &= \text{PVI} + 0.5L \\ &= (\text{sta } 21{+}50) + (0.5)(500 \text{ ft}) \\ &= (\text{sta } 21{+}50) + (\text{sta } 2{+}50) \\ &= \text{sta } 24{+}00\end{aligned}$$

Determine the tangent elevation.

$$\text{tangent elevation} = Y_{\text{PVC}} + g_1 x = Y_{\text{PVI}} + g_2\left(x - \frac{L}{2}\right)$$

$$\begin{aligned}Y_{\text{PVC}} &= Y_{\text{PVI}} - g_1\left(\frac{L}{2}\right) \\ &= 51.50 \text{ ft} - (-0.04)\left(\frac{500 \text{ ft}}{2}\right) \\ &= 61.50 \text{ ft}\end{aligned}$$

$$\begin{aligned}Y_{\text{PVT}} &= Y_{\text{PVI}} + g_2\left(x - \frac{L}{2}\right) \\ &= 51.50 \text{ ft} + (0.03)\left(500 \text{ ft} - \frac{500 \text{ ft}}{2}\right) \\ &= 59.00 \text{ ft}\end{aligned}$$

The answer is (A).

500. The rainfall-runoff is

$$\text{runoff} = \text{total precipitation} - \text{losses}$$

The NRCS rainfall-runoff method assumes 20% losses due to depression storage, evaporation, and interception in the runoff. This is reflected in the numerator of the equation shown.

$$Q = \frac{(P - 0.2S)^2}{P + 0.8S}$$

The term P denotes the total precipitation in inches, and S denotes the storage capacity.

The answer is (C).

501. The diameter of the pipe is 36 inches, and the depth of flow is 27 inches. Determine the diameter ratio as

$$\frac{d}{D} = \frac{3}{4} = 0.75$$

step 1: See the Fluid Mechanics section of the *NCEES Handbook*. The hydraulic radius is

$$R_H = \frac{\text{cross-sectional area}}{\text{wetted perimeter}}$$

In this case, the wetted perimeter, P, is an arc length. The cross-sectional area of the flow is the area of a partial circle. Use the Hydraulic-Elements Graph for Circular Sewers given in the Hydrology/Water Resources section of the *NCEES Handbook*. For a diameter ratio of 0.75, the corresponding radii ratio is

$$\frac{R}{R_f} = 1.21$$
$$R = (1.21)(R_f)$$

step 2: The factor, R_f, is the hydraulic radius of the pipe running full. Find the hydraulic radius.

$$\begin{aligned}R_{H,\text{full}} &= \frac{\text{cross-sectional area}}{\text{wetted perimeter}} \\ &= \frac{\pi r^2}{2\pi r} \\ &= \frac{r}{2} \\ &= \frac{D}{4} \\ &= \frac{36 \text{ in}}{4} \\ &= 9 \text{ in}\end{aligned}$$

step 3: Determine the hydraulic radius for the pipe running ¾ full.

$$R_H = (1.21)(9 \text{ in}) = 10.89 \text{ in} \quad (11 \text{ in})$$

The answer is (B).

502. Find the storage to be depleted.

$$\text{storage depleted} = (0.95)(200{,}000 \text{ gal}) = 190{,}000 \text{ gal}$$

The gross outflow rate is 800 gpm and the inflow rate is 300 gpm. Find the net outflow rate.

$$\text{outflow rate} = 800 \frac{\text{gal}}{\text{min}} - 300 \frac{\text{gal}}{\text{min}} = 500 \frac{\text{gal}}{\text{min}}$$

Determine the time needed to empty 95% of the tank.

$$\begin{aligned}t &= \frac{\text{volume of water}}{\text{outflow rate}} \\ &= \frac{190{,}000 \text{ gal}}{500 \frac{\text{gal}}{\text{min}}} \\ &= (380 \text{ min})\left(\frac{1 \text{ hr}}{60 \text{ min}}\right) \\ &= 6.33 \text{ hr} \quad (6 \text{ hr})\end{aligned}$$

The answer is (C).

503. The energy equation for incompressible flow is

$$\frac{p_1}{\gamma} + \frac{v_1^2}{2g} + z_1 = \frac{p_2}{\gamma} + \frac{v_2^2}{2g} + z_2 + h_f + h_{f,\text{fitting}}$$

The difference in elevation of two reservoirs connected by a pipe equals the losses in the pipe line and in the entry and exit fittings. List the parameters.

$$\begin{aligned} h_f &= 36 \text{ ft} \\ h_{f,\text{fitting}} &= 1.5 \text{ ft} + 4 \text{ ft} + 7 \text{ ft} = 12.5 \text{ ft} \\ p_A &= v_A = 0 \\ z_A &= 180 \text{ ft} \\ p_B &= v_B = 0 \end{aligned}$$

At both points A and B, the water velocity is zero, and the pressure heads are equal (i.e., open to air).

Substitute the values in the energy equation.

$$\begin{aligned} 0 \text{ ft} + 0 \text{ ft} + 180 \text{ ft} &= 0 \text{ ft} + 0 \text{ ft} + z_B + 36 \text{ ft} + 12.5 \text{ ft} \\ z_B &= 131.5 \text{ ft} \quad (130 \text{ ft}) \end{aligned}$$

The answer is (B).

504. The maximum height that can be used in siphoning is limited by the absolute pressure needed to avoid separation of dissolved gases. In this case, separation occurs at an absolute pressure of 8 ft of water head, which is less than the atmospheric pressure of 34 ft of water. The difference can be used to raise the pipe elevation above point A. Apply Bernoulli's energy equation to points A and C.

$$\begin{aligned} \text{energy at point A} &= \text{energy at point C} \\ &\quad + \text{losses between A and C} \end{aligned}$$

$$\frac{p_A}{\gamma} + \frac{v_A^2}{2g} + z_A = \frac{p_C}{\gamma} + \frac{v_C^2}{2g} + z_C + h_f + h_{f,\text{fitting}}$$

Express the height as the difference between the heights of point C and A.

$$h = z_C - z_A$$

In order to maximize the siphoning height, the water pressure at point C should be reduced as much as possible without causing separation of the gases from the water. The water velocity at point A is practically zero and at point C it is 3 ft/sec.

Substitute the values into the energy equation.

$$\begin{aligned} 34 \text{ ft} &+ 0 \text{ ft} + z_A \\ &= 8 \text{ ft} + \frac{\left(3 \frac{\text{ft}}{\text{sec}}\right)^2}{(2)\left(32.2 \frac{\text{ft}}{\text{sec}^2}\right)} + z_C + 15 \text{ ft} \\ &\quad + 1 \text{ ft} + 0.1 \text{ ft} \\ h &= z_C - z_A \\ &= 9.76 \text{ ft} \quad (10 \text{ ft}) \end{aligned}$$

The answer is (A).

505. Determine the flow in two parallel pipes. Two conditions to meet are

1. Friction loss in the 6 in diameter pipe should be equal to the friction loss in the 10 in diameter pipe.
2. The total flow of the 6 in and 10 in pipes should add up to equal the flow of the 12 in pipe.

These conditions are mathematically described in the Fluid Mechanics section of the *NCEES Handbook*. The friction loss in a pipe is given by the Darcy-Weisbach equation.

$$h_f = f\frac{L}{D}\frac{v^2}{2g}$$

Write the equation after equating the friction losses in both pipes.

$$h_f = f\frac{(L_{10})(v_{10})^2}{(D_{10})(2)(g)} = f\frac{(L_6)(v_6)^2}{(D_6)(2)(g)}$$

Because the friction coefficient is constant, the Moody diagram is not needed to solve this problem. Since the lengths of both branches of the pipe are equal, the friction loss equations can be simplified.

$$\begin{aligned} \frac{v_{10}^2}{D_{10}} &= \frac{v_6^2}{D_6} \\ v_{10} &= \sqrt{\frac{10}{6}}\, v_6 \\ &= 1.29 v_6 \end{aligned}$$

Substitute the values in the parallel pipe flow conditions.

$$\begin{aligned} &\left(\frac{\pi}{4}\right)\left(\frac{6 \text{ in}}{12}\right)^2 v_6 \\ &+ \left(\frac{\pi}{4}\right)\left(\frac{10 \text{ in}}{12}\right)^2 v_{10} = 18 \text{ ft}^3/\text{sec} \end{aligned}$$

$$0.2v_6 + (0.55)(1.29)v_6 = 18 \text{ ft}^3/\text{sec}$$
$$v_6 = 19.8 \text{ ft}^3/\text{sec}$$
$$Q_6 = v_6 A_6$$
$$= \left(19.8 \frac{\text{ft}^3}{\text{sec}}\right)(0.2)$$
$$= 3.96 \text{ ft}^3/\text{sec} \quad (4 \text{ ft}^3/\text{sec})$$

The answer is (B).

506. Statements A, B, and C are correct. Statement D is incorrect because the hydraulic radius is one-fourth of the hydraulic diameter of a pipe running full.

$$R_H = \frac{\text{hydraulic diameter}}{4}$$

The answer is (D).

507. Use the log equation to find the noise pollution.

$$\text{SPL}_{\text{total}} = 10\log_{10}\sum 10^{\text{SPL}/10}$$
$$= 10\log_{10}\sum\left(10^{\frac{60}{10}} + 10^{\frac{70}{10}} + 10^{\frac{80}{10}}\right)$$
$$= 10\log_{10}\sum(10^6 + 10^7 + 10^8)$$
$$= 10\log_{10}(1.11 \times 10^8)$$
$$= (10)(8.045)$$
$$= 80.45 \text{ dB} \quad (80 \text{ dB})$$

The answer is (B).

508. See the *NCEES Handbook* section Units and Conversion Factors. Evaporation is measured in reduction of water depth in a pan, container, or reservoir per unit time per unit surface area of the water.

Put some water in a wide pan and measure the drop in its depth over a specified period of time. Find the area of the pan.

$$A = \frac{\pi D^2}{4}$$
$$= \frac{\pi(3 \text{ ft})^2}{4}$$
$$= 7.07 \text{ ft}^2$$

Find the rate of evaporation.

$$\text{rate of evaporation} = \frac{\text{surface evaporation volume}}{\text{time in hours}}$$
$$= \frac{4 \text{ in} + 2 \text{ in}}{12 \text{ hr}}$$
$$= 0.5 \text{ in/hr}$$

Determine evaporation per unit surface area.

$$\text{evaporation rate per unit surface area} = \frac{0.5 \text{ in/hr}}{7.07 \text{ ft}^2}$$
$$= 0.0707 \text{ in/hr/ft}^2$$

The answer is 0.0707 in/hr/ft².

509. A meander, in general, is a bend in a sinuous watercourse or river. A meander forms when moving water in a stream erodes the outer banks (the concave side) and widens its valley. The inner part of the river (the convex side) has less energy and deposits silt, as shown.

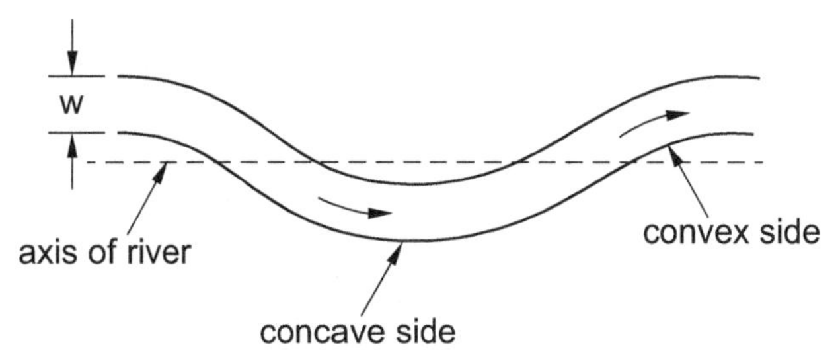

Option A is correct. The lower Mississippi is an example of a meandering alluvial river.

The answer is (A).

510. Use the *NCEES Handbook*, Engineering Probability and Statistics section. Use the laws of probability. The first law of probability states that "the probability $P(E)$ of an event E is a real number in the range of 0 to 1. The probability of an impossible event is 0 and that of an event certain to occur is 1."

step 1: Let, the probability of recurrence of an event in one year $= p$.

step 2: The probability of nonrecurrence of the event in that year $= 1 - p$.

step 3: It is known that flood events are independent. The event in one year is not dependent on events in another year. Therefore, the probability of nonrecurrence of the event in 100 years $= (1 - p)^{100}$.

step 4: Equate the probability of nonrecurrence with the design criteria of 90% probability and solve the equation.

$$(1 - p)^{100} = 90\% = 0.9$$
$$(1 - p) = (0.9)^{\frac{1}{100}}$$
$$= (0.9)^{0.01}$$
$$= 0.99895$$
$$p = 1 - 0.99895$$
$$= 0.00105$$

step 5: If the probability is p, then the recurrence interval of the event is its inverse.

$$\frac{1}{p} = \frac{1}{0.00105} = 950 \text{ years}$$

Therefore, for the dam to survive 100 years with 90% probability, the design flood recurrence should be 1 in 950 years.

The answer is 1 in 950 years.

511. The member forces in both members AC and BC are compressive. Consider the equilibrium at joint C.

$$\sum F_x = F_{\text{CB}} \cos 60° + F_{\text{CA}} \cos 30° = 0$$
$$\sum F_y = F_{\text{CB}} \sin 60° + F_{\text{CA}} \sin 30° = 30$$

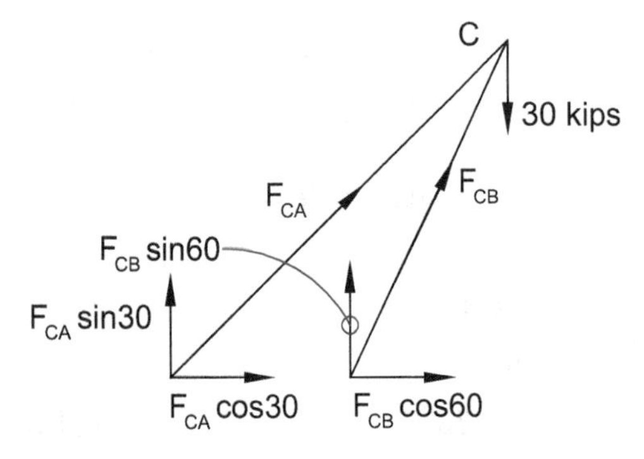

Isolate F_{CB} from the F_x equation.

$$F_{\text{CB}} \cos 60° = -\ F_{\text{CA}} \cos 30°$$
$$F_{\text{CB}} = \frac{-F_{\text{CA}}(0.866)}{0.5}$$
$$= -1.73\ F_{\text{CA}}$$

Substitute the value of F_{CB} into the F_y equation.

$$-1.73 F_{\text{CA}} \sin 60° + F_{\text{CA}} \sin 30° = 30$$
$$(-1.73)(0.866) F_{\text{CA}} + (0.5) F_{\text{CA}} = 30$$
$$F_{\text{CA}} = -30 \text{ kips}$$

The negative sign indicates that the member CA is in tension. Find F_{CB}.

$$F_{\text{CB}} = -(1.73)(-30 \text{ kips})$$
$$= 51.9 \text{ kips} \quad (52 \text{ kips, compression})$$

The answer is (D).

512. See the table Simply Supported Beam Slopes and Deflections in the Mechanics of Materials section of the *NCEES Handbook*. The equation for the deflection of a beam carrying a uniformly distributed load (UDL) is

$$y_{\max} = \frac{5wL^4}{384EI} = \frac{(5)\left(480 \frac{\text{lbf}}{\text{ft}}\right)(180 \text{ in})^4}{\left(12 \frac{\text{in}}{\text{ft}}\right)(384)\left(2 \times 10^6 \frac{\text{lbf}}{\text{in}^2}\right) I} = 1 \text{ in}$$

$$I = 273.4 \text{ in}^4$$

For a rectangular section, the moment of inertia is

$$I = \frac{bd^3}{12}$$
$$\frac{(3.5 \text{ in})\, d^3}{12} = 273.4 \text{ in}^4$$
$$d = \frac{(273.4 \text{ in}^4)(12)}{3.5 \text{ in}}$$
$$= 9.79 \text{ in}$$

In order to control the deflection, select the beam depth that is larger than the computed depth. Use a 12 in nominal depth beam, which has 11.5 in actual depth.

The answer is (C).

513. See the *NCEES Handbook*, Design of Steel Components, W Shapes Dimensions and Properties table. The area of W10 × 49 is 14.4 in^2.

Its slenderness ratio as given in the table is

$$\frac{KL}{r} = 120$$

Use the critical stress table to find the available stress.

$$\phi_c F_{\text{cr}} = 15.7 \text{ ksi}$$

Determine the compressive strength of the column.

$$\phi P_n = \phi_c F_{\text{cr}} A = (15.7 \text{ ksi})(14.4 \text{ in}^2) = 226 \text{ kips}$$

Using the factored load combination, find the live load.

$$U = 1.2D + 1.6L$$
$$226 \text{ kips} = (1.2)(100 \text{ kips}) + 1.6L$$
$$L = 66 \text{ kips} \quad (65 \text{ kips})$$

The answer is (A).

514. The propping force is an upward load that is applied at the shaft's midspan, as shown. It is in the direction opposite to the applied load. The upward deflection is

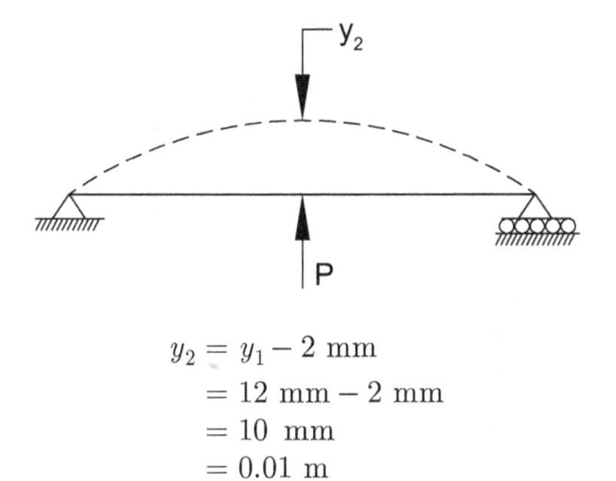

$$\begin{aligned} y_2 &= y_1 - 2 \text{ mm} \\ &= 12 \text{ mm} - 2 \text{ mm} \\ &= 10 \text{ mm} \\ &= 0.01 \text{ m} \end{aligned}$$

Midspan deflection of a beam subjected to a point load at midspan is

$$\begin{aligned} y_2 &= \frac{PL^3}{48EI} \\ P &= \frac{48EIy_2}{L^3} \\ &= \frac{(48EI)(0.01 \text{ m})}{L^3} \end{aligned}$$

From the Typical Material Properties table in the *NCEES Handbook*, the modulus of elasticity, E, for steel is 200 GPa. The force needed for propping up the beam is

$$\begin{aligned} P &= \frac{(48)\left(200 \times 10^9 \ \frac{\text{N}}{\text{m}^2}\right)(0.55 \times 10^{-3} \text{ m}^4)(0.01 \text{ m})}{(12 \text{ m})^3} \\ &= 30.56 \times 10^3 \text{ N} \quad (30 \text{ kN}) \end{aligned}$$

The answer is (B).

515. See the moving concentrated load set principle stated in the Structural Analysis section of the *NCEES Handbook*. The maximum moment in the girder occurs when the resultant loads and an adjacent load from the set are equidistant to the beam centerline.

The resultant load, R, is 40 kips, and its location relative to other loads is shown.

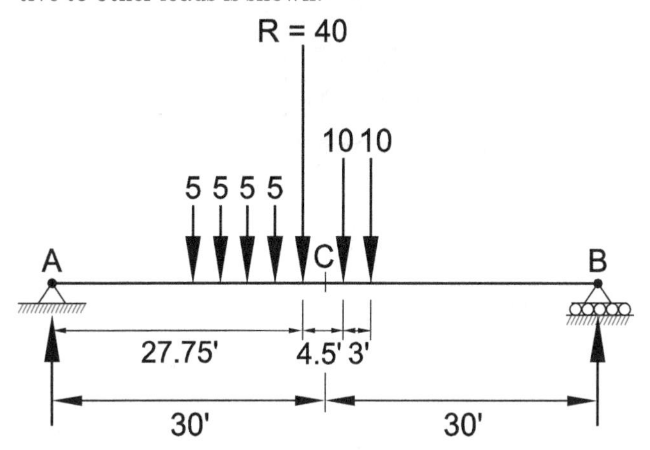

There are two loads adjacent to the resultant force: a 5 kip load and a 10 kip load. The maximum moment in the girder occurs when the resultant load and an adjacent load from the set are equidistant to the beam centerline. Therefore, load R and the 10 kip load to its right should be positioned at

$$30 \text{ ft} + \frac{4.5 \text{ ft}}{2} = 32.25 \text{ ft}$$

Thus, the distance from the outer 10 kip wheel to support A is

$$32.25 \text{ ft} + 3 \text{ ft} = 35.25 \text{ ft} \quad (35 \text{ ft})$$

The answer is (D).

516. See the principle of moving concentrated load sets in the Structural Analysis section of the *NCEES Handbook*. The absolute maximum bending moment in the girder occurs under one of the wheel loads adjacent to the resultant (i.e., either the second 10 kip wheel load or the first 5 kip wheel load). Since the load resultant, midspan, and the 10 kip wheel load on the right are not adjacent wheel load locations, options A, B, and C are incorrect. The 10 kip wheel load on the left is the only adjacent load listed in the options. Therefore, the absolute maximum bending moment occurs under the 10 kip wheel load on the left.

The answer is (D).

517. Determine the tributary area of the column at each floor level. The tributary area of a column is the area adjoining the column, which sheds its load to the column, as shown.

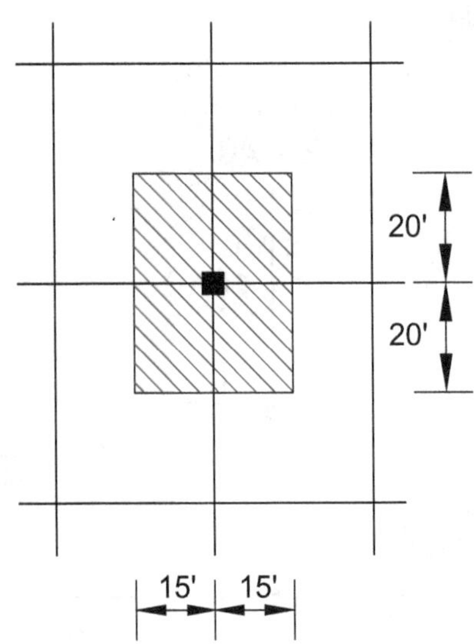

The tributary area of an interior column is

$$(30 \text{ ft})(40 \text{ ft}) = 1200 \text{ ft}^2$$

Find the service dead load, D, on the interior column.

$$\begin{aligned} D &= (\text{floor unit DL})(\text{tributary area})(\text{no. of floors}) \\ &= \left(120 \ \frac{\text{lbf}}{\text{ft}^2}\right)(1200 \text{ ft}^2)(3) \\ &= 432{,}000 \text{ lbf} \\ &= 432 \text{ kips} \quad (400 \text{ kips}) \end{aligned}$$

The answer is (B).

518. Use AISC Table 3-10 (W Shapes Available Moment vs. Unbraced Length) to find the available moment.

$$\phi M_n = 366 \text{ ft-kips}$$

Since the table is based on the factor $C_b = 1$, and the given $C_b = 1.67$, the design strength should be multiplied by 1.67.

$$\begin{aligned} \phi M_n &= (1.67)(366 \text{ ft-kips}) \\ &= 611 \text{ ft-kips} \end{aligned}$$

The answer is (B).

519. The capacity of the connection depends on the plates being connected and the bolts connecting them. For three bolts in single shear, the available capacity is

$$\phi r_n = (3)(7.95 \text{ kips}) = 23.9 \text{ kips}$$

Since the capacity of the bolts is less than the plate capacity of 126.3 kips, the bolt capacity controls the design.

The answer is (B).

520. See the *NCEES Handbook*, Civil Engineering Design of Reinforced Concrete Components (ACI 318-14) section. The concrete's 28 day strength is 5 ksi, the steel yield stress is 60 ksi, and the beam width is 14 in.

Find the effective depth.

$$d_t = 32 \text{ in} - 3 \text{ in} = 29 \text{ in}$$

Determine the ratio of depth of rectangular stress block a to depth to neutral axis c.

$$\begin{aligned} \beta_1 &= 0.85 - (0.05)\left(\frac{f'_c - 4000}{1000}\right) \\ &= 0.85 - (0.05)\left(\frac{5000 \text{ ksi} - 4000 \text{ ksi}}{1000}\right) \\ &= 0.8 \end{aligned}$$

The maximum allowable steel area is

$$\begin{aligned} A_{s,\,\text{max}} &= \left(\frac{(3)(29 \text{ in})}{7}\right)\left(\frac{(0.85)(5 \text{ ksi})(0.8)(14 \text{ in})}{60 \text{ ksi}}\right) \\ &= 9.86 \text{ in}^2 \quad (10 \text{ in}^2 \text{ or ten \#9 bars}) \end{aligned}$$

The answer is (C).

521. Use the formulas given in the Design of Reinforced Concrete Components subsection of the Civil Engineering section of the *NCEES Handbook.*

$$\begin{aligned} \beta_1 &= 0.85 - (0.05)\left(\frac{f'_c - 4000}{1000}\right) \\ &= 0.85 - (0.05)\left(\frac{6000 - 4000}{1000}\right) \\ &= 0.75 \\ a &= \frac{A_s f_y}{0.85 b f'_c} \\ &= \frac{(3 \text{ in}^2)(60 \text{ ksi})}{(0.85)(14 \text{ in})(6 \text{ ksi})} \\ &= 2.52 \text{ in} \end{aligned}$$

The equation relating the depth of the equivalent rectangular block, and the distance from the extreme compression fiber to the neutral axis, is

$$a = \beta_1 c$$
$$c = \frac{a}{\beta_1} = \frac{2.52 \text{ in}}{0.75} = 3.36 \text{ in}$$

Calculate the steel tensile strain.

$$\varepsilon_t = \frac{0.003(d_t - c)}{c} = \frac{(0.003)(29 \text{ in} - 3.36 \text{ in})}{3.36 \text{ in}} = 0.023$$

The strain diagram is

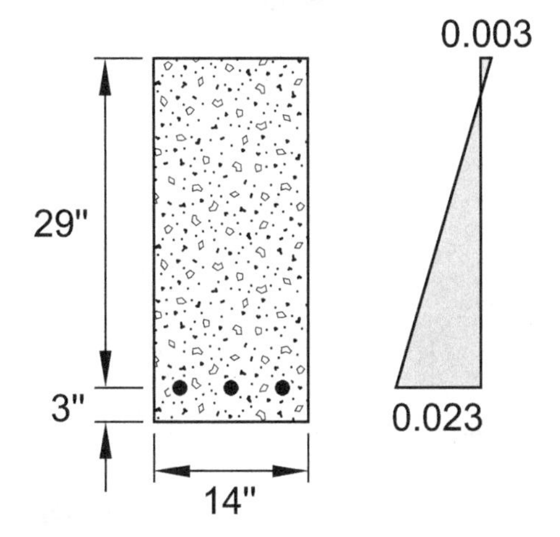

The answer is (C).

522. The deflection of a flexural member is inversely proportional to the modulus of elasticity of its material. The larger the modulus, the smaller the deflection. The ACI concrete modulus of elasticity formula given in the Design of Reinforced Concrete Components subsection of the Civil Engineering section of the *NCEES Handbook* is

$$E_c = 33 w_c^{1.5} \sqrt{f'_c}$$
$$= (33)\left(150 \ \frac{\text{lbf}}{\text{ft}^3}\right)^{1.5}\left(\sqrt{6000 \ \frac{\text{lbf}}{\text{ft}^2}}\right)$$
$$= 4{,}696{,}000 \text{ psi} \quad (4696 \text{ ksi})$$

Similarly, the modulus of elasticity for 3000 psi is

$$E_c = (33)\left(150 \ \frac{\text{lbf}}{\text{ft}^3}\right)^{1.5}\left(\sqrt{3000 \ \frac{\text{lbf}}{\text{ft}^2}}\right)$$
$$= 3{,}321{,}000 \text{ psi} \quad (3321 \text{ ksi})$$

The beams have identical size and moments of inertia. Therefore, the deflection ratio is

$$\text{deflection ratio} = \frac{\text{deflection of as-built beam}}{\text{deflection of designed beam}} = \frac{E_{\text{design}}}{E_{\text{as-built}}} = \frac{4{,}696 \ \frac{\text{lbf}}{\text{ft}^2}}{3{,}321 \ \frac{\text{lbf}}{\text{ft}^2}} = 1.41 \quad (1.5)$$

The answer is (C).

523. See the *NCEES Handbook*, Civil Engineering section, Geotechnical subsection.

$$\text{sand (\%)} = \text{(percent passing through sieve no. 10)} - \text{(percent passing through sieve no. 200)}$$

$$\text{percent passing sieve no. 10} = 100\% - (0\% + 15\%) = 85\%$$
$$\text{percent passing sieve no. 200} = 100\% - (15\% + 22\% + 25\% + 10\%) = 28\%$$
$$\text{sand} = 85\% - 28\% = 57\%$$

The answer is (C).

524. The statements in options A, B, and C are correct. The higher the group index (GI), the less suitable is the soil. For example, A-6 with a GI equal to 15 is less suitable than A-6 with a GI equal to 10. In other words, the higher the GI of a soil, the thicker the sub-base required to carry a given traffic loading. Option D is incorrect.

The answer is (D).

525. The void ratio is the ratio of the volume of voids to the volume of solids in a given volume of material. The soil grains are incompressible. Therefore, their mass and volume remain the same at any void ratio. The porosity is the ratio of the volume of voids to the total volume of a soil mass.

$$\text{porosity} = n = \frac{V_V}{V} = \frac{e}{e+1}$$
$$\text{void ratio, } e = \frac{V_V}{V_S}$$

Since porosity is always less than the void ratio, options C and D are incorrect.

step 1: Find the weight of water in the sample that weighs 10 lbf when wet and 9.25 lbf when dried out.

$$W_w = 10.00 \text{ lbf} - 9.25 \text{ lbf} = 0.75 \text{ lbf}$$

The volume of solids is

$$\begin{aligned} V_S &= \frac{\text{weight of solids}}{\text{density of solids}} = \frac{W_S}{\gamma_S} \\ &= \frac{W_S}{(G_S)(\gamma_W)} \\ &= \frac{9.25 \text{ lbf}}{(2.70)\left(62.4 \ \frac{\text{lbf}}{\text{ft}^3}\right)} \\ &= 0.055 \ \text{ft}^3 \end{aligned}$$

step 2: Find the volume of the voids, V_V, that is the volume of all voids in the sample, whether filled with water or air. Use the definition of total volume.

$$\begin{aligned} V &= V_S + (V_W + V_A) \\ V &= V_S + V_V \\ V_V &= V - V_S \\ &= 0.085 \text{ ft}^3 - 0.055 \text{ ft}^3 \\ &= 0.03 \ \text{ft}^3 \end{aligned}$$

step 3: Find the porosity and the void ratio, using the definitions and applicable values determined in this problem.

$$n = \frac{V_V}{V} = \frac{0.03 \text{ ft}^3}{0.085 \text{ ft}^3} = 0.35$$

$$e = \frac{V_V}{V_S} = \frac{0.03 \text{ ft}^3}{0.055 \text{ ft}^3} = 0.55$$

The answer is (A).

526. The problem involves using the relationships and formulas given in the *NCEES Handbook*, Civil Engineering section, to determine the unknown properties.

step 1: Determine the porosity, n, of the soil sample.

The void ratio, $e = 0.55$, was given in the problem statement.

$$\begin{aligned} n &= \frac{V_V}{V} = \frac{e}{e+1} \\ &= \frac{0.55}{0.55+1} \\ &= 0.35 \end{aligned}$$

step 2: Find the degree of saturation.

$$S = \text{degree of saturation (\%)} = \frac{V_W}{V_V} (100)$$

Both volume parameters, V_W and V_V, are unknown. To determine the ratio of the two, assume

$$\text{volume of solids, } V_s = 1.0 \text{ ft}^3$$

$$W_s = \text{weight of solids} = \text{volume} \times \text{density}$$

$$W_s = V_s G_s \gamma_w = (1.0 \text{ ft}^3)(2.70)(62.4 \text{ pcf}) = 168.5 \text{ lbf}$$

$$W_w = \text{weight of water}$$

$$W_w = wW_s = (0.18)(168.5 \text{ lbf}) = 30.33 \text{ lbf}$$

$$V_W = \text{volume of water} = \frac{W_w}{\gamma_w} = \frac{30.33 \text{ lbf}}{62.4 \text{ pcf}} = 0.486 \text{ ft}^3$$

By definition,

$$\begin{aligned} e &= \frac{V_V}{V_S} = \frac{V_V}{1.0} = 0.55 \\ V_V &= 0.55 \end{aligned}$$

The degree of saturation is

$$S = \frac{V_W}{V_V}(100) = \frac{0.486}{0.55}(100) = 88\%$$

The answer is (C).

527. Using the constant head method with Darcy's equation, the coefficient of permeability is

$$\begin{aligned} k &= \frac{Q}{(iAt_E)} \\ &= \frac{1000 \text{ cm}^3}{\left(\frac{200 \text{ cm}}{100 \text{ cm}}\right)\left(\left(\frac{\pi}{4}\right)(4 \text{ cm})^2\right)\left((15 \text{ min})\left(60 \ \frac{\text{s}}{\text{min}}\right)\right)} \\ &= 0.04 \text{ cm/s} \end{aligned}$$

The answer is (A).

528. The Reynolds number is used to characterize flow.

$$\text{Re} = \frac{vD\rho}{\mu}$$

For 20°C water, water density, ρ, is 998.2 kg/m^3.

The dynamic viscosity is

$$\mu = 10^{-3}\ \frac{\text{kg}}{\text{m·s}}$$

For soil seepage or permeability, the parameter D is the average diameter of the soil grains (D = 0.35 mm = 0.00035 m).

$$\text{flow velocity, } v = \frac{\text{discharge rate}}{\text{area of sample X} - \text{section}}$$

Note that Q is not the discharge, but the quantity of water collected. (The total quantity of water, Q, is 1000 ml).

$$A = \frac{\pi d^2}{4} = \frac{3.14}{4}(4\ \text{cm})^2 = 12.56\ \text{cm}^2$$

$$\begin{aligned} v &= \frac{Q}{At_e} \\ &= \frac{1000\ \text{ml}}{\left(\frac{\pi}{4}(4\ \text{cm})^2\right)\left((15\ \text{min})\left(60\ \frac{\text{s}}{\text{min}}\right)\right)} \\ &= 0.089\ \text{cm/s} \quad (0.00089\ \text{m/s}) \end{aligned}$$

Find the Reynolds number.

$$\begin{aligned} \text{Re} &= \frac{vD\rho}{\mu} \\ &= \frac{\left(0.00089\ \frac{\text{m}}{\text{sec}}\right)(0.00035\ \text{m})\left(998.2\ \frac{\text{kg}}{\text{m}^3}\right)}{10^{-3}} \\ &= 0.31 < 2100 \end{aligned}$$

A flow with a Reynolds number less than 2100 is laminar.

The answer is (A).

529. Consider a unit length of the strip footing.

The overturning moment is defined as moment = load × eccentricity.

An equivalent loading diagram for the footing subjected to an axial load and bending moment is shown.

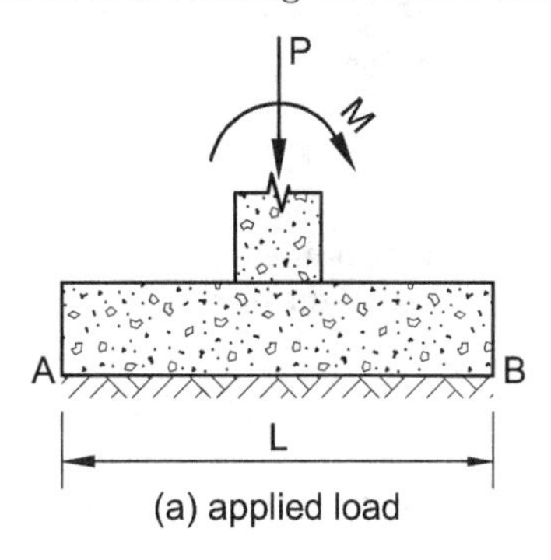

(a) applied load

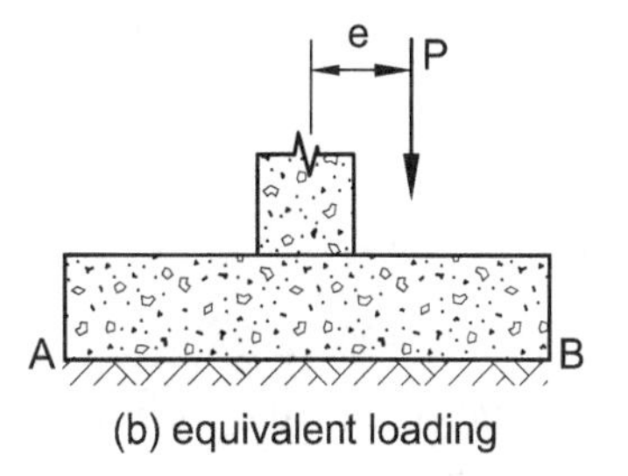

(b) equivalent loading

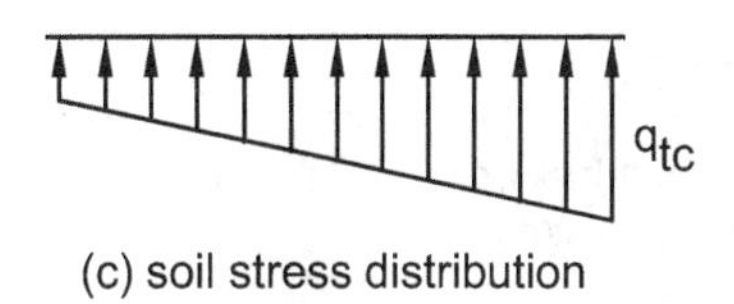

(c) soil stress distribution

The moment causes the soil stress distribution under the footing to increase on one side and decrease on the other side of the load, as shown in the illustration.

For wall strip footing, consider a unit length of the wall. The load eccentricity is

$$e = \frac{M}{P} = \frac{10\ \text{ft-kips}}{10\ \text{kips}} = 1\ \text{ft} \quad (12\ \text{in})$$

The answer is (C).

530. Given that F = 55%, the group index is given by

$$\begin{aligned} \text{GI} &= (F - 35)\big(0.2 + (0.005)(\text{LL} - 40)\big) \\ &\quad + (0.01)(F - 15)(\text{PI} - 10) \\ &= (55 - 35)\big(0.2 + (0.005)(46 - 40)\big) \\ &\quad + (0.01)(55 - 15)(20 - 10) \\ &= (20)\big(0.2 + (0.005)(6)\big) + (0.01)(40)(10) \\ &= 4.6\ \text{in} + 4\ \text{in} = 8.6\ \text{in} \end{aligned}$$

The total thickness is

$$t_{\text{total}} = 8.5 + \text{GI} = 8.5 \text{ in} + 8.6 \text{ in} = 17.1 \text{ in} \quad (17 \text{ in})$$

The answer is (B).

531. See the *NCEES Handbook*, Civil Engineering section, Geotechnical subsection for the shear equation.

$$\tau_F = c + \sigma_N \tan\phi$$

The equation represents a linear relationship. Therefore, it is an equation of a straight line; angle $\phi = 31°$ and tan $\phi = 0.6$.

Rearrange the equation and find the y-intercept.

$$c = \tau_F - \sigma_N \tan \phi$$

The standard form of a straight-line equation is

$$y = mx + b$$

Use the test data from sample 1.

$$\begin{aligned} c &= 700 - 1000\tan\phi \\ &= 700 \text{ psf} - (1000)(\tan 31°) \\ &= 100 \text{ psf} \end{aligned}$$

The answer is (B).

532. As a tensile force is applied, two types of failure mechanisms can occur:

(1) The anchor rod mechanism can fracture, or

(2) the mechanism can pull out along with some soil.

The failure surface of a mechanical anchor is an inverted cone, as shown.

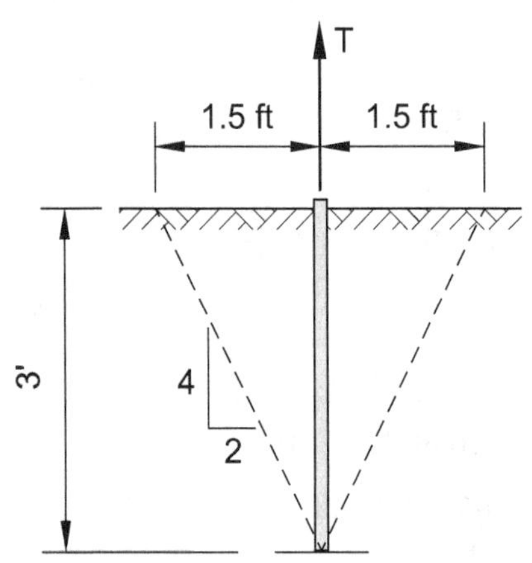

The resistance to pullout is derived from the weight of soil in the soil cone and the cohesion at the cone's fracture surface.

All information needed to compute the pullout resistance is given. See the *NCEES Handbook*, Mathematics section for the volume of the cone.

$$\begin{aligned} V &= \frac{1}{3}\pi r^2 h \\ &= \frac{1}{3}\pi(1.5 \text{ ft})^2(3 \text{ ft}) \\ &= 7.07 \text{ ft}^3 \end{aligned}$$

Find the surface area.

$$\begin{aligned} A &= \pi r\left(\sqrt{r^2 + h^2}\right) \\ &= \pi(1.5 \text{ ft})\left(\sqrt{(1.5 \text{ ft})^2 + (3 \text{ ft})^2}\right) \\ &= 15.8 \text{ ft}^2 \end{aligned}$$

Determine the uplift capacity.

$$\begin{aligned} T_{\text{ult}} &= \text{weight of soil cone} + \text{soil cohesion} \\ &= \gamma V + cA \\ &= (120 \text{ pcf})(7.07 \text{ ft}^3) + (70 \text{ psf})(15.8 \text{ ft}^2) \\ &= 1954 \text{ lbf} \quad (1.95 \text{ kips}) \end{aligned}$$

The answer is (C).

533. To have minimum footing size, the applied load should be distributed uniformly over the entire footing area. The total vertical load is 200 kips + 400 kips = 600 kips.

Find the footing area.

$$\begin{aligned} A_{\text{ftg}} &= \frac{\text{total load}}{\text{allowable soil capacity}} \\ &= \frac{600 \text{ kips}}{6 \text{ ksf}} \\ &= 100 \text{ ft}^2 \end{aligned}$$

Determine the footing length.

$$\begin{aligned} A_{\text{ftg}} &= LB_{\text{ftg}} \\ L &= \frac{A_{\text{ftg}}}{B_{\text{ftg}}} \\ &= \frac{100 \text{ ft}^2}{7 \text{ ft}} \\ &= 14.33 \text{ ft} \quad (15 \text{ ft}) \end{aligned}$$

The selection of the minimum length requires that the footing be placed so that the column load is concentric.

The answer is (C).

534. See the *NCEES Handbook*, Dynamics and Civil Engineering: Transportation sections.

The units of speed and time must be consistent. Use the *NCEES Handbook*, Units and Conversion Factors section. In this problem, the speed is in miles per hour and time is given in seconds. Convert the speed from miles per hour to feet per second.

$$\begin{aligned} D &= (\text{speed})(\text{time}) \\ &= \left(\left(50\ \frac{\text{mi}}{\text{hr}}\right)\left(1.47\ \frac{\dfrac{\text{hr}}{\text{mi-sec}}}{\text{ft}}\right)\right)(2\ \text{sec}) \\ &= 147\ \text{ft} \quad (150\ \text{ft}) \end{aligned}$$

The answer is (D).

535. See the *NCEES Handbook*, Civil Engineering: Transportation section. Transition or spiral curves provide a gradual change from a tangent section to a circular curve. The minimum spiral transition length is

$$\begin{aligned} L_s &= \frac{3.15V^3}{RC} \\ C &= \text{rate of increase of lateral acceleration} \\ &= 1\ \text{ft/sec}^3\text{, unless otherwise stated} \\ &= \frac{(3.15)\left(35\ \frac{\text{mi}}{\text{hr}}\right)^3}{(200\ \text{ft})\left(1\ \frac{\text{ft}}{\text{sec}^3}\right)} \\ &= 675\ \text{ft} \end{aligned}$$

The answer is (D).

536. The minimum curve length depends upon the sight distance. For this there are two criteria.

step 1: Use the stopping sight distance to find the difference in the grades.

$$A = |g_1 - g_2| = |-5\% - (-1\%)| = 4\%$$

step 2: Determine whether the curve length is more or less than the stopping sight distance. First assume that the curve is longer that the sight distance.

For $S \leq L$, the curve length is

$$\begin{aligned} L &= \frac{AS^2}{400 + 3.5S} \\ &= \frac{(4)(424\ \text{ft})^2}{400 + (3.5)(424\ \text{ft})} \\ &= 382\ \text{ft} \end{aligned}$$

This condition is not met. The curve is shorter than the given stopping sight distance of 424 ft.

step 3: Use the alternate condition $S > L$ to determine the curve length.

$$\begin{aligned} L &= 2S - \frac{400 + 3.5S}{A} \\ &= (2)(424\ \text{ft}) - \frac{400 + (3.5)(424\ \text{ft})}{4} \\ &= 377\ \text{ft} \quad (400\ \text{ft}) \end{aligned}$$

The condition is met. The minimum length required is less than given stopping sight distance of 424 ft. Since $S > L$, the curve length should be approximately 400 ft.

The answer is (D).

537. The total number of transit trips per day for the population is

$$\begin{aligned} \text{total transit trips} &= (40\%)(5000\ \text{people})\left(0.51\ \frac{\text{trips}}{\text{day}}\right) \\ &\quad + (60\%)(5000\ \text{people})\left(0.25\ \frac{\text{trips}}{\text{day}}\right) \\ &= 1770\ \text{trips/day} \end{aligned}$$

The answer is (D).

538. See the *NCEES Handbook*, Civil Engineering/Basic Freeway Segment Highway Capacity section.

BFFS pertains to the basic section of the freeway that is outside of the influence of exit ramps and weaving area. It is an ideal speed and does not include the effects of heavy traffic. It is based on the lane width of 12 ft. BFFS is given by the expression

$$\text{BFFS} = \text{FFS} + f_{\text{LW}} + f_{\text{RLC}} + 3.22\ \text{TRD}^{0.84}$$

BFFS is always equal to or more than FFS, depending on applicable adjustments.

The answer is (A).

539. This is a dilemma zone issue. In order to eliminate the dilemma zone, the minimum duration for the yellow phase should equal the time taken to clear the intersection without acceleration by a vehicle that otherwise cannot stop without encroaching into the intersection. It is expressed as

$$\begin{aligned} T_{\text{min}} &= y + r \\ T_{\text{min}} &= t + \frac{v}{2a \pm 64.4G} + \frac{W + l}{v} \end{aligned}$$

In this case, the length of the yellow light interval appears to be too short, which the driver can show by calculations.

The answer is (B).

540. The traffic characteristics are determined in terms of the repetitions of an 18,000 lbf (80 kN) single-axle load applied to the pavement. This is referred to as the equivalent single-axle load (ESAL). ESAL is based on the truck factor determined for different types of vehicles using the roadway. In this case, the fully loaded trucks comprise the entire traffic.

Use the load equivalency factors from the Highway Pavement Design table in the *NCEES Handbook*. The truck factor for a fully loaded truck leaving the stone pit is

$$\begin{aligned}\text{truck factor} &= 1.0 + (2)(0.658) + 1.51\\ &= 3.83\end{aligned}$$

The daily ESAL is

$$\begin{aligned}\text{daily ESAL} &= (\text{no. of trucks})\\ &\quad\times(\text{no. of trips/day})(\text{truck factor})\\ &= (5)(20)(3.83)\\ &= 383\ \text{ESAL/day}\end{aligned}$$

Find the weekly ESAL.

$$\begin{aligned}\text{weekly ESAL} &= \left(5\ \frac{\text{days}}{\text{week}}\right)\left(383\ \frac{\text{ESAL}}{\text{day}}\right)\\ &= 1915\ \text{ESAL/week}\end{aligned}$$

Find the 10 year ESAL.

$$\begin{aligned}\text{10 year ESAL} &= \left(50\ \frac{\text{week}}{\text{year}}\right)\left(1915\ \frac{\text{ESAL}}{\text{week}}\right)(10\ \text{years})\\ &= 957{,}500\ \text{ESAL}\quad (1{,}000{,}000\ \text{ESAL})\end{aligned}$$

The answer is (C).

541. The wear and tear to the roadway caused by a vehicle is proportional to its truck factor and ESAL. It can be calculated using the Load Equivalency Factors (LEF) table in the *NCEES Handbook*, Civil Engineering (Transportation) section. Find the truck factor of light vehicles and HS-20 trucks.

$$\begin{aligned}\text{truck factor of a light vehicle} &= 0.00018 + 0.00209\\ &= 0.00227\\ \text{truck factor of HS-20 truck} &= 0.0343 + 8.88 + 8.88\\ &= 17.7943\end{aligned}$$

Find the number of light vehicles equaling one HS-20 truck.

$$\begin{aligned}N &= \frac{\text{truck factor of HS-20}}{\text{truck factor of a light vehicle}}\\ &= \frac{17.7943}{0.00227}\\ &= 7839\ \text{light vehicles}\quad (8000\ \text{vehicles})\end{aligned}$$

The answer is (C).

542. The expression to determine the free-flow speed (FFS) is

$$\text{FFS} = \text{BFFS} - f_{\text{LW}} - f_{\text{RLC}} - 3.22\ \text{TRD}^{0.84}$$

Highway capacity tables are given in the *NCEES Handbook*. Since the lane width is less than 12 ft, apply the lane width adjustment.

$$\begin{aligned}&\text{adjustment for 3 ft right-side lateral clearance,}\\ &\qquad f_{\text{RLC}} = 0.6\ \text{mi/hr}\\ &\text{adjustment for 11 ft lane width,}\\ &\qquad f_{\text{LW}} = 1.9\ \text{mi/hr}\end{aligned}$$

$$\text{total ramp density, TRD} = 1\ \text{ramp/mile}$$

Substitute the values in the free-flow speed equation.

$$\begin{aligned}\text{FFS} &= 70\ \frac{\text{mi}}{\text{hr}} - 1.9\ \frac{\text{mi}}{\text{hr}}\\ &\quad - 0.6\ \frac{\text{mi}}{\text{hr}} - (3.22)\left(1\ \frac{\text{ramp}}{\text{mile}}\right)^{0.84}\\ &= 67.5\ \frac{\text{mi}}{\text{hr}} - 3.22\ \frac{\text{mi}}{\text{hr}}\\ &= 64.28\ \text{mi/hr}\quad (64\ \text{mi/hr})\end{aligned}$$

The answer is (B).

543. A CPM arrow diagram is also called an activity-on-arrow network. In this diagram, the nodes are events, and the arrows represent activity. To complete an event, all activities culminating at the node must be completed. The dotted lines represent dummy activities, which consume no time or resource. There are two paths to initiate activity 4-6.

path 1: 1-2-3-4 = 3 days + 5 days + 0 days = 8 days

path 2: 1-2-4 = 3 days + 2 days = 5 days

Dummy activity 3-4 imposes a restriction so that activity 4-6 cannot occur until event 3 can occur. Dummy activity 3-4 is on the critical path, so there are 3 float days available to complete activity 4. The earliest activity 4-6 can start is 8 days from the start.

The answer is 8 days.

544. Use Simpson's one-third rule. This is an accurate method to compute an area; however, it applies to an area where (1) the offsets are at regular intervals and (2) the number of offsets is odd. The offsets in this problem meet the requirement. See the *NCEES Handbook*, Civil Engineering section, Earthwork Formulas subsection. For 7-offset data, the formula is

$$\text{area} = \frac{w}{3}[h_1 + 2 \times (h_3 + h_5) + 4(h_2 + h_4 + h_6) + h_7]$$

$$\begin{aligned}\text{area} &= \frac{20}{3}[16 + 2(23 + 23) \\ &\quad + 4(12 + 26 + 22) + 12] \\ &= \frac{20}{3}(360) = 2400 \text{ ft}^2\end{aligned}$$

The answer is 2400 ft².

545. Use the *NCEES Handbook*, Units and Conversion Factors section. The weight of the concrete mix is

$$\begin{aligned}W &= 1 \text{ lbf of cement produces} \\ &= 0.5 \text{ lbf} + 1 \text{ lbf} + 2 \text{ lbf} + 4 \text{ lbf} \\ &= 7.5 \text{ lbf} \\ W_{\text{total}} &= \left(\frac{7.5 \text{ lbf concrete}}{1 \text{ lbf cement}}\right)\left(94 \frac{\text{lbf}}{\text{sack}}\right)(4998 \text{ sacks}) \\ &= 3{,}523{,}590 \text{ lbf}\end{aligned}$$

Based on the weight and volume, find the amount payable to the contractor.

$$\begin{aligned}V &= (3{,}523{,}590 \text{ lbf})\left(\frac{1 \text{ ft}^3}{145 \text{ lbf}}\right)\left(\frac{1 \text{ yd}^3}{27 \text{ ft}^3}\right) \\ &= 900.02 \text{ yd}^3 \quad (900 \text{ yd}^3) \\ \text{amount} &= (900 \text{ yd}^3)\left(\frac{\$100}{1 \text{ yd}^3}\right) \\ &= \$90{,}000\end{aligned}$$

The answer is $90,000.

546. A fence installed as a part of temporary erosion control is called a silt fence. It should be installed along or parallel to a contour and before upslope land disturbance begins, as shown. As the drainage is perpendicular to the contour lines, the silt fence reduces, slows, or stops the sediment-laden runoff.

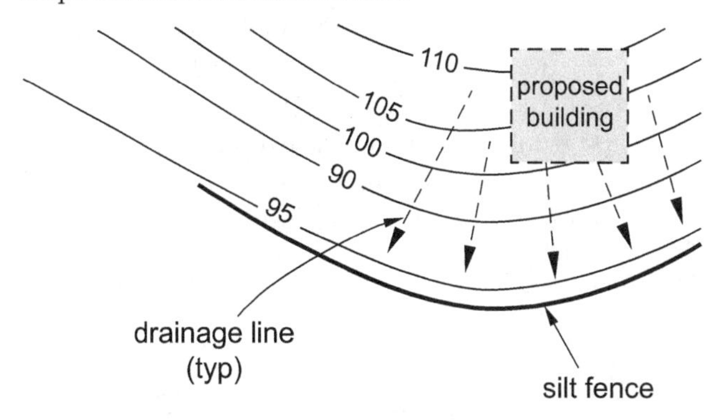

The length of the posts is generally 32 in long, out of which about 16 in should be above original ground. Silt fence has a life of 6 months to a year when installed. An inspection is required weekly and after each storm event and repairing as needed. When the fence height is half full with sediments, it should be cleaned out.

The answer is (C).

547. The water flows to a lower elevation along the steepest slope. Grid point D-3 is located at approximately EL. 66. The lowest elevation in the area is approximately EL. 40, located at or close to grid point C-1. The water drop would travel along the steepest slope starting from D-3 to the lowest point, C-1. Point to grid point C-1.

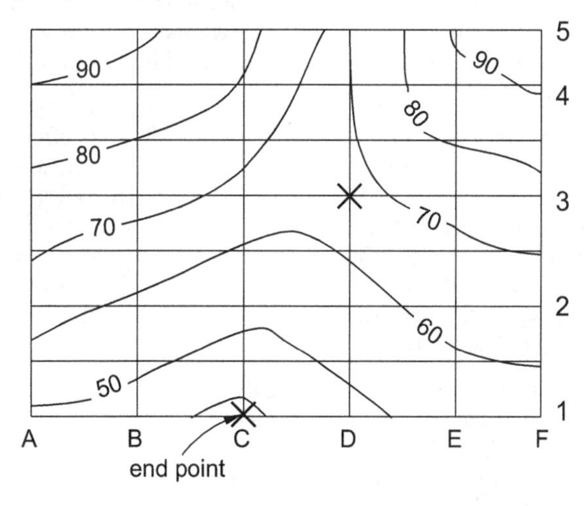

The answer is point C-1.

548. See the *NCEES Handbook*, Civil Engineering/Construction section for earned-value analysis parameters. The three parameters are

- BCWS, the budgeted cost of work scheduled (planned)
- ACWP, the actual cost of work performed (actual)
- BCWP, the budgeted cost of work performed (earned)

For the parameters, the two variances are

$$\text{CV} = \text{BCWP} - \text{ACWP}$$
$$(\text{cost variance} = \text{earned} - \text{actual})$$
$$\text{SV} = \text{BCWP} - \text{BCWS}$$
$$(\text{schedule variance} = \text{earned} - \text{planned})$$

The variance shows that the project is running a deficit since the earned value is less than the budgeted value. The variance CV is noted by C.

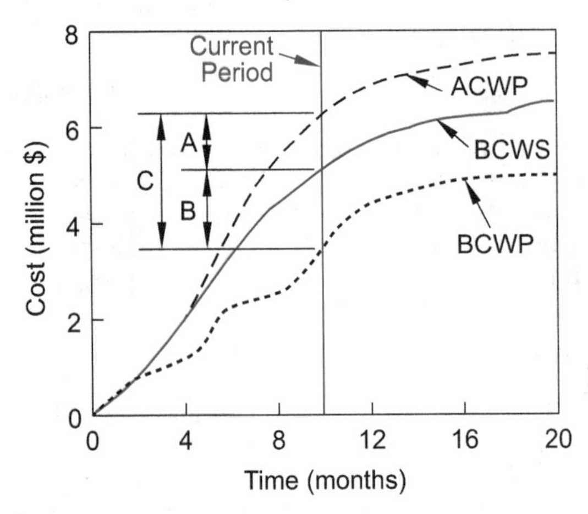

The answer is C.

549. The backbone of a safety program is not to have a firm handshake understanding with the construction superintendent. The backbone of a safety program is to have a written plan and adhere to it. Therefore, option B is incorrect.

The experience modification rating (EMR) offers a reflection of a company's safety program in relation to other similar employers. The lower the EMR, the better the safety program. A rating of 1.00 is considered average and correlates to achieving a "C" on a safety report card. An EMR rating less than 1.00 is considered better than average, and EMRs above 1.00 are worse than average. Therefore, option A is incorrect.

In most jurisdictions, it is illegal for an employer to pay medical bills related to work injury. Rather, it is required that all work-related claims should be submitted to your workers' compensation carrier. Therefore, option C is incorrect.

The safety of employees is a paramount concern of every employer. It is a component of any project as important as estimating, scheduling, and equipment maintenance. Therefore, option D is correct.

The answer is (D).

550. For unit conversion, see the *NCEES Handbook*, Units and Conversion Factors section. Since the loader capacity is given in cubic yards, it is important to use unit weight of the loose soil. The problem provides the information, and the swell factor is not needed.

The loader production per cycle is

$$3\ \text{yd}^3 = (3\ \text{yd}^3)\left(3000\ \frac{\text{lbf}}{\text{yd}^3}\right) = 9000\ \text{lbf}$$

The loader production/min is

$$\left(\frac{9000\ \text{lbf}}{1\ \text{cycle}}\right)\left(\frac{1\ \text{cycle}}{30\ \text{sec}}\right)\left(\frac{60\ \text{sec}}{1\ \text{min}}\right) = 18{,}000\ \text{lbf}$$

$$\begin{aligned}\text{production/min} &= \left(18{,}000\ \frac{\text{lbf}}{\text{min}}\right)\left(\frac{1\ \text{ton}}{2000\ \text{lbf}}\right)\\ &= 9\ \text{ton/min}\\ \text{production/50 min} &= \left(\frac{9\ \text{tons}}{1\ \text{min}}\right)(50\ \text{min})\\ &= 450\ \text{tons}\end{aligned}$$

The loader's efficiency is 50 min/hour. Therefore, its hourly production is 450 tons.

The answer is 450 tons per hour.